Ken Ring's

IRELAND
WEATHER ALMANAC

2024

BY THE SAME AUTHOR

New Zealand Weather Almanac (1999-Present)
Australia Weather Almanac (2006-Present)
Ireland Weather Almanac (2010-Present)
How to Get Your Kid to Like Maths
Predicting Weather by The Moon
New Meteorological Techniques
We Used to Call It Summer
The Romany's Apprentice
Moon and WeatherLore
Secrets of the Moon
Notes From the Pit
The Lunar Code
Fingermaths
Supertramp
A Fiery Tale

ISBN: 978-1-73862-301-3

Email: enquiries@predictweather.com
Web: www.predictweather.com

CONTENTS

INTRODUCTION

DISCLAIMER

The forecasts in this work are best-endeavour opinions of the authors and associates. No guarantees of 100% accuracy is claimed. Neither can the author and/or publisher be held accountable for decisions made on the basis of information in these pages.

ABBREVIATIONS

MOON SYMBOLS

1st Q	Moon in first-quarter phase.
3rd Q	Moon in third-quarter phase.
^	Northern declination Moon at furthest point north for month.
A	Apogee Moon furthest from earth for the month.
F	Full moon.
N	New moon.
P	Perigee Moon at closest point to the earth out of the 27 day cycle.
P(n)	Priority closest e.g. P3 means third closest for year.
V	Southern declination Moon at furthest point south for the month.
XhN	Moon crossing the equator heading north.
XhS	Moon crossing the equator heading south.

RAIN SUMMARY TABLES

d	Dry most of the day.
h	Heavy rain.
l	Light, intermittent showers.
n	Fog, mist, dew, or non-recordable showers.
p	Passing Showers
r	Rain.
s	Showery.

SUNSHINE SUMMARY TABLES

F	Fine, sunny (8+ hours sunshine)
pc	Partly cloudy (4-7 hrs sunshine)
c	Cloudy (1-3 hrs sunshine)
o	Overcast (0 hrs sunshine)

TEMPERATURE SUMMARY TABLES

clr	Cooler than average temperatures.
a-c	Average to cooler temperatures.
av	Average temperatures.
a-w	Average to warmer temperatures.
wrm	Warmer than average temperatures.

WEATHER GLOSSARY

To get the most out of the forecasts, it helps to be able to read a weather map and to understand the language used. A high air pressure – a fair-weather sign, is when the isobaric values increase inwards towards the center. A low-pressure zone or depression, indicates worsening conditions, and when isobaric values decrease towards the center. Closer isobars = stronger winds.

Apogee	Moon furthest from Earth for month	Winds generally light. Calm conditions.
Perigee	Moon closest to Earth for month	Spells trouble Earthquake risk in some countries.
Cold front	Cold air advancing, pushing away warm air	Changeable
Depression	Area of low barometric pressure	Unpredictable, bad weather, strong winds
Ridge of high pressure	Edge of anticyclone	Good weather approaching, cooler nights in winter
Stationary front	Slow-moving weather	No change for 2-3 days
Northern declination	Moon at northernmost point for month. Moon rises in the north.	Slow moving weather system, northerlies, rain in north, colder conditions
Southern declination	Moon at southernmost point for month	Slow moving system, southerlies, precipitation in south, generally milder.
Moon crossing equator	Faster moving weather system, moonrise is due east.	Turbulent, changeable, often electrical storms, gales or winds expected.

WEATHER FORECASTING IS AN INEXACT SCIENCE

The morning skies generally will indicate the state of the atmosphere, and not much will change during the course of the day. The situation can be compared to the visit to the doctor. One might wake up with a flu that has been a couple of days in the forming, and seek advice for how long it may continue. The doctor might say, 'it may clear up today, in a day or two, or in a few days, but if it hasn't cleared up in a week then come back and see me'. Most patients will be happy with that, and gladly pay the visit fee. But imagine if a weather forecaster said that; that it may clear up today or in a couple of days and if it hasn't cleared up in a week's time then come back for another report. The public would complain loudly, unaware of the double standard.

Weather forecasting is not more precise than medicine. But like many other sciences, prediction about weather determines potential and trends, whether it will be a wet year or a dry year - so do we prepare for flooding, will the next month require haymaking contractors to be booked or do we wait a month, and hospitals need to know which will be the coldest week in winter. Will the summer be a good one for camping? What week of autumn should rams be put in the field, because adding gestation time one can calculate when spring lambs will be born, and if it is still snowing by then the newborn lambs may suffer. These are questions that long-range forecasting can address.

POTENTIAL

It may rain if there is potential, or, with potential in place, for various reasons, it may not rain at all. But it will not rain without potential. Weather forecasting is an inexact science, like medicine, geology, psychology, and economics. In medicine, often a second 'opinion' might be sought. Prediction is informed opinion, nothing more. No one would call the authors of commercial tide tables 'satanic soothsayers', and yet tides will be found in real life to vary from previously published heights and times. Allow for a leeway of 1-3 days and as weather is general to an area of 50-80kms radius, that is the accepted range of applicability and error potential.

WEATHER FROM THE OCEAN

Weather may be thought of as the tide of the atmosphere. Because the sea and air interface over nearly three quarters of the earth's surface, and because most rain comes from the sea and then drops back into it, the world's weather systems originate and usually end in the oceans. Tidal currents in the sea transfer energy to the ocean surface and create surface wind systems which travel from ocean to ocean, crossing strips of land that get in their way.

There is nothing on the land that generates weather – when it hits the atmosphere it is already half obsolete. Storms generate deep within the ocean. Otherwise, where does the massive amounts of water for a thunderstorm or cyclone come from? Swell always comes *before* wind and changes direction *before* wind change. The air tide which is joined to and therefore is part of the ocean tide provides changes in temperatures to air layers which in turn provides conditions for condensation.

For that reason, neither the rain forests nor any land masses have anything to do with the generation of global weather – the Amazon rain forests could be all cut down in one day and the world's weather patterns would be unaltered.

Neither do vehicle emissions, which are on land, have any power to change what happens to ocean currents in the far-down ocean depths which are a function of orbits of the moon. The weather will be generated anyway, and when it arrives it instantly deals deal to the emissions, not vice versa. It is not that rain forests cause rain, it is the rain that provides conditions for a forest. Forests will tend to grow where it rains often. For example, a line of green hills or a patch of green trees near the coast is a sign that it often rains there, compared to a wide area of brown land where rain is very scarce.

It is the lack of rain that creates a desert. The rain only comes from the storehouse of water - the oceans, or nearby lakes. For ocean tide strength, timing and turbulence, the prevailing factors that encourage evaporation come from millions of kms beyond the earth, culminating in ocean currents, waves, rain and storms, and these forces can finally be seen as wavelets running up a beach.

WHAT CONTROLS WEATHER

Weather is determined by astronomical and physical reasons which are under relatively fixed control of cycles. It is not determined by any activities by Man. Climate-scientists erroneously claim that the heavier-than-air "trace" gas, carbon dioxide, which is absent in 99.96% of the atmosphere; and methane, which is absent in 99.9999% of it, have a controlling effect on the climate.

Approximately 80-85% of the weather is caused by the moon, changing and repeating its position above the earth. Planning for major droughts or floods is therefore possible, but not by measuring any emissions of a negligible gas.

Here is the weighting system. There is space weather, climatic weather and everyday weather.

1.CHANGES TO WHOLE PLANET:

(Irrespective of geography)

• Deviation of Earth's rotational axis to the vertical.
• Variation of Earth's orbit.
• Magnetosphere
• Heliosphere

2. CLIMATE AT LOCATION:

(long-term trends in weather-patterns)

• Proximity to equator (angle of the Sun's rays – latitude)
• Direction of prevailing winds
• Topography – shape of land, proximity to other land masses, elevation and distance from coast.

3. WEATHER AT LOCATION:

(trends and daily occurrences, observable and recordable)

• Influence of Moon on tides (including air tides), cycle of rain from ocean, and back to ocean. (80-85%), also ocean current cycles e.g. El Nino.
• Level and intensity of cloud cover (keeps heat in, determines what, where, when, of rain (subject to ocean and air tides)
• Solar winds, cycles of sunspots (temperature of seasons), hemispherical and seasonal factors.

4. NOT IN THE MIX AT ALL, EITHER FOR CLIMATE OR WEATHER

• Land use changes, deforestation and greenification of areas of the earth.
• Volcanic activity ongoing.
• Ice-reflectivity feedback, constant.
• Industrial and commercial pollution, and emissions.
• other activities of Man e.g. diet.

Immediate weather is modified by cycles of solar activity (position and number of occurring sunspots), and the cyclic position of the Moon relative to the earth. Solar cycles determine floods and drought periods.

Sunspots have a cycle of 9-14 years, averaging about 11 years or 23 years. A minimal year of no sunspots is called the sunspot minimum. 2020 was the last minimum, bottoming out in December, but with effects continuing for a couple of years. Colder conditions worldwide have been a feature of the last 3 years.

After solar minimum, 4 or 5 years elapse until sunspot maximum, with spots nearer the Sun's equator. Then they gradually decline in number until minimum again is reached. This decline is about 7 years. Intensity has connection to the sunspot cycle. The next solar maximum is expected 2025-6, and the next solar minimum in 2031.

The Moon is a planet like earth, also orbiting the Sun, such that Earth holds the Moon to its own orbit, so Earth and Moon orbit the Sun together. Because it is a planet, the Moon is really orbiting the Sun, but it may be said that the earth keeps getting in its way.

Therefore, the Moon is tied to both earth and the Sun, and alternately absorbs and shields solar electro-charges, according to Moon phase. By this process the Moon controls timing of weather events. An analogy is that the Sun is the engine and the Moon is the driver.

THE JUPITER/SATURN EFFECT

Sunspots have been shown to be electromagnetic disturbances produced by external charges which act on the Sun's internal structure. Just as Sun and Moon, either both on one side of Earth (new Moon) or on opposite sides (full Moon) cause internal stress on Earth to raise tides in land, sea and air, by the same process the largest of our solar system's planetary bodies on one side of the Sun or on opposite sides cause exaggerations in Sun-tides manifesting as sunspots/solar wind.

When the planet Jupiter draws the Sun outwards from its normal line, sunspot maxima occurs, and when the Sun is drawn inwards, the minima are the result. When the gas giants Jupiter and Saturn pass across the line of the Sun's advance (Solar Apex), sunspot minimum is very close to that date. Jupiter, Saturn, Uranus and Neptune, in that order, affect the Sun the most, with a total period of 178 years, roughly 10 lunar cycles. The 11-12-year average sunspot period is roughly the same as the orbit of Jupiter, and any variations are due to the varying positions of the other three great planets.

In the prolonged drought of 1862, Jupiter and Saturn lined up on one side of the Sun in that year. Both straddled the Sun in 1871, were on one side of the Sun 1900-2, then on opposite sides of the Sun again in 1931. Jupiter/Saturn together go through this cycle every 35-36 years, with the other two planets, Uranus and Neptune, of much slower periodicity. Jup/Sat were together again on one side of the Sun in 2020.

In 1982, Saturn/Jupiter were on the same side of the Sun, causing severe drought to flat terrain of the northern hemisphere. In 2001 Jupiter crossed the vertex, bringing further drought. When Jupiter crosses the path of the Sun there is usually a drought just beforehand. Jupiter was crossing the Sun in 2007, when world-wide droughts brought a food crisis. No sunspots can result in little or no rain.

Jupiter has been at high peaks in 1766-69, 1779-81, 1790-93, 1802-4 (drought Jup+Sat), 1814-17 (drought), 1827-30 (drought), 1838-40 (Jup+Sat drought), 1850-52, 1862-64, 1873-76, 1885-88, 1887-1900, 1909-12, 1921-23, 1933-35, 1945-47, 1957-59, 1968-71, 1980-83, 1992-95, 2004-6, 2016-19 (all drought years). Low Jupiter peaks have been 1774, 1786, 1798, 1809, 1821, 1833, 1845, 1857, 1881, 1904, 1928, 1939-40, 1952, 1964, 1976, 1987, 1999, 2011 and will be in 2024.

Jupiter/Saturn peaks and conjunctions have been blamed for low yields and economic panic. Examples have been 1873, 1884, 1893, 1900-2, 1907, 1910, 1958-60, 1987, 1990, 1999, and the recession of 2011-12. The worst have been 1901 and 1959. The next has been 2016-2024, when solar minimum created potential for worldwide droughts through low rainfall and low agricultural yields in 2020. Closer perigees in minimum declination years typically bring longer lasting extreme events.

In 2020 Jupiter and Saturn were in "conjunction". That means they were occupying the same area of sky as seen from earth. When you looked up at them, they appeared very close together, especially in December 2020. Conjunction correlates with drought years. By 2024-5 these two planets will be at right angles to each other, called a "square", and wetter years are expected. Finally, in 2030 they will be in "opposition", being 180° apart, and typically this will be a year of floods. Half the cycle is then over. This is a worldwide phenomenon. Each country experiences its own particular cycle of events, according to its unique climate.

The last previous year of equivalent conjunction was 2001. In 2005 was the last square of these planets, and their last equivalent opposition time was 2012. Following that, the last previous square was in 2015. Since 2015 it has been noticeably getting drier.

THE ORDER IS:

1. J/S conjunction=drought years (2001, 2020),
2. J/S square=coming out of drought years, wetter (2005, 2025)
3. J/S opposition=flood years (2012, 2030)
4. J/S square=getting drier, approaching the next drought (2015, 2036).

Jupiter and Saturn together control the grand cycle of drought/flood. Jupiter orbits the Sun around every 12 years, and Saturn every 29. Jupiter and Saturn are the largest planets in the cosmos, and together have an effect on the Sun, such that every 35-38 years the SSB (solar system barycenter) sees an event-repeat. Thus 2024 may see repeats of events that last occurred in 1987. A scan into Ireland's weather history may give some clue.

LUNAR FACTORS

The "perigee" of the Moon is when it comes close to earth once per month. Taking note of averages of perigees over a year yields how in a particular year the Moon is averagely closer to earth. The cycle of perigees is 8.85 years. The closest averagely shorter earth-Moon distance over the past century or so have been 1905, 1922, 1940, 1958, 1976, 1993, 2011, and 2029.

These closest perigee-years happen typically when the Moon is at or near minimum "declination" (also called minor standstill). This is the slow trekking of the moon, over 18-19 years, between earth latitude lines. It is around minimum declination years that the Moon's inter-latitudinal range is narrower.

In the southern hemisphere, a phenomenon known as the El Niño occurs twice every decade. The ocean currents reverse along the equator, restabilise the difference in single level, and then resume again. The effect on the weather is greatest in the Pacific area, but the Gulf stream tends to eventually be affected as well. These are the years that the Gulf Sream tends to stall, slow down, and countries susceptible to its usual insulation are less affected by it. This will not happen in 2024 but is expected to in 2025.

The El Niño alternates with La Nina, but the La Nina is twice as long. Mariners have always known about the 'trade' winds, blowing strongly westward along the equatorial band, which is the old name for La Nina. La Nina can be viewed as the planks on a deck, with the El Nino akin to the gaps between them. La Nina is the stronger of the two systems, being the normal situation of strong easterly flows. Although there are twice as many weak La Nina years as there are weak El Nino years, there are about equal moderate to strong La Nina and El Nino years.

On or just after averagely closer perigee years, El Nino has occurred on 1905/6, 1923-6, 1940-2, 1957-8, 1976-7, 1994-5 and 2014-15. It is not expected again until 2032.

As well there is a monthly N-S declination. In minimum declination years the speed of changing hemispheres in the course of each monthly declination is less. Such years tend more towards the slackening of sea currents due to a more sluggish lunar lateral pull, and therefore more tendency towards El Nino systems.

EVERY 19 YEARS; SUN AND MOON

19 years, the lunar Metonic Cycle, is approximately the period between the conjunctions (occurrence side-by-side of Jupiter and Saturn). That is to say, between the times when they are nearest to one another as they revolve round the Sun.

A cycle of about 76 years shows rather conspicuously in the records of the Nile annual flood levels, but this also is a multiple of the Jupiter-Saturn conjunction period 19 x 4.76, and is slightly in excess of six revolutions of Jupiter's 71¼ years.

Solar cycles peak when earth-Moon distance is on the decrease, when the Moon is coming "in from the cold". Solar peaks are closer to La Nina than El Nino, which has some bearing on the phenomena of ocean currents and therefore sea surface temperatures. Strongest El Ninos form either just after solar minima or on the rising side of sunspot cycles.

The last El Nino conditions came about in July 2015. The next lunar declination minimum was 2015. The next solar minimum began about 2015 and peaked in 2020. From 1957 to 1998 spanned 40 years and 5 strong, 5 moderate and 4 weak El Ninos. This makes for an oscillation alert each 2.5 years. If we take the 5 strong and 5 moderate events we get an average of significant El Nino-type cycle of 4-5 years, which is a multiple of the lunar declination cycle. Because it is an average, this can vary to a couple of years either side.

THE CASE FOR THE MOON

The bottom of the air and the surface of the sea are invisibly joined. Things that are joined are part of a joined system. Both do indeed comprise one interactive and connected interface, covering a surface area of 73% of the earth. Lunar forces begin with the Earth (or Land) tide, in which the land surface rises vertically around about 20cm per day in NZ and Ireland, 50cm per day in Australia and Spain, and 55cm on the equator, towards the moon, and recedes again in regular daily tidal rhythm, and which in some locations gives rise to earthquakes. The Land Tide supports the tide of the ocean which then gives rise (and fall) to a daily Air (or atmospheric) Tide. All tides are repeatable, to some extent measurable, and therefore predictable.

Matching the timing of the air-tide to oceanic tide-tables at a location yields a repeating frequency of weather extremes. Nobody questions the predictability of sea tide-tables. The more time spent outdoors, the more it will be realized that nothing in nature is random, but governed by repeating cycles. The rotation of the earth and planets, the recurrence of day and night, summer and winter, animal, bird, insect and fish breeding, feeding and migration habits, and the cycles of the growing seasons can all be predicted, not exactly, but to a degree of usefulness. Weather is the no different.

There are 3 main cycles of the moon that we use for predicting weather, gardening and fishing. They each have influences, and is easier to describe them separately to determine their effects.

1. (PHASE)

The monthly cycle that we can all see, from New moon to Full and back to New again is called phase. That includes the intermediate phases Waxing Crescent, First Quarter, Waxing Gibbous, Waning Gibbous (Disseminating), Last Quarter, and Waning Crescent (Balsamic). For weather, a change to full or new or quarter moons very often brings a weather pattern change. If raining before a phase change it may stop on the day of changeover, but if dry before and rain is about, it may choose the day of the phase to manifest the new trend.

An old mariners' saying is that the full moon 'eats' clouds, because the night of a full moon is generally clear. That is why you can see the full moon. The full moon actually clears the sky, especially around midnight. Maximum temperature of the day will correspond with tides; and at the beach, temperature is likely to be highest when the tide is lowest.

For gardening, plant above ground with attention to leafy things when the moon is climbing up in the sky in the daytime, which is before and up to full moon. After full moon, the moon is descending in the daytime because it is rising on the other side of the earth, so give attention to root things, e.g. bulbs, carrots, potatoes. For fishing, the new and full moon tides are the best, around 12 noon and midnight, and the quarter moons are good around dawn and dusk.

2. (DECLINATION)

The second cycle to consider is when the moon changes hemispheres because of the tilt of the earth. When it is in the northern hemisphere the tidal variation is greater because of greater gravitational pull, and higher tides mean more water to fill a bay which means fish compete for food in the faster currents, so bite-chances may be better, but also the water table rises in the land, and as well the sap tide in the trees. This means more insects appear, also more birds. It is why the Canadian Rockies Hunting calendar, the Maori Fishing Calendar and all ancient tribal teachings are similar. We all rotate under the same moon.

When the moon is in the northern hemisphere in winter, around full moon time, Ireland can get very cold winds, dragged southwards over Ireland from the North Pole. Northernmost full moons in summer time bring warm, moist northerly wind systems and warm rain. It is why if the winter full moon rises further in the north it will be a colder season, and if the summer full moon rises far in the south, it will be unusually warm. So, when the winter full moon is also when the moon is furthest north, one may plan for snow. This happened in January 2022. Unfortunately, Dublin may not see snow until 2027.

The moon traverses different parts of the sky as it orbits the earth in the course of a month. The "constellation" (or astrological *sign*) is the road map of the declination cycle. In folklore functions were assigned to signs. For instance Pisces moon was considered best for fishing, hence the fish zodiac symbol. Cancer full moons were said to bring more snow. In gardening, the constellations favour particular types of plant activity. For example those of a leafy, flowery, fruity, or rooty nature are said to have water (Scorpio, Pisces, Cancer), air (Libra, Aquarius, Gemini), fire (Leo, Sagittarius, Aries) or earth (Capricorn, Taurus, Virgo) signs. Air and fire signs are usually considered barren, and earth and water signs fertile. In weather, the constellations provide the template for changes in barometric pressure.

3. (APSIDAL LINE)

The third lunar cycle is called Perigee-Apogee, or Apsidal Line, which comes about because the moon's orbit around the earth is not a circle but an elliptical or egg shape. The point in the month when the moon is furthest from earth is called the Apogee. At perigee the moon is about 350,000 km from earth and at apogee it is about 400, 000 km from earth. This is a varying difference of about 50,000 km; and even that varies by about 10,000 kms over 12 months of the year.

The gravitational forces that apply between earth and moon are increased at perigee and decrease at apogee and account for extremes in tidal movement, particularly when coinciding with either a new moon or a full moon. In perigee the weather is more turbulent and strong winds are common. These may develop into gales or hurricanes and inflict property damage.

Fishermen use apogee because fish are less skittery at that time. On perigee the tides are bigger but fish don't like the breakers with their bigger swells, so surfcasting isn't the best that day. But estuaries, where fish hide from the turbulence of the shoreline, are good for fishing during perigees. Perigee is responsible for the increase in tidal height per month, the so-called king or spring tide, not, as is popularly believed, the proximity of full or new moons.

Just as the moon affects the oceans tides, so it affects the sap 'tide' in a plant's structure. Each plant, seed, and tree, and organism in the garden has a tidal flow or magnetic force that we see daily in the ocean's tidal flow. At perigee (and a couple of days either side) you would avoid planting seeds, as germination can be poor and consequent growth of the seedling may be leggy and weak. But it is an opportune time to maximize on feeding and fertilizing as the nutrient is more easily distributed in the soil. Everything is 'going for it', so to speak.

Perigee favors fungi and they can become prolific over a 3-day period. Perigee is a good time to put fermentation brews to work. Kelp can be gathered because the higher tides bring it onto beaches. At apogee, the magnetic force is weaker and the tidal pull also weaker, gentler and more consistent, so the ground is more settled for putting in things that need a calm environment to get going. While some plants may grow rapidly, others might run to seed quickly if planted in apogee. If the moon is descending and waning, good nourishment is coming to roots, and if an ascending and waxing moon, energies are traveling upwards away from the ground, tending to draw water upwards. Then again, if the apogee, the strongest influence would be lessened, but if perigee, exaggerated.

THE CASE FOR THE SUN

Solar cycles determine annual temperatures and amounts of annual rain. Moon cycles determine timing of weather events.

The sun's radiation, that we call heat, is a function of the solar cycle. For thousands of years the Chinese noticed drought years correlated to periods of intensity of visible black spots on the sun. They called these sunspots.

2020/2021 saw the end of Sunspot cycle #24.

Cycle #23 had begun in 1998, peaked around 2000/1 (solar maximum), and died off (solar minimum) from 2007-2009. After a 2-3 year period of relative inactivity Cycle #24 began around midway through 2010, and peaked in 2014. It had "bottomed-out" in 2020-21.

Cycle #25 began in 2022, will peak in 2026 and bottom out again around 2032.

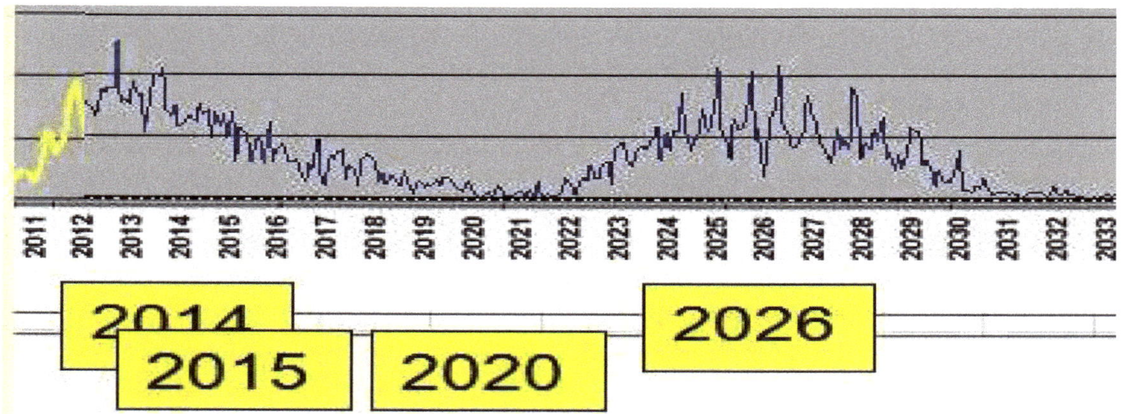

Just after a solar maximum (2015), there is usually a run of wetter years. The last solar minimum of 1996 was followed by the late 1990s which were wet years. So were the late 1970s after the previous solar minimum of 1975. There is also typically more rain between peak and minimum years than between solar minima and maxima. 2014 and 2015 weather trends were always going to resemble the later 1970s and the mid-1990s for event timings, and the mid-1950s for intensity and rain amounts. 2016 was notable for its mild wet winter. The way of longrange forecasting is to find matching and therefore equivalent years in the solar and lunar cycles.

2024 is a near match to 1962/63. By combining lunar and solar influences, temperature (solar) may lead to amounts of rainfall (due to evaporation rates), whereas timing of severe weather events may follow strict lunar cycles.

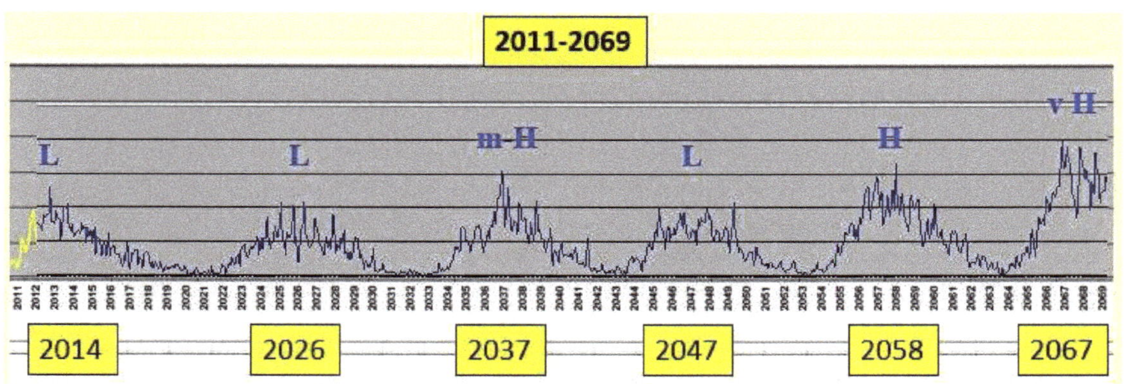

Looking at the trend in sunspot cycles the next 50 years will resemble the first 50 years of the 20th century. The next 3 or 4 solar cycles may be weaker than the last 3 or 4, with low maximum sunspot numbers and long periods with no sunspots at all, perhaps initiating a 30 - 50-year decline in global temperatures. 2007 may have been the start of this decline and global temperatures may bottom out around 2030, perhaps even staying cooler until regaining higher sunspots around 2070.

Warmer periods have typically less and weaker hurricanes, while cooler periods have more and stronger hurricanes. Solar activity may increase again from 2070 to the next century which will be the next cycle of global warming, to be followed early in the 22nd century with another extended solar activity decline. None of us will be here to witness this, but humans have in the past survived by adaptation, and will continue to do so.

It is the Sun's heat that causes evaporation which brings rain. The amount of heat that the Sun puts out determines the amount of evaporation, which must then fall within about 7 days, near to the water source. Coastal locations are generally wetter than places that are some distance inland. Without a source of nearby water, a place quickly dries out. The amount of rain falling is due to the amount of solar radiation which caused prior evaporation. Therefore, rain amounts depend on the sunspot cycle, varying over 11 and/or 22-24 years. The amount of rain falling is also dependent on the extent of the air tide, which can be likened to volume of water in each sea tide. As the solar cycle of radiation is cyclic, then so too are rain intensities.

Sea tide timing is predictable, subject to positions of Moon and Sun. The volume of water flowing into an estuary depends on wind direction, subject to shape and size of surrounding hills, wind forces increased or confused by surface currents, and less visible factors like close proximity of underwater eruptions, fissures, earthquakes, and time of the day due to temperature gradients. Amounts of rain are influenced by Sun's warmth for evaporation, and then how much cold air comes to the layers of the atmosphere containing clouds.

Sometimes colder air is at a higher or lower altitude than rain-filled clouds, and expected rains must wait for the marrying of the two factors. Until then a cloud may look like rain, but heat from the ground is sufficient to buoy up the rain-laden cloud and prevent rain falling. Analyzing sunspot cycles is at least better than having nothing at all, by looking to see which previous years correspond to same points of the cycle.

THE MOON IN PERIGEE

The moon in perigee exerts increased gravitational pull on fluids such as air and water (and land, influencing earthquakes), depending on the season, and by generating more turbulence, can cause weather to worsen.

Because the earth rotates beneath the moon once every 24 hours, unsettled conditions can happen in many countries on the same day. Up to three days before or after a perigee it is common to see the month's most rain, particularly in autumn and winter. Clear skies might return shortly afterwards as the Moon recedes from the earth. For 2024, new moon perigees will be until June, and the rest of the year see full moon perigees. The last year this happened was 2016.

As mentioned, perigees vary in closeness. In 2014 the closest perigee was in August, with #2 closest in January. September had the closest perigee in 2015, perigee #2 in February, and perigee #3 in March. Consequently, the closer 2014-15 summer moon was set to fuel increased cyclone and monsoonal activity (in countries prone to cyclones) and some above normal temperatures. But perigees were weak in December 2015 through January 2016, and cyclones were weak.

In 2016 the closest perigee was in November and tied to a full moon, and 2016-17 was a more intense cyclone season for countries not far from the equator. 2016 was also an averagely closer moon across all months, whereas overall, 2017 was an averagely further away moon, but 2018 again started to come closer, with 2019 continuing this trend. Closest perigee in 2024 happens in March and April.

Where 2020 was the 8th closest (in the 125 years from 1905 to 2030), 2021 was not amongst the closest perigee years. Neither was 2022, which was only the 90th closest. But the moon comes closer again in 2024 (12th) and some intense weather events in Ireland and the UK may be the result.

For most of 2022, perigees were level with the Southern declination. Previous years this has happened have been 2013 and 2004, both of which saw stronger than average cyclones in other countries. But in 2024 the perigees start to trek up towards the equator.

GENERAL TRENDS ARE:

- Closer perigees = more cyclonic activity (1997, 2002, 2007, 2011, 2016, 2024)
- Far away perigees = less cyclonic activity (1983, 1991, 2000, 2009, 2013, 2017, 2027)

GUIDE AS TO WHEN IT WILL RAIN THROUGH THE PHASES

The below is an approximation of rain timing, provided rain is about. If no rain, expect a drop in temperatures. Cut and paste this somewhere handy as your guide to year-around weather.

NEW MOON (INVISIBLE)
Most rain, if about, is overnight. If rain occurs during the new moon day, expect rain also the following night. If winter, expect temperature drop. Most new moon days are mainly pleasant and dry; if wet then only isolated dumps, i.e. no long-lasting storm activity.

FIRST-QUARTER MOON (D-SHAPED)
Most rain, if about, is between midnight and noon. If no rain before noon, expect it to stay dry during the afternoon. If wet in the morning, weather should clear in the afternoon.

FULL MOON (FULLY ROUND)
Most rain, if about, is during daylight hours. Otherwise it may have the look of rain all day. If rain does come overnight, which is rare, it may continue the next day. Mostly, clouds will be seen to drift away as the sun sets, and may stay away until dawn.

LAST QUARTER MOON (C-SHAPED)
Most rain, if about, arrives between noon and midnight. If rain arrives before noon, expect more rain between the noon and the evening. If the afternoon is dry, the following morning will usually stay clear.

PITFALLS OF FORECASTING

Long range forecasting is different from meteorology. Long range is more trends-based, even though it relies on archived records. Even then rain statistics from earlier years may not tell the full story.

Statistics taken from e.g., Dublin Airport, may tell a different story to what actually falls in the county, because the airport may not be always representative of the greater district.

A possibility is that whereas rain may arrive, winds around the rain gauge, especially at an airport, can push rain quickly away, which then gives a subsequent low readout and a possible misread for the region, even though the winds have already blown the rain in.

A region can easily get confused with its capital, because of population numbers. For that reason the reader is advised to consider notes pertaining to a whole region when planning ahead, and regarding any weather event within a 50-80mile range as a potential for their area also, even if it is listed for a neighboring district.

Often the amount of rain may be localized, and it is not uncommon for each hill valley to have a different microclimate to the next one over. In a forested and hilly terrain overall amounts of rain for one small area is practically impossible to determine accurately beyond a general trend. A flat region is much easier as rain is more likely to be widespread due to less quickly changing temperatures.

Quite another problem is that rain gauges often shift locations, to suit the convenience of meteorological staff. This makes it difficult for historian/long-range forecasters to obtain a valid continuous record for a cycle.

It would seem from what has been said that we should give up before we begin. But it is worth in the exercise because if this book is used correctly, to spot trends rather than look for literal amounts, then good planning decisions may be made.

A 24-hr error is the most obvious because we cannot know, from our archived data files of weather 20-50 years away, incorporating at least two lunar cycles, which side of midnight historical data was gathered. Most gathering stations display a rain reading taken in the 24hrs previous to 9am, but did the rain on the day happen just prior to 9am, at say, 8.55am and hence on the day in the listing, or, say, 7pm some 14 hours before, qualifying as previous day's rain?

The scientist Benjamin Franklin (1706 - 1790) the inventor of the lightning conductor was also a lunar weather forecaster. His regular *Poor Richard's Almanac* ran for 25 years and he had 10,000 subscribers. Franklin hedged his bets slightly by stating: 'I should ask the indulgence of the reader for a day of grace on either side of the date for a specific weather prediction.' He was rightly judicious. The one-day error is universal.

Similarly, temperatures are gathered as maximums and minimums, one of each per day per town. Suppose a reading is taken at 4pm (often the practice) but let's say on a particularly hot day a very high maximum is reached at 5pm. The hotter value will miss the cut, and the 4pm reading will be recorded as the day's maximum. But if the next day brings a freezing cold maximum due to an anomalous and temporary southern wind shift, nevertheless the previous day's 4.30pm reading will be the relevant maximum figure when taken over the previous 24-hrs. The error in temperatures can be large because over a month, this error in average maximums can be magnified. Before long the "normal" is invalid, and the breaking of records simply fiction media hype to sell newspapers.

Also, as cold air layers have to match with rain-laden clouds for rain to actually fall, there can sometimes be up to about a 24-hr error in rain arrival. When such errors for every town every day every year are collated, a so-called trend can be compromised in favor of whatever bias an operator may have. Weather can be generated between 200ft and 8 miles up, which can incur the potential overshoot of a rain system over a radius of up to 50-60 miles (80-100km). So if rain is forecast for a coastal location but drops 100kms out to sea, it may arguably still qualify as a successful forecast. It is not uncommon for one forecast report heard on radio or TV to cover half a county in one sentence, yet there will be many microclimatic events within even one small area, especially in hilly terrain.

Most historical data is taken at airports, and these are usually well-sited for wind which makes it easier for planes to take off and land safely, and as windier places dry out more rapidly, it means they are typically drier areas than the surrounding hills. Unless your location is at or near the airport the weather may turn out slightly different to the area forecast in this book. This is not a problem in an unchanging landscape but may be one in a varying terrain whereby a farmer living high on a hill or in a valley can't understand why the airport-derived data he finds in this book doesn't tally with what weather occurred on his farm, yet in the same breath he may be aware of his own small microclimates around particular mounds, hills, valleys and flats.

The almanac process may be the meteorology of the future. There is a respected place now in our society for differing viewpoints. We already have alternative music and alternative medicine. From the old man on the hill who could tell from the spitting of logs in his fire when snow was coming, to the way the sailors hauled-in sail believing (correctly) that the appearance of dolphins heralded approaching storms, there has always been room for alternative weather. Because weather is so much a part of our lives, we need all the tools, modern and ancient, and all the past observations about cycles that we can acquire.

The climate of Ireland has been the same for thousands of years. We know this by the types of natural species that have long adapted to inhabiting this country and the vegetation. We can also see, by the rivers and streams, where floods from heavy rains have carved the landscape, giving us a dossier for rainfall history. Climate is always weather from the past, and hopefully it is a guide to the future. Pollution has nothing to do with weather. If it did, bad weather would be seen over pigsties or rubbish dumps. It is recognized in weather science that it requires 30-years to establish a weather trend average, suitable to comment on the following year. It therefore would take 30 centuries of archived data to be able to comment on the weather one century from now.

GRAPHS

MAX/MIN TEMPERATURES FOR COUNTIES

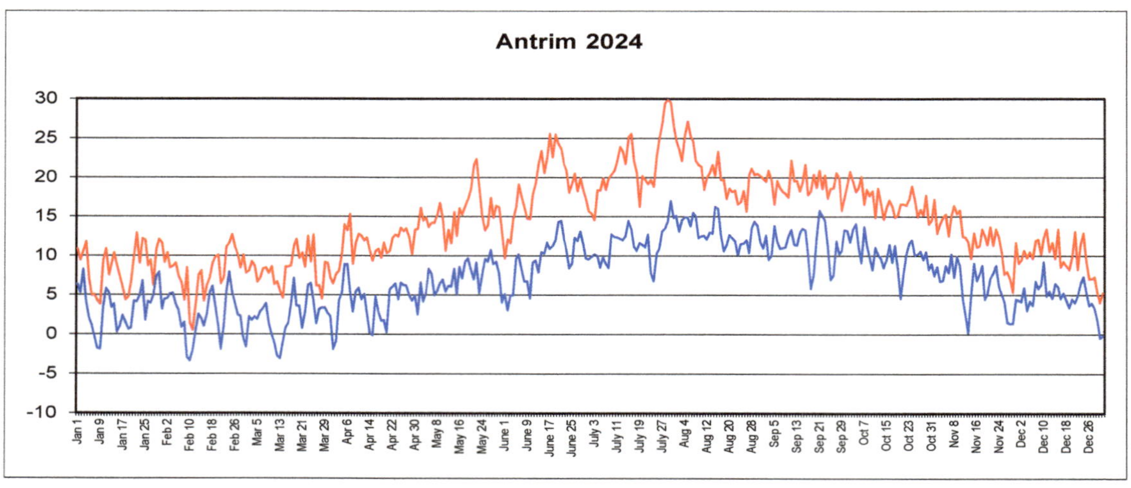

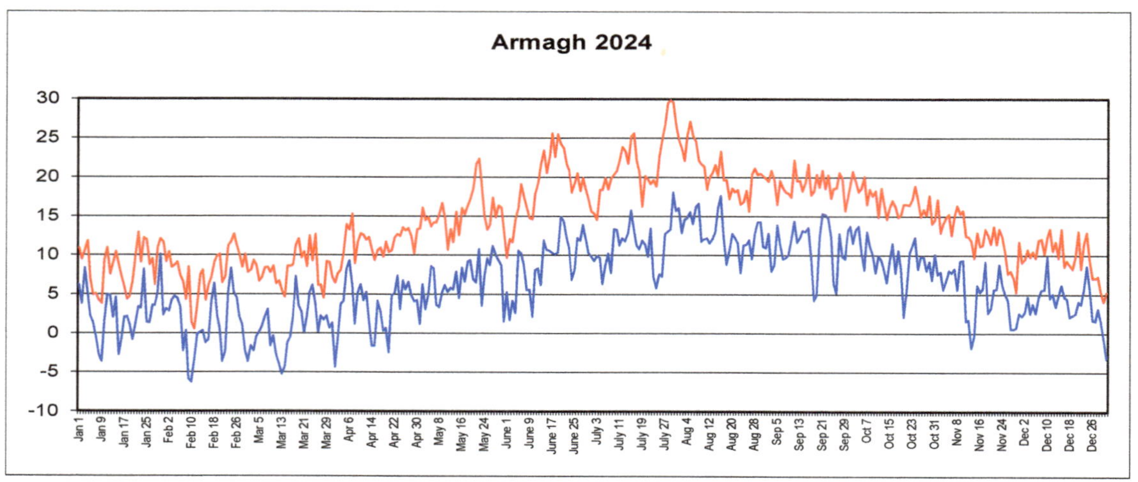

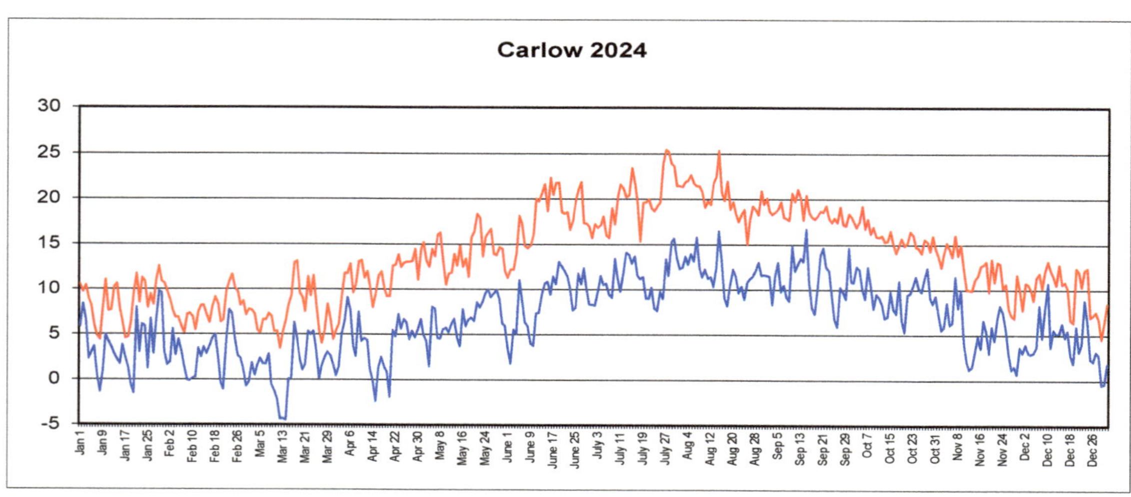

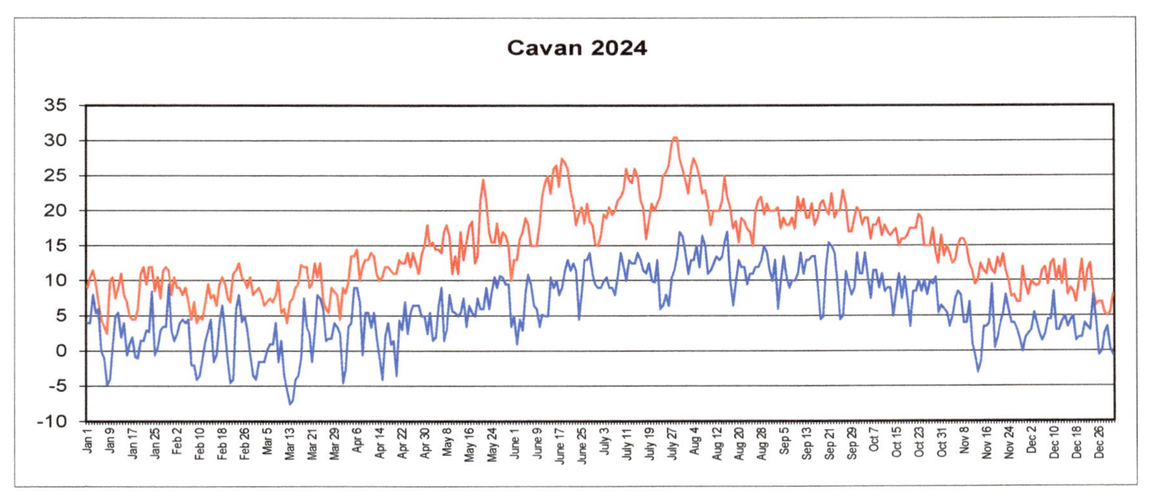
Cavan 2024
35
30
25
20
15
10
5
0
-5
-10
Jan 1
Jan 9
Jan 17
Jan 25
Feb 2
Feb 10
Feb 18
Feb 26
Mar 5
Mar 13
Mar 21
Mar 29
Apr 6
Apr 14
Apr 22
Apr 30
May 8
May 16
May 24
June 1
June 9
June 17
June 25
July 3
July 11
July 19
July 27
Aug 4
Aug 12
Aug 20
Aug 28
Sep 5
Sep 13
Sep 21
Sep 29
Oct 7
Oct 15
Oct 23
Oct 31
Nov 8
Nov 16
Nov 24
Dec 2
Dec 10
Dec 18
Dec 26

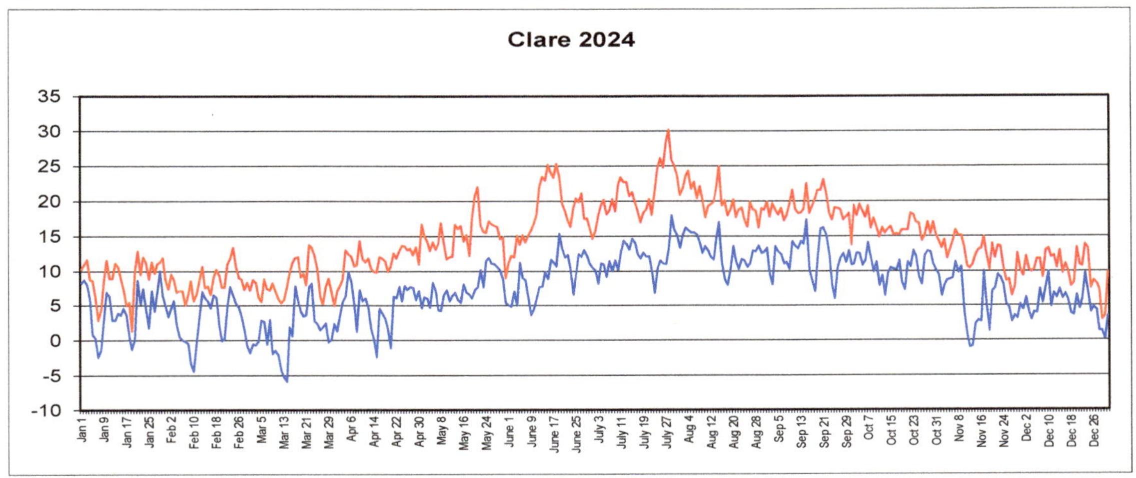
Clare 2024
35
30
25
20
15
10
5
0
-5
-10
Jan 1
Jan 9
Jan 17
Jan 25
Feb 2
Feb 10
Feb 18
Feb 26
Mar 5
Mar 13
Mar 21
Mar 29
Apr 6
Apr 14
Apr 22
Apr 30
May 8
May 16
May 24
June 1
June 9
June 17
June 25
July 3
July 11
July 19
July 27
Aug 4
Aug 12
Aug 20
Aug 28
Sep 5
Sep 13
Sep 21
Sep 29
Oct 7
Oct 15
Oct 23
Oct 31
Nov 8
Nov 16
Nov 24
Dec 2
Dec 10
Dec 18
Dec 26

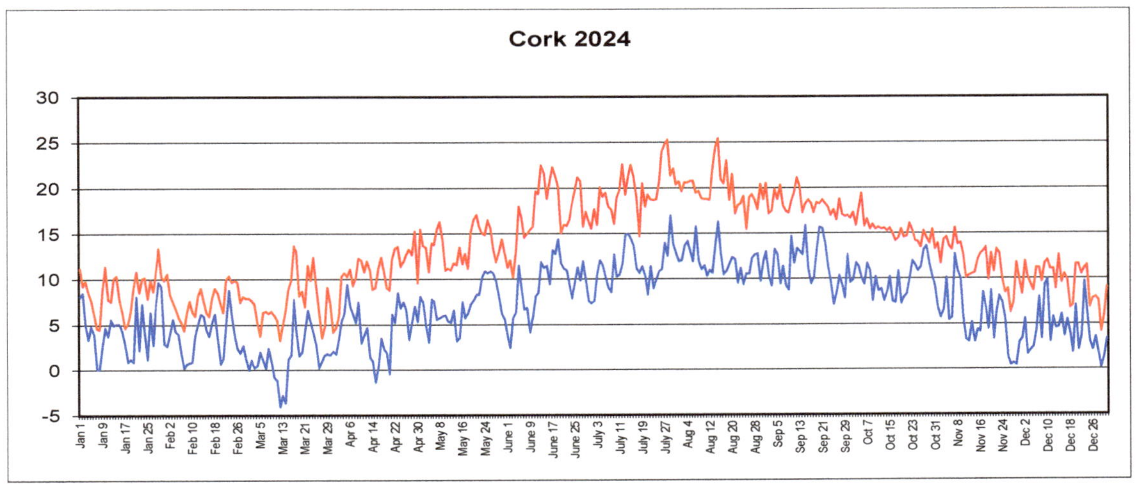
Cork 2024
30
25
20
15
10
5
0
-5
Jan 1
Jan 9
Jan 17
Jan 25
Feb 2
Feb 10
Feb 18
Feb 26
Mar 5
Mar 13
Mar 21
Mar 29
Apr 6
Apr 14
Apr 22
Apr 30
May 8
May 16
May 24
June 1
June 9
June 17
June 25
July 3
July 11
July 19
July 27
Aug 4
Aug 12
Aug 20
Aug 28
Sep 5
Sep 13
Sep 21
Sep 29
Oct 7
Oct 15
Oct 23
Oct 31
Nov 8
Nov 16
Nov 24
Dec 2
Dec 10
Dec 18
Dec 26

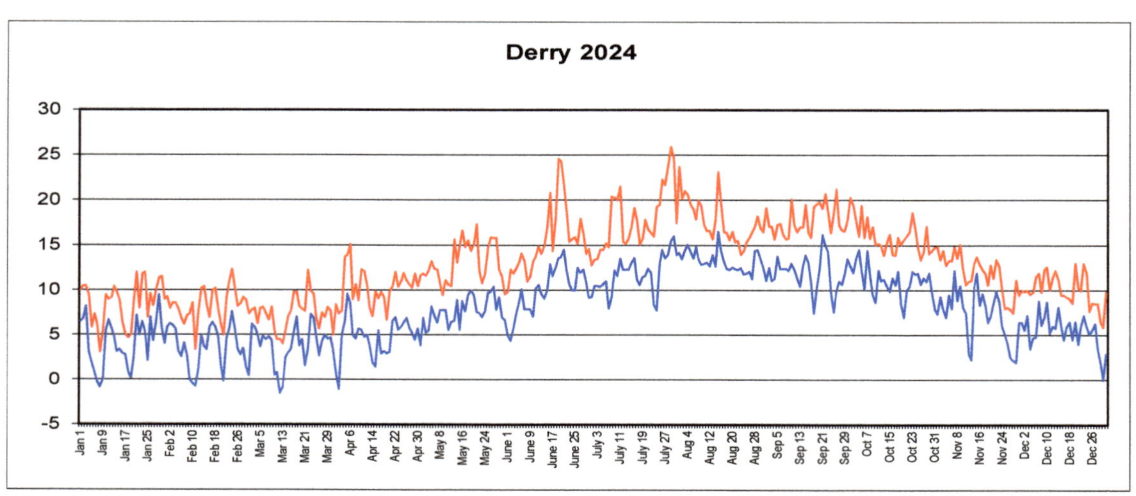
Derry 2024
30
25
20
15
10
5
0
-5
Jan 1
Jan 9
Jan 17
Jan 25
Feb 2
Feb 10
Feb 18
Feb 26
Mar 5
Mar 13
Mar 21
Mar 29
Apr 6
Apr 14
Apr 22
Apr 30
May 8
May 16
May 24
June 1
June 9
June 17
June 25
July 3
July 11
July 19
July 27
Aug 4
Aug 12
Aug 20
Aug 28
Sep 5
Sep 13
Sep 21
Sep 29
Oct 7
Oct 15
Oct 23
Oct 31
Nov 8
Nov 16
Nov 24
Dec 2
Dec 10
Dec 18
Dec 26

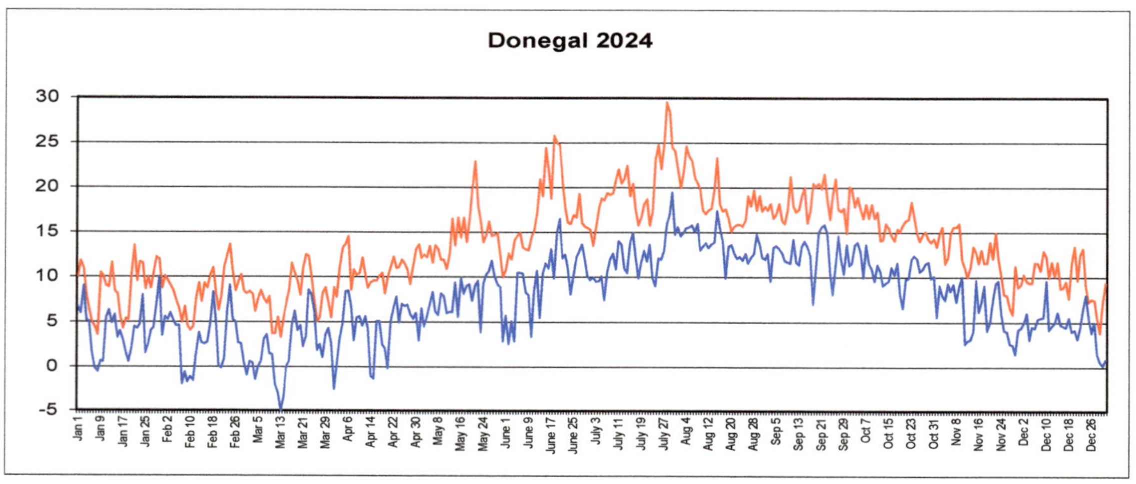
Donegal 2024
30
25
20
15
10
5
0
-5
Jan 1
Jan 9
Jan 17
Jan 25
Feb 2
Feb 10
Feb 18
Feb 26
Mar 5
Mar 13
Mar 21
Mar 29
Apr 6
Apr 14
Apr 22
Apr 30
May 8
May 16
May 24
June 1
June 9
June 17
June 25
July 3
July 11
July 19
July 27
Aug 4
Aug 12
Aug 20
Aug 28
Sep 5
Sep 13
Sep 21
Sep 29
Oct 7
Oct 15
Oct 23
Oct 31
Nov 8
Nov 16
Nov 24
Dec 2
Dec 10
Dec 18
Dec 26

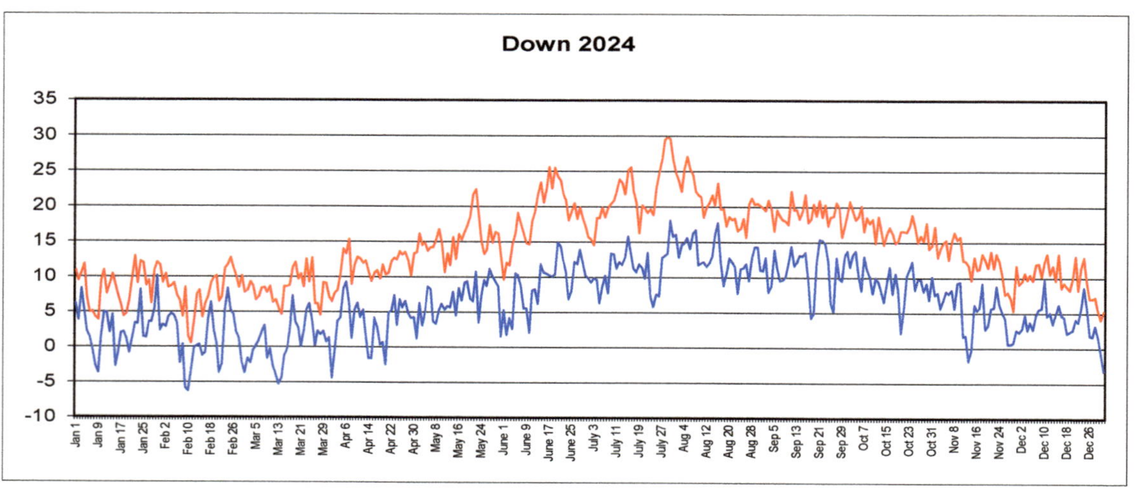
Down 2024
35
30
25
20
15
10
5
0
-5
-10
Jan 1
Jan 9
Jan 17
Jan 25
Feb 2
Feb 10
Feb 18
Feb 26
Mar 5
Mar 13
Mar 21
Mar 29
Apr 6
Apr 14
Apr 22
Apr 30
May 8
May 16
May 24
June 1
June 9
June 17
June 25
July 3
July 11
July 19
July 27
Aug 4
Aug 12
Aug 20
Aug 28
Sep 5
Sep 13
Sep 21
Sep 29
Oct 7
Oct 15
Oct 23
Oct 31
Nov 8
Nov 16
Nov 24
Dec 2
Dec 10
Dec 18
Dec 26

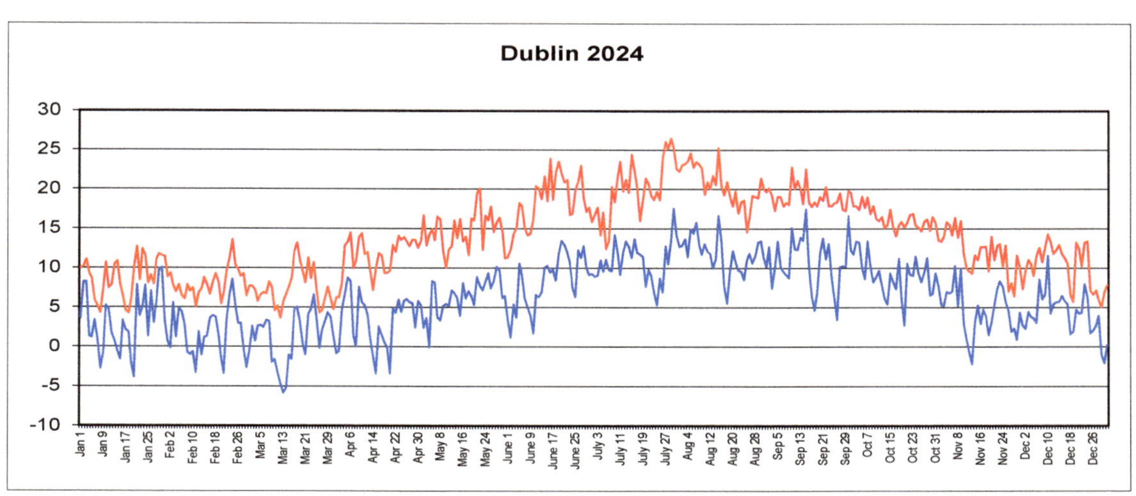
Dublin 2024
30
25
20
15
10
5
0
-5
-10
Jan 1 Jan 9 Jan 17 Jan 25 Feb 2 Feb 10 Feb 18 Feb 26 Mar 5 Mar 13 Mar 21 Mar 29 Apr 6 Apr 14 Apr 22 Apr 30 May 8 May 16 May 24 June 1 June 9 June 17 June 25 July 3 July 11 July 19 July 27 Aug 4 Aug 12 Aug 20 Aug 28 Sep 5 Sep 13 Sep 21 Sep 29 Oct 7 Oct 15 Oct 23 Oct 31 Nov 8 Nov 16 Nov 24 Dec 2 Dec 10 Dec 18 Dec 26

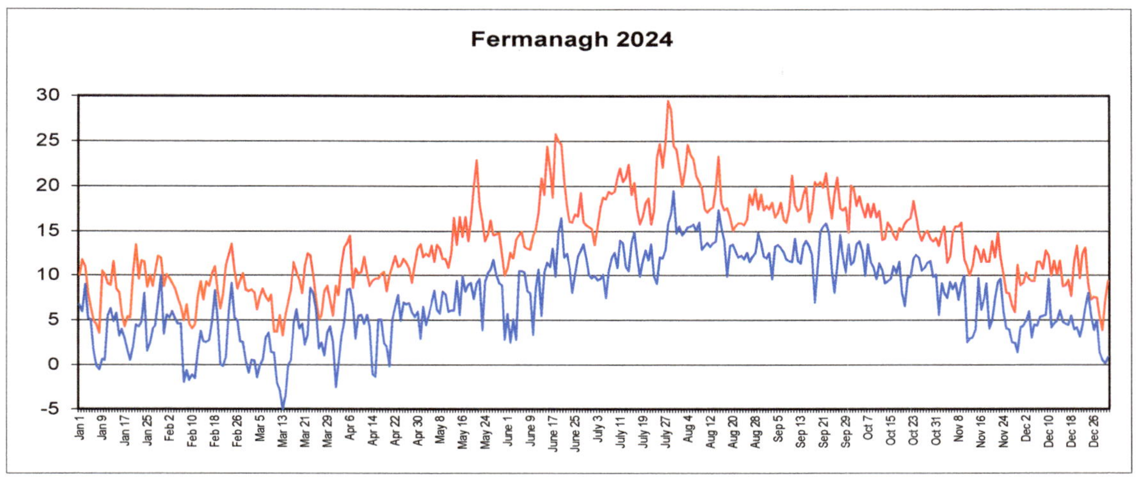
Fermanagh 2024
30
25
20
15
10
5
0
-5
Jan 1 Jan 9 Jan 17 Jan 25 Feb 2 Feb 10 Feb 18 Feb 26 Mar 5 Mar 13 Mar 21 Mar 29 Apr 6 Apr 14 Apr 22 Apr 30 May 8 May 16 May 24 June 1 June 9 June 17 June 25 July 3 July 11 July 19 July 27 Aug 4 Aug 12 Aug 20 Aug 28 Sep 5 Sep 13 Sep 21 Sep 29 Oct 7 Oct 15 Oct 23 Oct 31 Nov 8 Nov 16 Nov 24 Dec 2 Dec 10 Dec 18 Dec 26

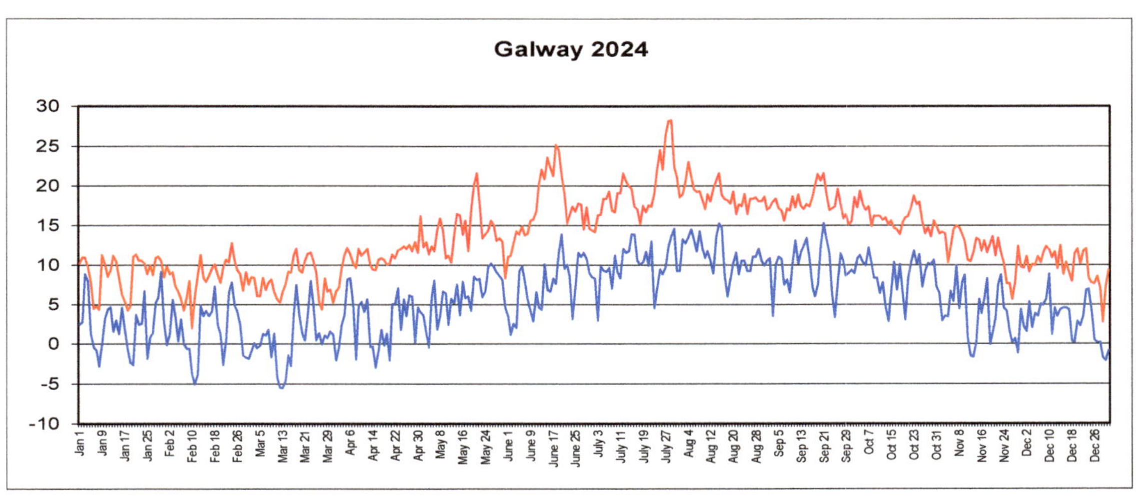
Galway 2024
30
25
20
15
10
5
0
-5
-10
Jan 1 Jan 9 Jan 17 Jan 25 Feb 2 Feb 10 Feb 18 Feb 26 Mar 5 Mar 13 Mar 21 Mar 29 Apr 6 Apr 14 Apr 22 Apr 30 May 8 May 16 May 24 June 1 June 9 June 17 June 25 July 3 July 11 July 19 July 27 Aug 4 Aug 12 Aug 20 Aug 28 Sep 5 Sep 13 Sep 21 Sep 29 Oct 7 Oct 15 Oct 23 Oct 31 Nov 8 Nov 16 Nov 24 Dec 2 Dec 10 Dec 18 Dec 26

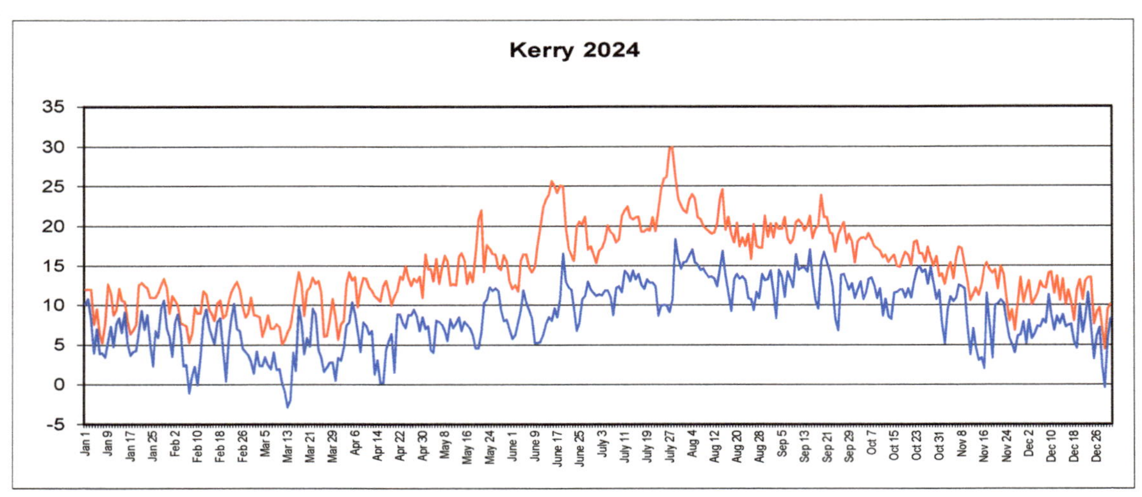
Kerry 2024
35
30
25
20
15
10
5
0
-5
Jan 1
Jan 9
Jan 17
Jan 25
Feb 2
Feb 10
Feb 18
Feb 26
Mar 5
Mar 13
Mar 21
Mar 29
Apr 6
Apr 14
Apr 22
Apr 30
May 8
May 16
May 24
June 1
June 9
June 17
June 25
July 3
July 11
July 19
July 27
Aug 4
Aug 12
Aug 20
Aug 28
Sep 5
Sep 13
Sep 21
Sep 29
Oct 7
Oct 15
Oct 23
Oct 31
Nov 8
Nov 16
Nov 24
Dec 2
Dec 10
Dec 18
Dec 26

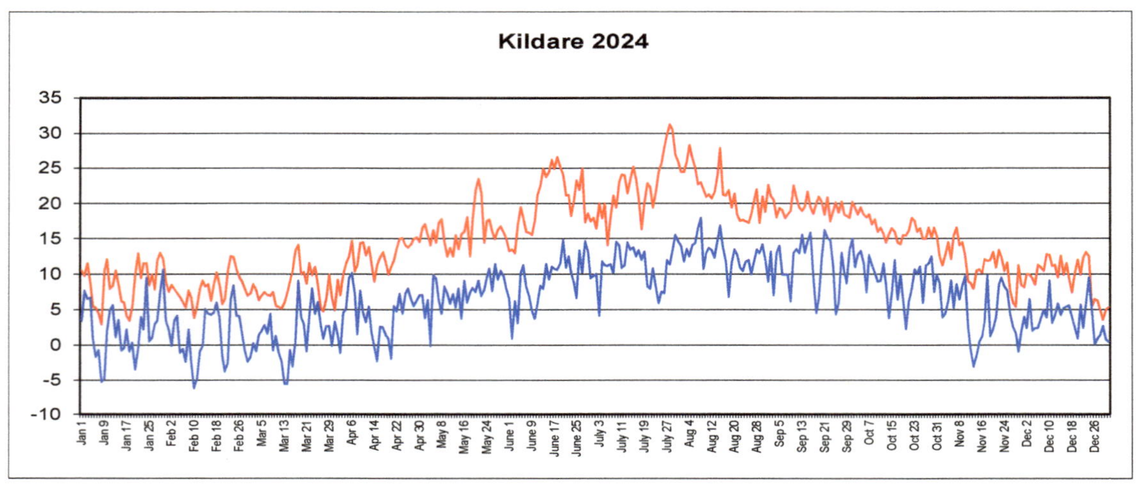
Kildare 2024
35
30
25
20
15
10
5
0
-5
-10
Jan 1
Jan 9
Jan 17
Jan 25
Feb 2
Feb 10
Feb 18
Feb 26
Mar 5
Mar 13
Mar 21
Mar 29
Apr 6
Apr 14
Apr 22
Apr 30
May 8
May 16
May 24
June 1
June 9
June 17
June 25
July 3
July 11
July 19
July 27
Aug 4
Aug 12
Aug 20
Aug 28
Sep 5
Sep 13
Sep 21
Sep 29
Oct 7
Oct 15
Oct 23
Oct 31
Nov 8
Nov 16
Nov 24
Dec 2
Dec 10
Dec 18
Dec 26

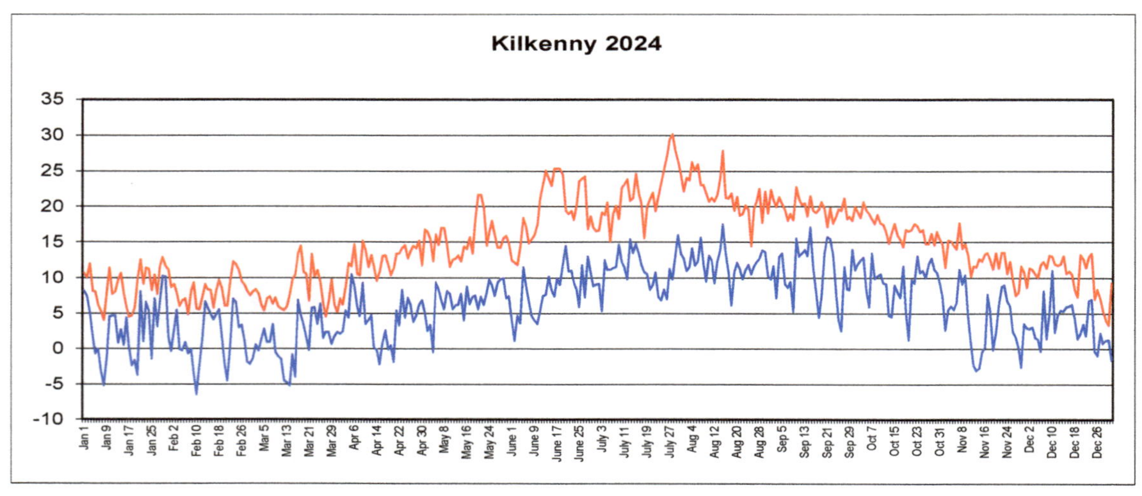
Kilkenny 2024
35
30
25
20
15
10
5
0
-5
-10
Jan 1
Jan 9
Jan 17
Jan 25
Feb 2
Feb 10
Feb 18
Feb 26
Mar 5
Mar 13
Mar 21
Mar 29
Apr 6
Apr 14
Apr 22
Apr 30
May 8
May 16
May 24
June 1
June 9
June 17
June 25
July 3
July 11
July 19
July 27
Aug 4
Aug 12
Aug 20
Aug 28
Sep 5
Sep 13
Sep 21
Sep 29
Oct 7
Oct 15
Oct 23
Oct 31
Nov 8
Nov 16
Nov 24
Dec 2
Dec 10
Dec 18
Dec 26

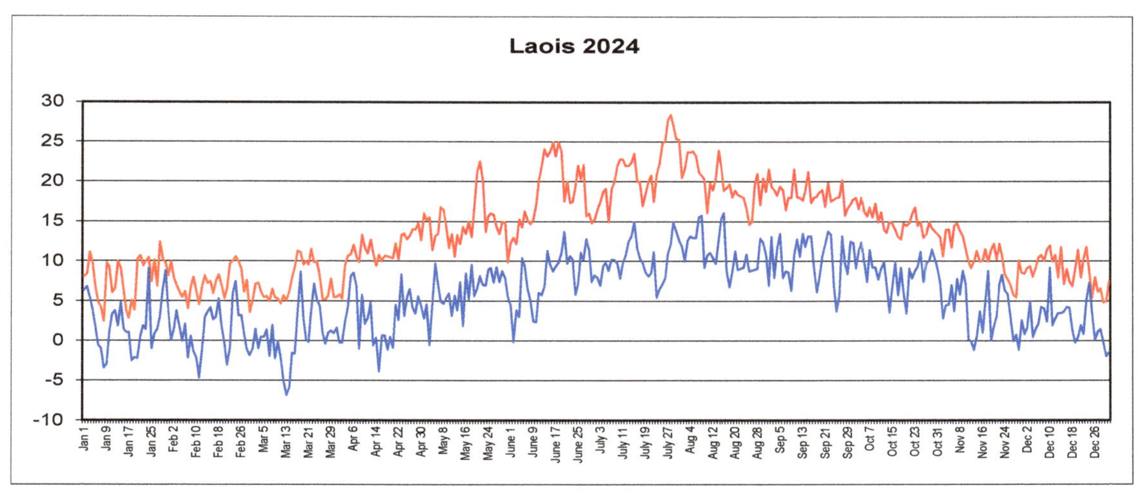
Laois 2024
30
25
20
15
10
5
0
-5
-10
Jan 1
Jan 9
Jan 17
Jan 25
Feb 2
Feb 10
Feb 18
Feb 26
Mar 5
Mar 13
Mar 21
Mar 29
Apr 6
Apr 14
Apr 22
Apr 30
May 8
May 16
May 24
June 1
June 9
June 17
June 25
July 3
July 11
July 19
July 27
Aug 4
Aug 12
Aug 20
Aug 28
Sep 5
Sep 13
Sep 21
Sep 29
Oct 7
Oct 15
Oct 23
Oct 31
Nov 8
Nov 16
Nov 24
Dec 2
Dec 10
Dec 18
Dec 26

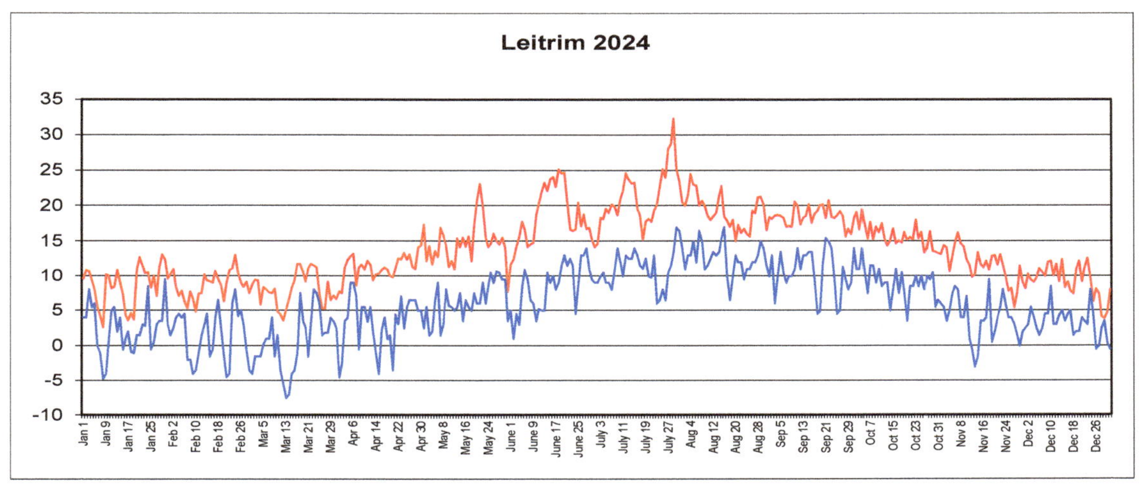
Leitrim 2024
35
30
25
20
15
10
5
0
-5
-10
Jan 1
Jan 9
Jan 17
Jan 25
Feb 2
Feb 10
Feb 18
Feb 26
Mar 5
Mar 13
Mar 21
Mar 29
Apr 6
Apr 14
Apr 22
Apr 30
May 8
May 16
May 24
June 1
June 9
June 17
June 25
July 3
July 11
July 19
July 27
Aug 4
Aug 12
Aug 20
Aug 28
Sep 5
Sep 13
Sep 21
Sep 29
Oct 7
Oct 15
Oct 23
Oct 31
Nov 8
Nov 16
Nov 24
Dec 2
Dec 10
Dec 18
Dec 26

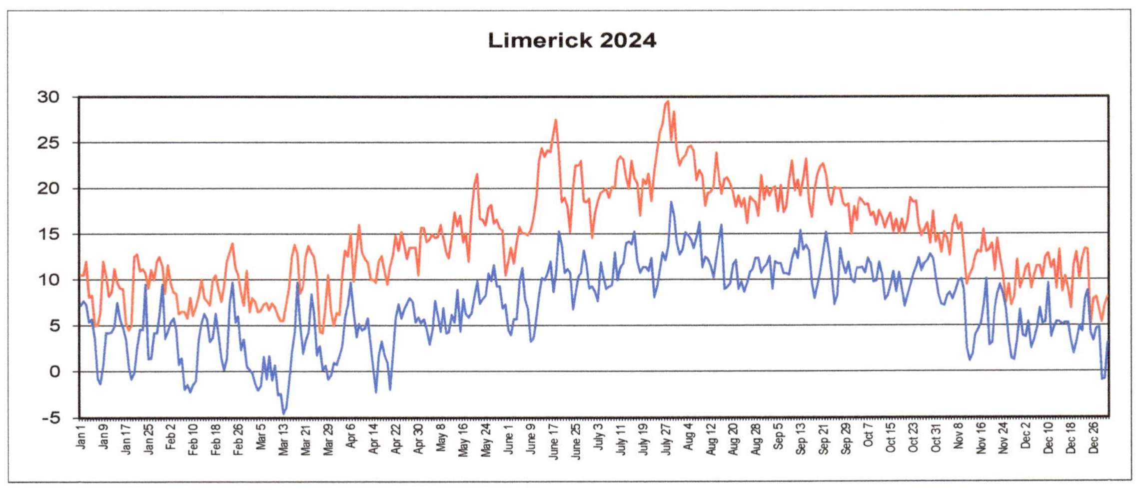
Limerick 2024
30
25
20
15
10
5
0
-5
Jan 1
Jan 9
Jan 17
Jan 25
Feb 2
Feb 10
Feb 18
Feb 26
Mar 5
Mar 13
Mar 21
Mar 29
Apr 6
Apr 14
Apr 22
Apr 30
May 8
May 16
May 24
June 1
June 9
June 17
June 25
July 3
July 11
July 19
July 27
Aug 4
Aug 12
Aug 20
Aug 28
Sep 5
Sep 13
Sep 21
Sep 29
Oct 7
Oct 15
Oct 23
Oct 31
Nov 8
Nov 16
Nov 24
Dec 2
Dec 10
Dec 18
Dec 26

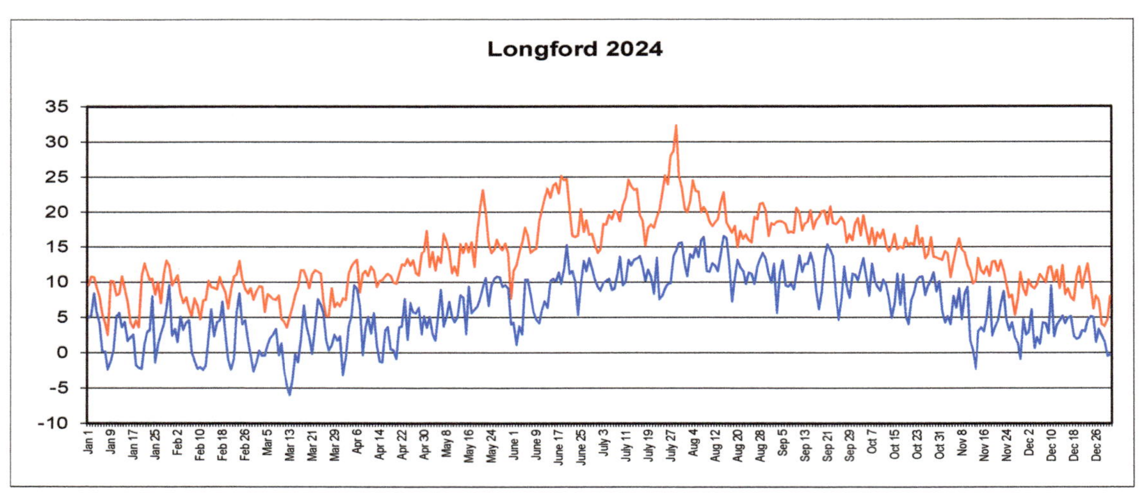
Longford 2024
35
30
25
20
15
10
5
0
-5
-10
Jan 1
Jan 9
Jan 17
Jan 25
Feb 2
Feb 10
Feb 18
Feb 26
Mar 5
Mar 13
Mar 21
Mar 29
Apr 6
Apr 14
Apr 22
Apr 30
May 8
May 16
May 24
June 1
June 9
June 17
June 25
July 3
July 11
July 19
July 27
Aug 4
Aug 12
Aug 20
Aug 28
Sep 5
Sep 13
Sep 21
Sep 29
Oct 7
Oct 15
Oct 23
Oct 31
Nov 8
Nov 16
Nov 24
Dec 2
Dec 10
Dec 18
Dec 26

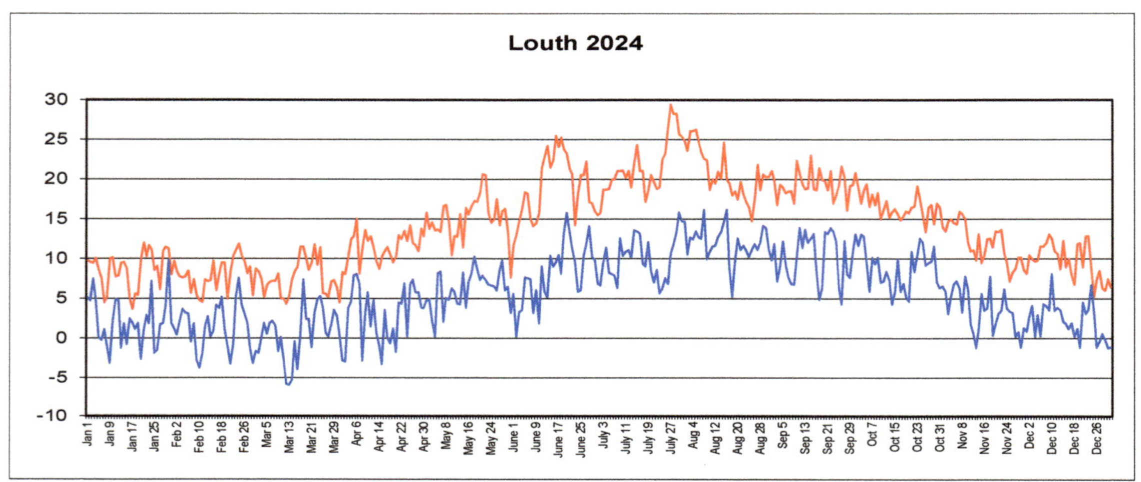
Louth 2024
30
25
20
15
10
5
0
-5
-10

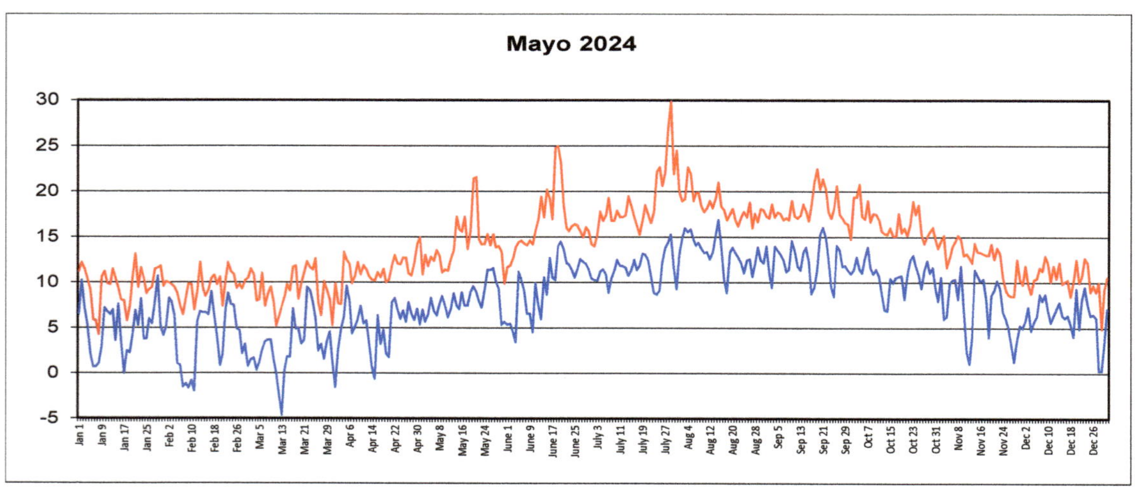
Mayo 2024
30
25
20
15
10
5
0
-5

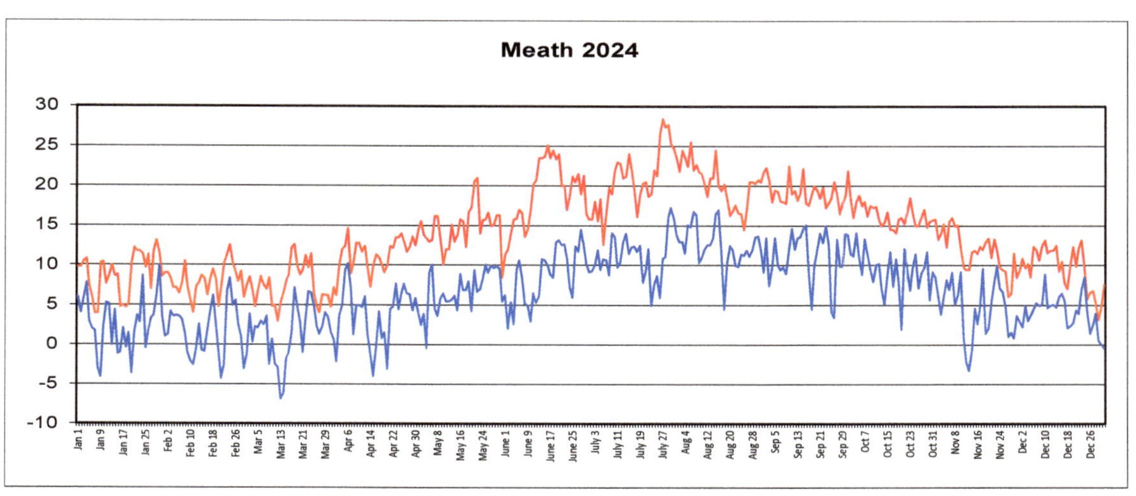

Meath 2024
30
25
20
15
10
5
0
-5
-10
Jan 1
Jan 9
Jan 17
Jan 25
Feb 2
Feb 10
Feb 18
Feb 26
Mar 5
Mar 13
Mar 21
Mar 29
Apr 6
Apr 14
Apr 22
Apr 30
May 8
May 16
May 24
June 1
June 9
June 17
June 25
July 3
July 11
July 19
July 27
Aug 4
Aug 12
Aug 20
Aug 28
Sep 5
Sep 13
Sep 21
Sep 29
Oct 7
Oct 15
Oct 23
Oct 31
Nov 8
Nov 16
Nov 24
Dec 2
Dec 10
Dec 18
Dec 26

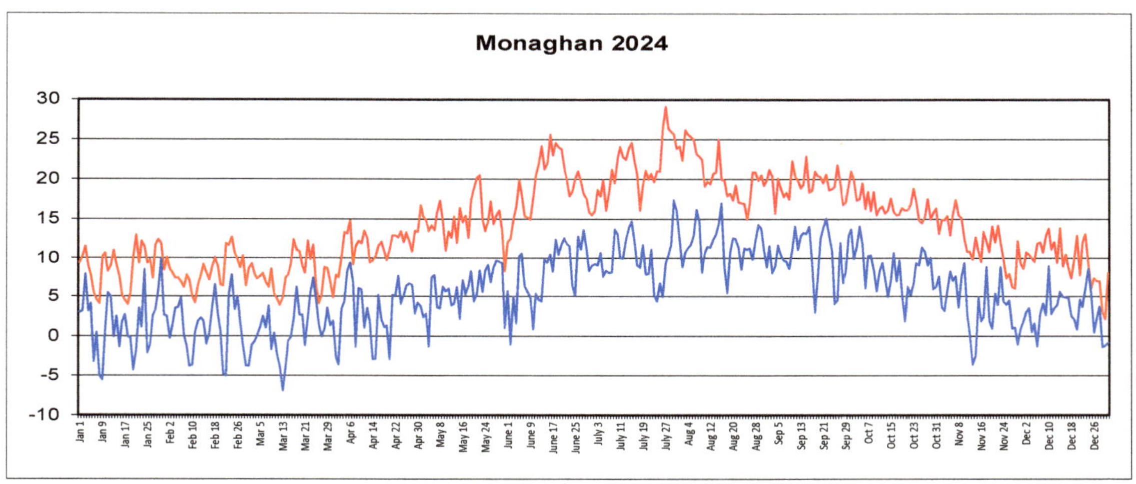

Monaghan 2024
30
25
20
15
10
5
0
-5
-10
Jan 1
Jan 9
Jan 17
Jan 25
Feb 2
Feb 10
Feb 18
Feb 26
Mar 5
Mar 13
Mar 21
Mar 29
Apr 6
Apr 14
Apr 22
Apr 30
May 8
May 16
May 24
June 1
June 9
June 17
June 25
July 3
July 11
July 19
July 27
Aug 4
Aug 12
Aug 20
Aug 28
Sep 5
Sep 13
Sep 21
Sep 29
Oct 7
Oct 15
Oct 23
Oct 31
Nov 8
Nov 16
Nov 24
Dec 2
Dec 10
Dec 18
Dec 26

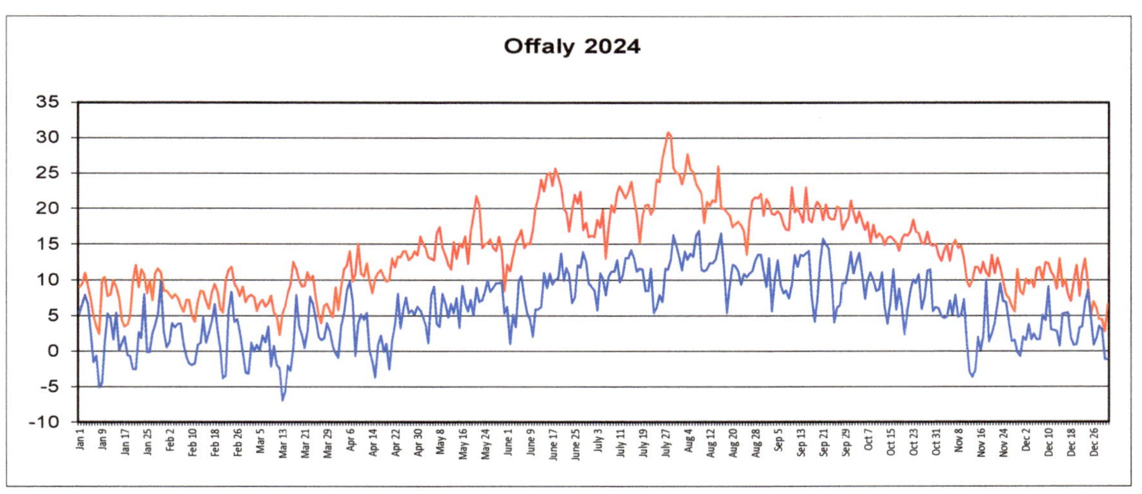

Offaly 2024
35
30
25
20
15
10
5
0
-5
-10
Jan 1
Jan 9
Jan 17
Jan 25
Feb 2
Feb 10
Feb 18
Feb 26
Mar 5
Mar 13
Mar 21
Mar 29
Apr 6
Apr 14
Apr 22
Apr 30
May 8
May 16
May 24
June 1
June 9
June 17
June 25
July 3
July 11
July 19
July 27
Aug 4
Aug 12
Aug 20
Aug 28
Sep 5
Sep 13
Sep 21
Sep 29
Oct 7
Oct 15
Oct 23
Oct 31
Nov 8
Nov 16
Nov 24
Dec 2
Dec 10
Dec 18
Dec 26

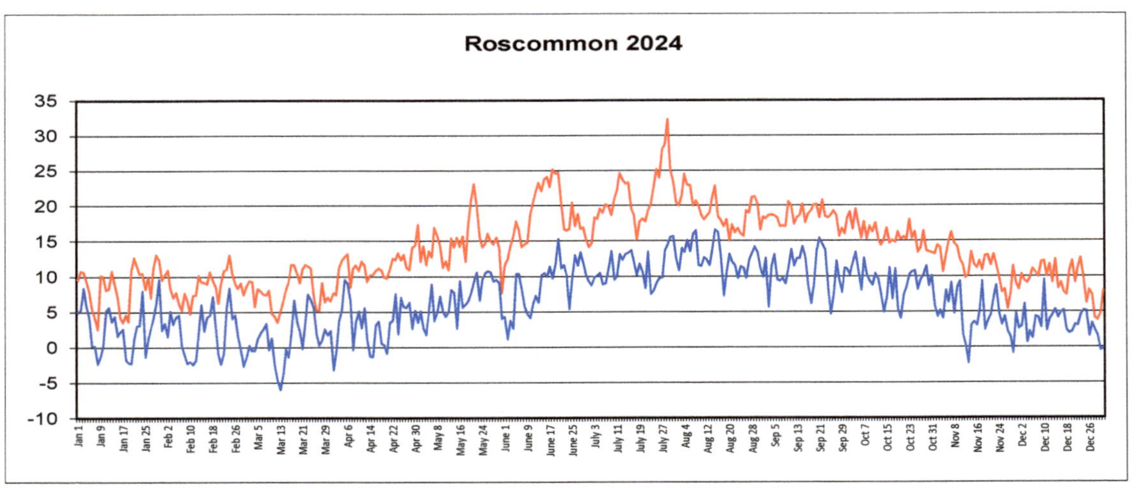
Roscommon 2024
35
30
25
20
15
10
5
0
-5
-10
Jan 1
Jan 9
Jan 17
Jan 25
Feb 2
Feb 10
Feb 18
Feb 26
Mar 5
Mar 13
Mar 21
Mar 29
Apr 6
Apr 14
Apr 22
Apr 30
May 8
May 16
May 24
June 1
June 9
June 17
June 25
July 3
July 11
July 19
July 27
Aug 4
Aug 12
Aug 20
Aug 28
Sep 5
Sep 13
Sep 21
Sep 29
Oct 7
Oct 15
Oct 23
Oct 31
Nov 8
Nov 16
Nov 24
Dec 2
Dec 10
Dec 18
Dec 26

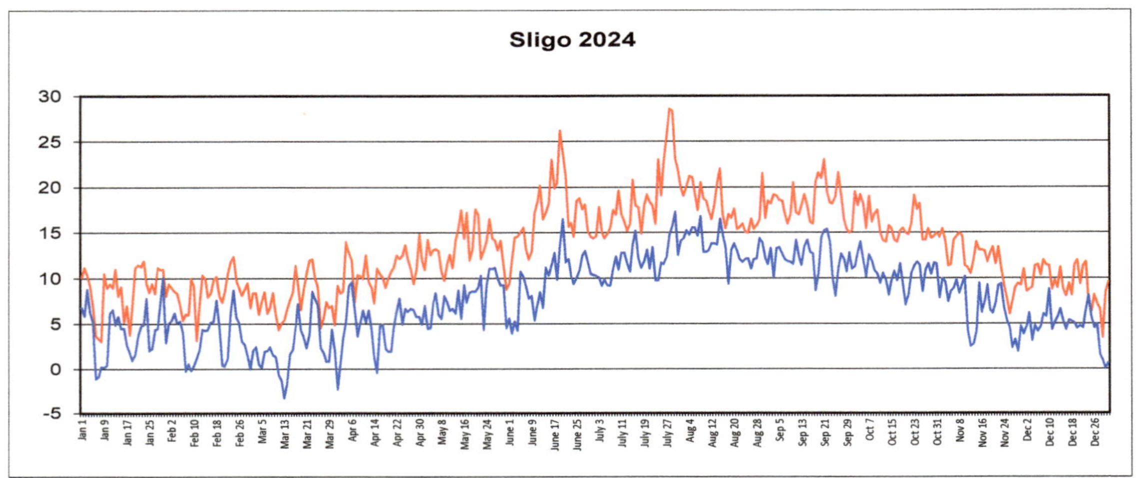
Sligo 2024
30
25
20
15
10
5
0
-5
Jan 1
Jan 9
Jan 17
Jan 25
Feb 2
Feb 10
Feb 18
Feb 26
Mar 5
Mar 13
Mar 21
Mar 29
Apr 6
Apr 14
Apr 22
Apr 30
May 8
May 16
May 24
June 1
June 9
June 17
June 25
July 3
July 11
July 19
July 27
Aug 4
Aug 12
Aug 20
Aug 28
Sep 5
Sep 13
Sep 21
Sep 29
Oct 7
Oct 15
Oct 23
Oct 31
Nov 8
Nov 16
Nov 24
Dec 2
Dec 10
Dec 18
Dec 26

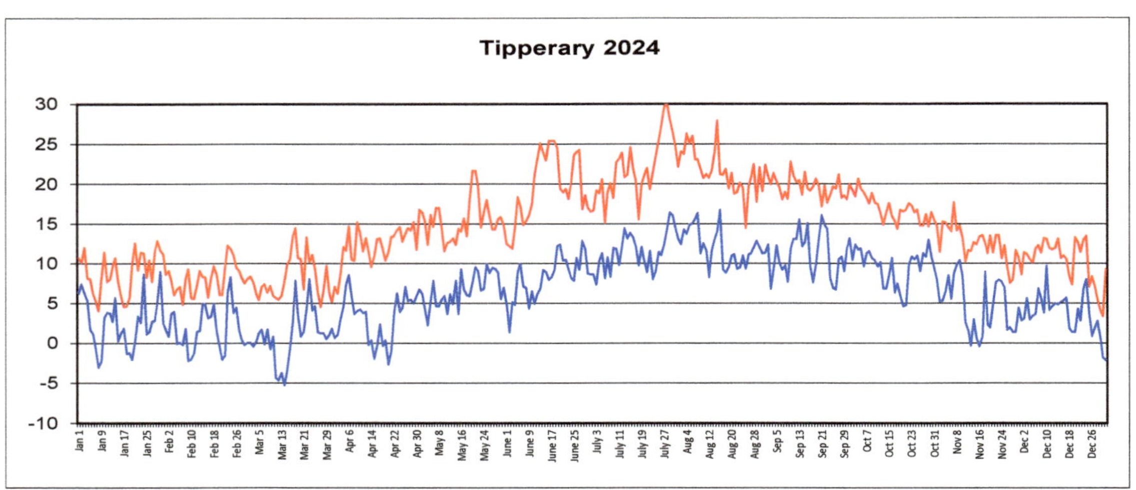
Tipperary 2024
30
25
20
15
10
5
0
-5
-10
Jan 1
Jan 9
Jan 17
Jan 25
Feb 2
Feb 10
Feb 18
Feb 26
Mar 5
Mar 13
Mar 21
Mar 29
Apr 6
Apr 14
Apr 22
Apr 30
May 8
May 16
May 24
June 1
June 9
June 17
June 25
July 3
July 11
July 19
July 27
Aug 4
Aug 12
Aug 20
Aug 28
Sep 5
Sep 13
Sep 21
Sep 29
Oct 7
Oct 15
Oct 23
Oct 31
Nov 8
Nov 16
Nov 24
Dec 2
Dec 10
Dec 18
Dec 26

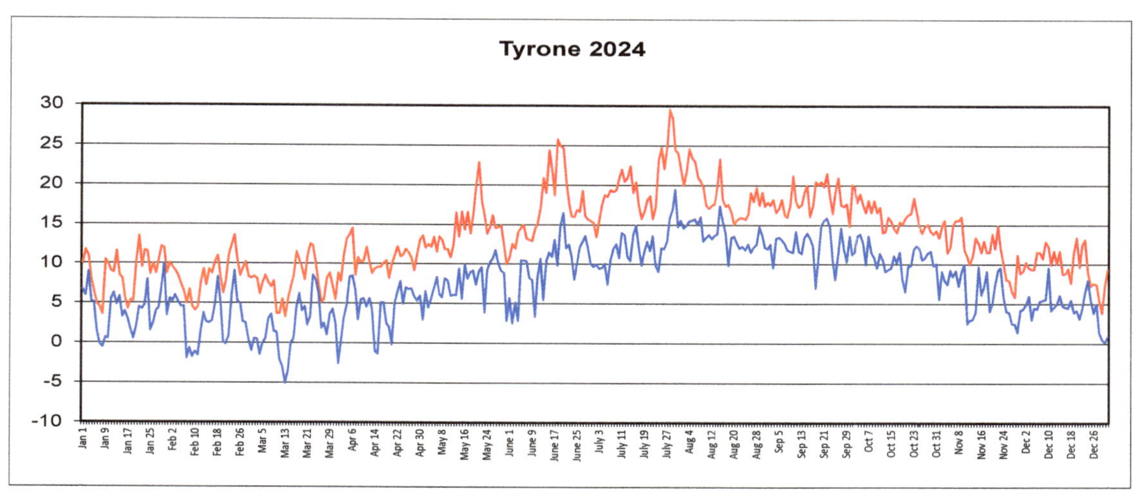
Tyrone 2024
30
25
20
15
10
5
0
-5
-10
Jan 1
Jan 9
Jan 17
Jan 25
Feb 2
Feb 10
Feb 18
Feb 26
Mar 5
Mar 13
Mar 21
Mar 29
Apr 6
Apr 14
Apr 22
Apr 30
May 8
May 16
May 24
June 1
June 9
June 17
June 25
July 3
July 11
July 19
July 27
Aug 4
Aug 12
Aug 20
Aug 28
Sep 5
Sep 13
Sep 21
Sep 29
Oct 7
Oct 15
Oct 23
Oct 31
Nov 8
Nov 16
Nov 24
Dec 2
Dec 10
Dec 18
Dec 26

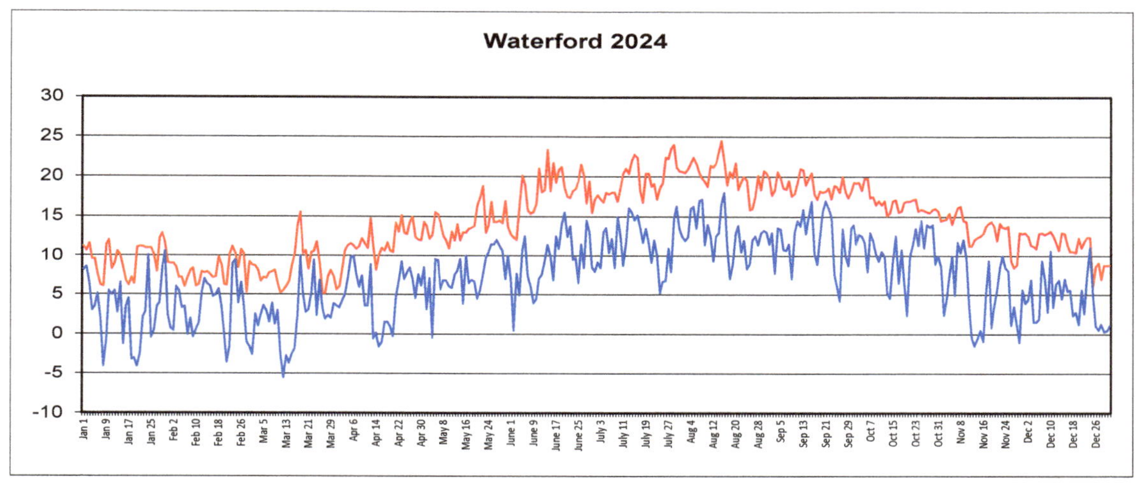
Waterford 2024
30
25
20
15
10
5
0
-5
-10
Jan 1
Jan 9
Jan 17
Jan 25
Feb 2
Feb 10
Feb 18
Feb 26
Mar 5
Mar 13
Mar 21
Mar 29
Apr 6
Apr 14
Apr 22
Apr 30
May 8
May 16
May 24
June 1
June 9
June 17
June 25
July 3
July 11
July 19
July 27
Aug 4
Aug 12
Aug 20
Aug 28
Sep 5
Sep 13
Sep 21
Sep 29
Oct 7
Oct 15
Oct 23
Oct 31
Nov 8
Nov 16
Nov 24
Dec 2
Dec 10
Dec 18
Dec 26

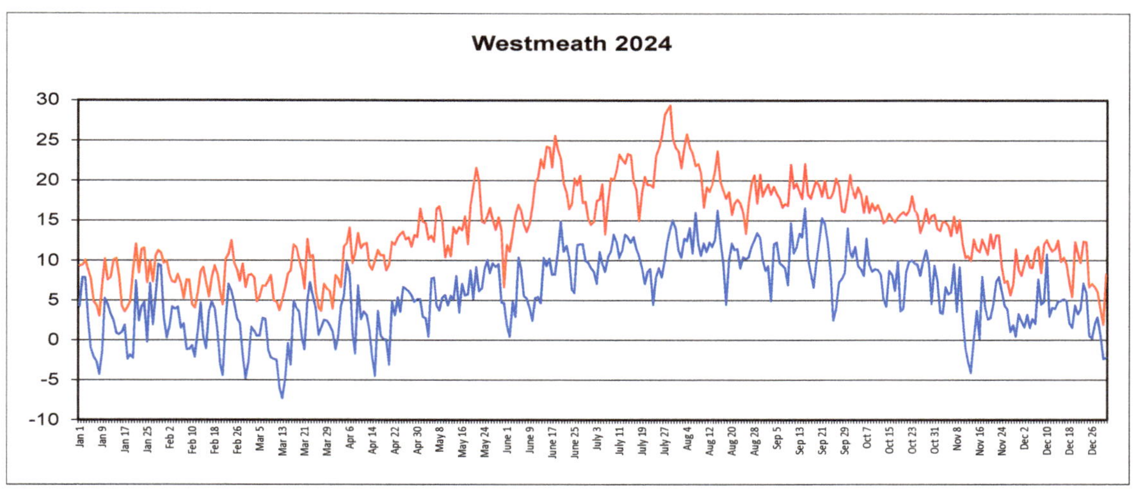
Westmeath 2024
30
25
20
15
10
5
0
-5
-10
Jan 1
Jan 9
Jan 17
Jan 25
Feb 2
Feb 10
Feb 18
Feb 26
Mar 5
Mar 13
Mar 21
Mar 29
Apr 6
Apr 14
Apr 22
Apr 30
May 8
May 16
May 24
June 1
June 9
June 17
June 25
July 3
July 11
July 19
July 27
Aug 4
Aug 12
Aug 20
Aug 28
Sep 5
Sep 13
Sep 21
Sep 29
Oct 7
Oct 15
Oct 23
Oct 31
Nov 8
Nov 16
Nov 24
Dec 2
Dec 10
Dec 18
Dec 26

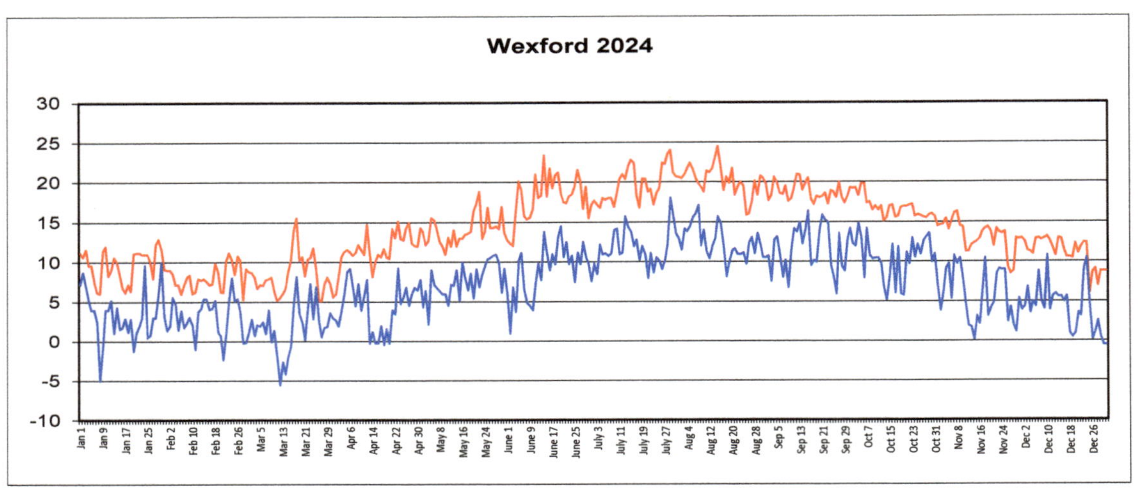

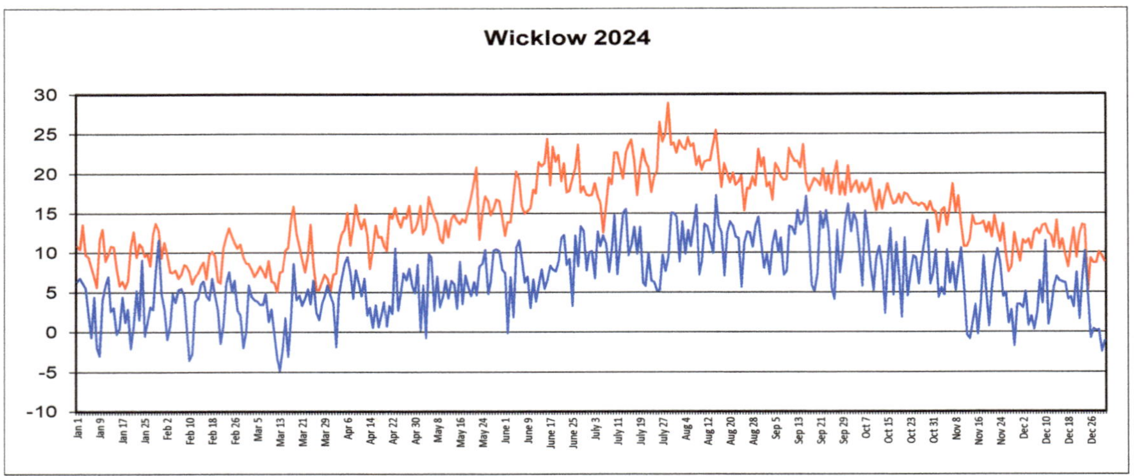

TEMPERATURE TRENDS

AVERAGED ACROSS THE WHOLE COUNTRY

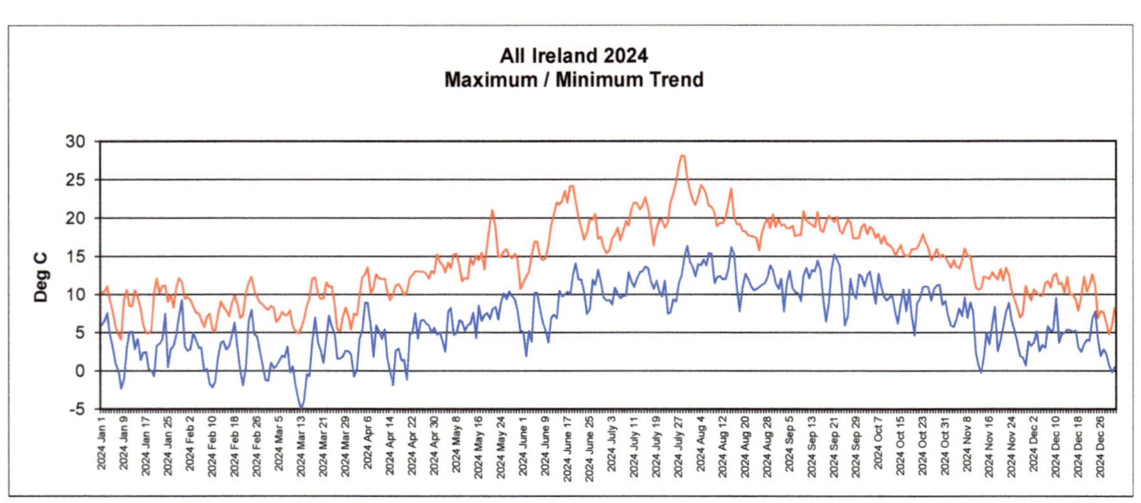

WIND

The graph below shows estimated daily wind force trends, for 2024, as calculated for Knock Airport, County Mayo. This western county has been chosen because of the availability of continuous historical data, and because most wind systems in Ireland come from the west. Other areas may show corresponding trends even if dissimilar force values.

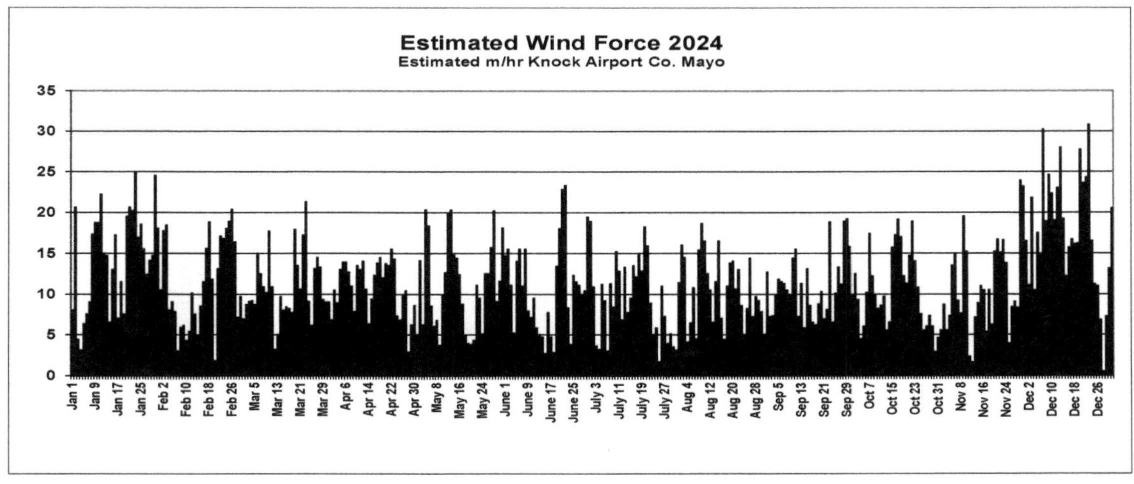

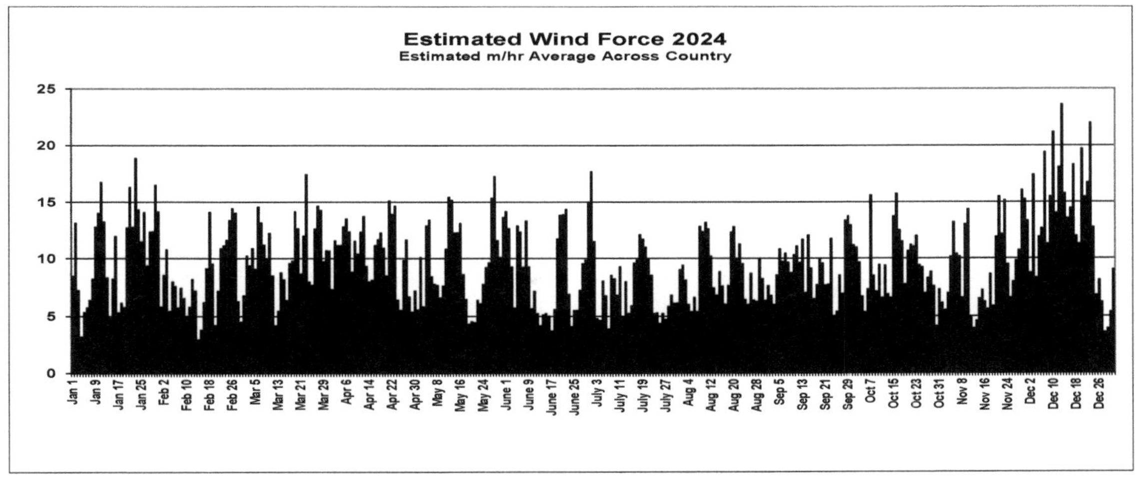

Averaging wind force values across the months reveals that December may be the windiest followed by January, March, and April, and June, August, February and July the calmest. The order of overall average windiest months is estimated to be December, January, March, April, May, October, November, September, June, August, February and July.

This makes the average windiest season, in order of windiest, spring, winter, autumn, then summer.

PRECIPITATION (RAIN/SNOW)

WHEN WILL RAIN AND LOW TEMPERATURES PRODUCE SNOW?

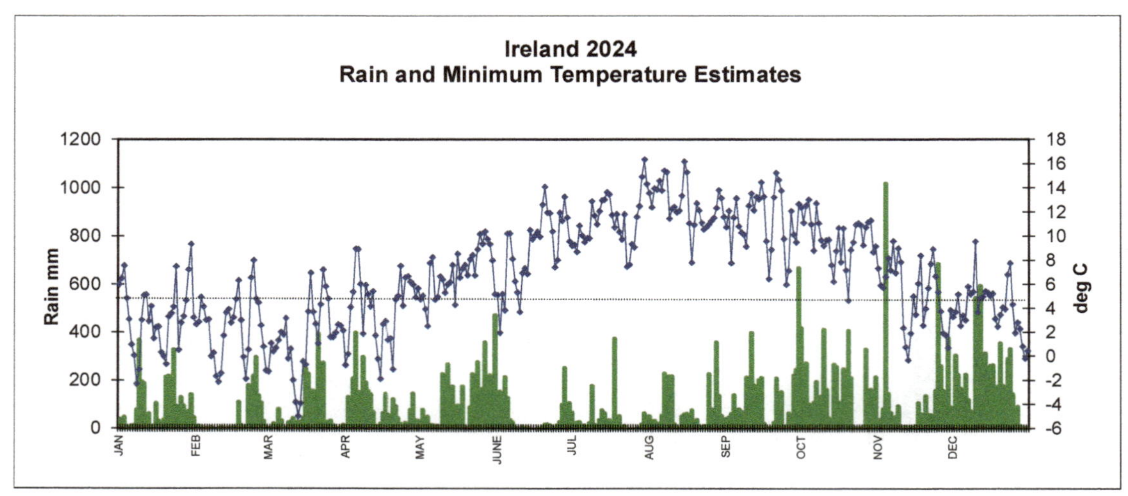

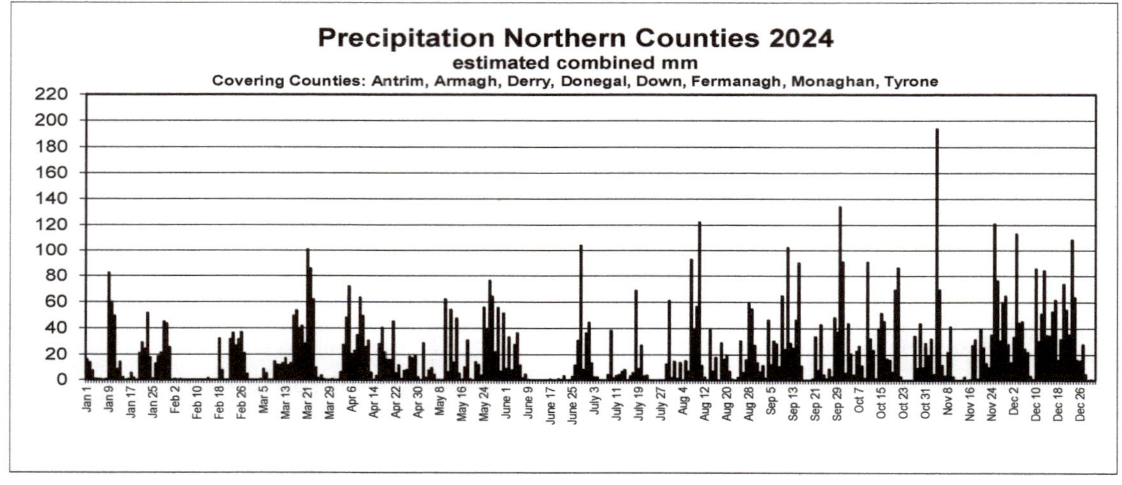

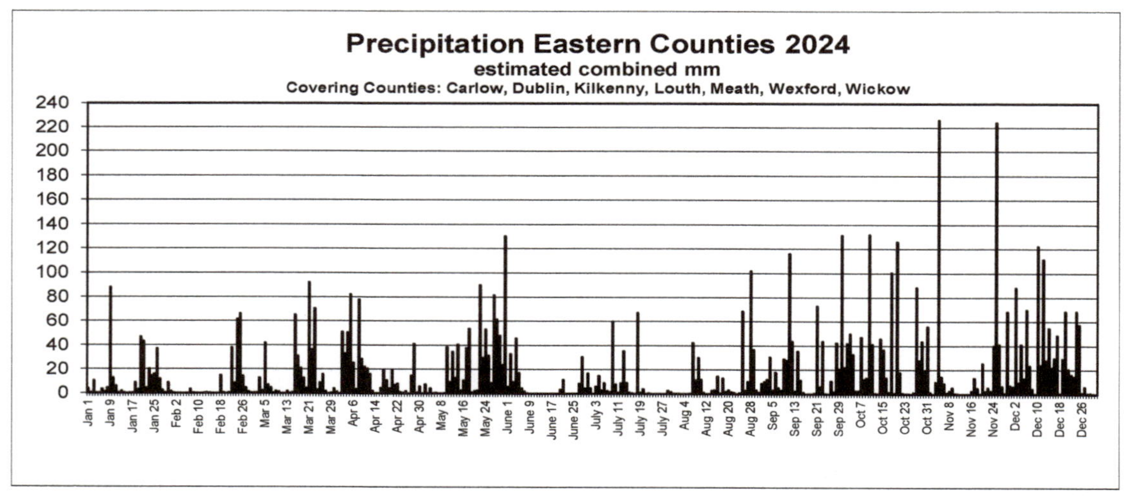

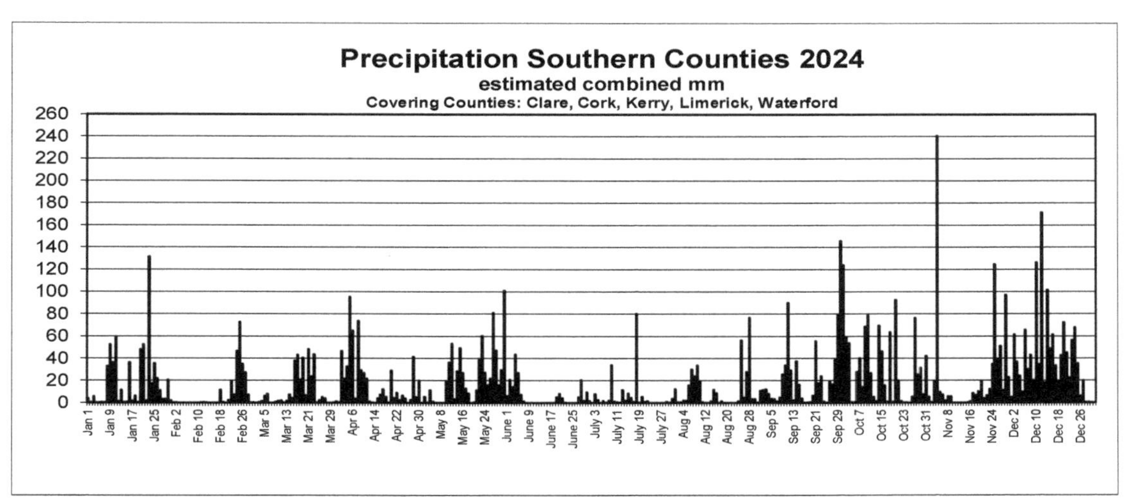
Precipitation Southern Counties 2024
estimated combined mm
Covering Counties: Clare, Cork, Kerry, Limerick, Waterford

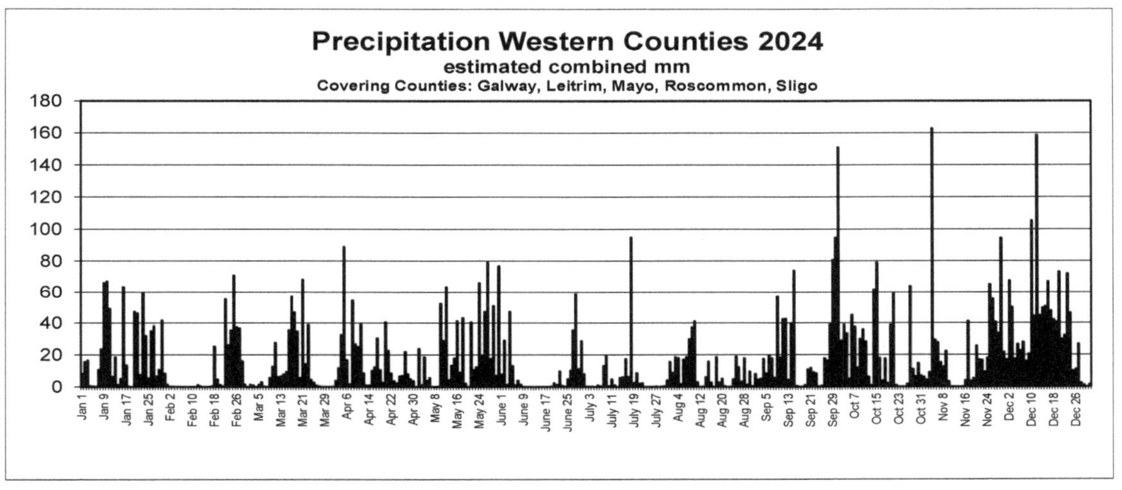
Precipitation Western Counties 2024
estimated combined mm
Covering Counties: Galway, Leitrim, Mayo, Roscommon, Sligo

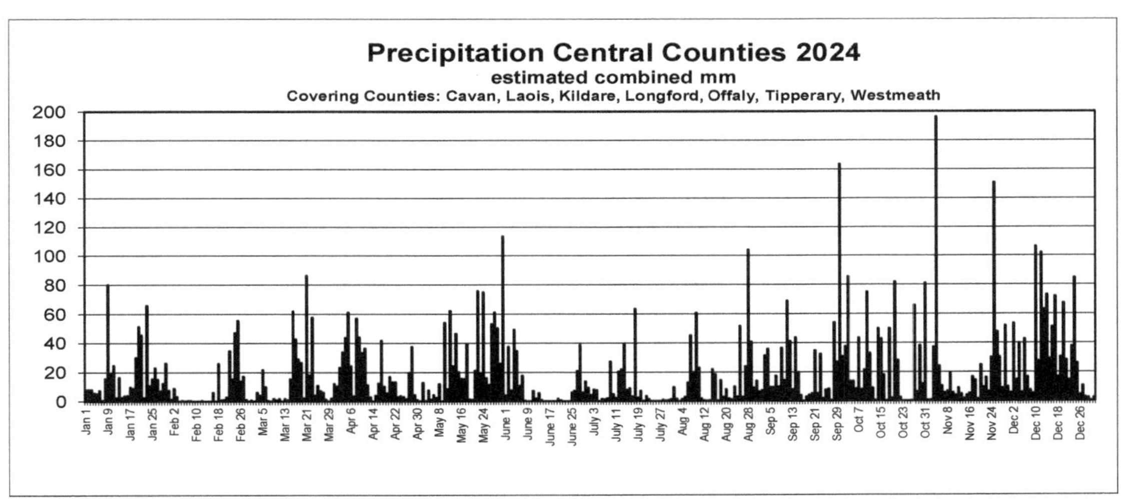
Precipitation Central Counties 2024
estimated combined mm
Covering Counties: Cavan, Laois, Kildare, Longford, Offaly, Tipperary, Westmeath

RAIN TRENDS 2024 -2025

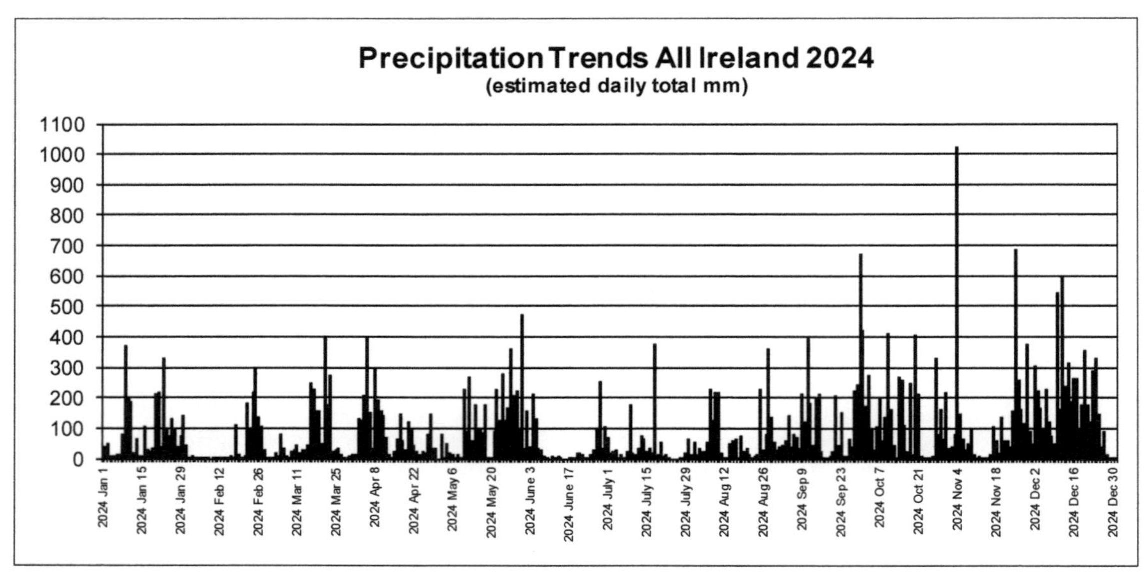

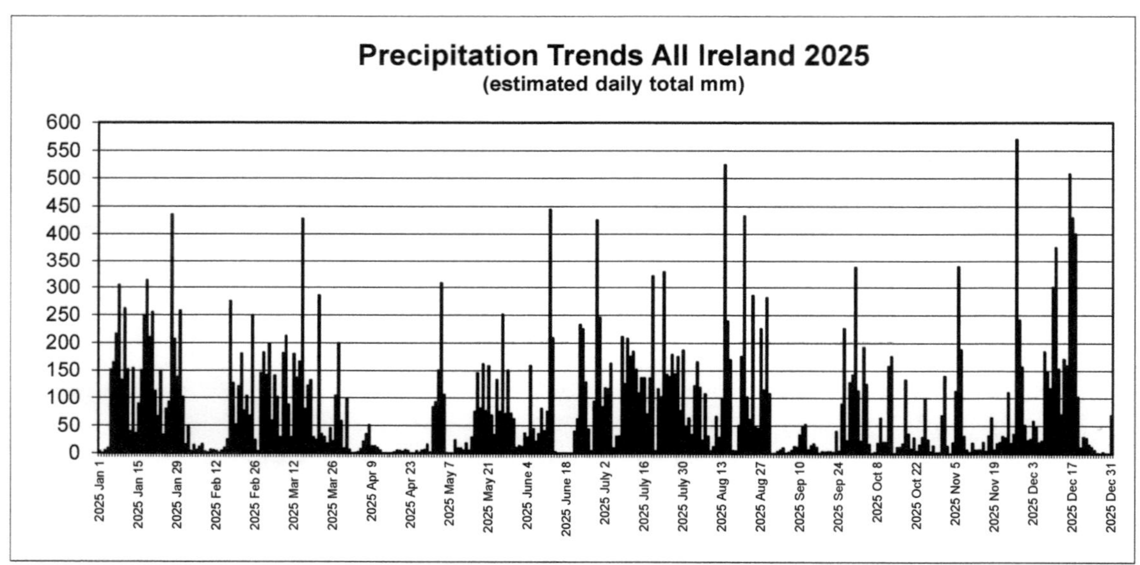

WHEN WILL THE SUN BE OUT?
AVERAGED ACROSS THE WHOLE COUNTRY.

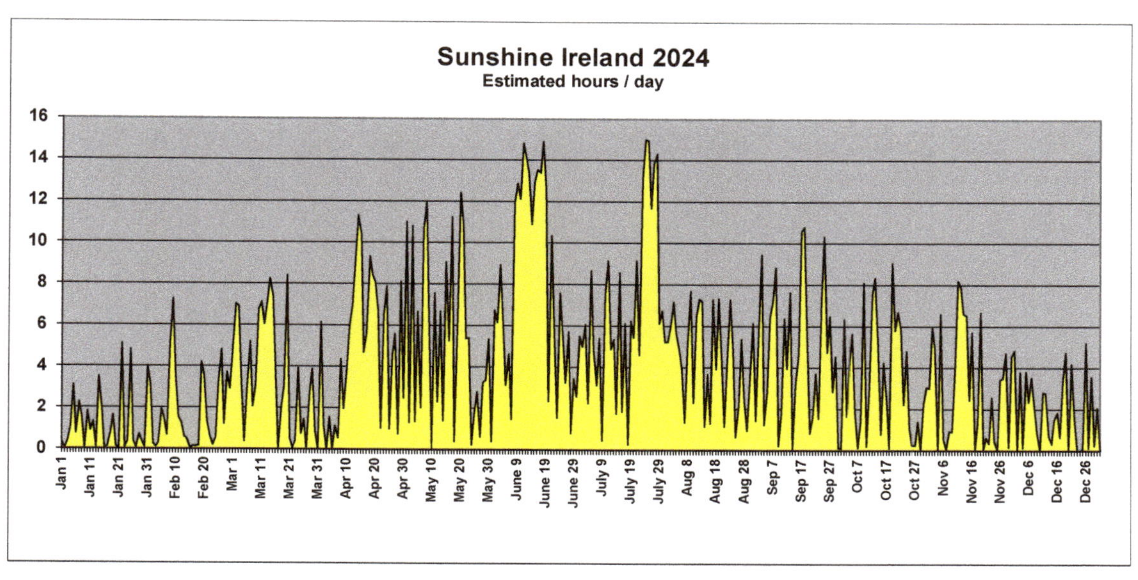

OTHER GRAPHS, NOTES, AND TRENDS

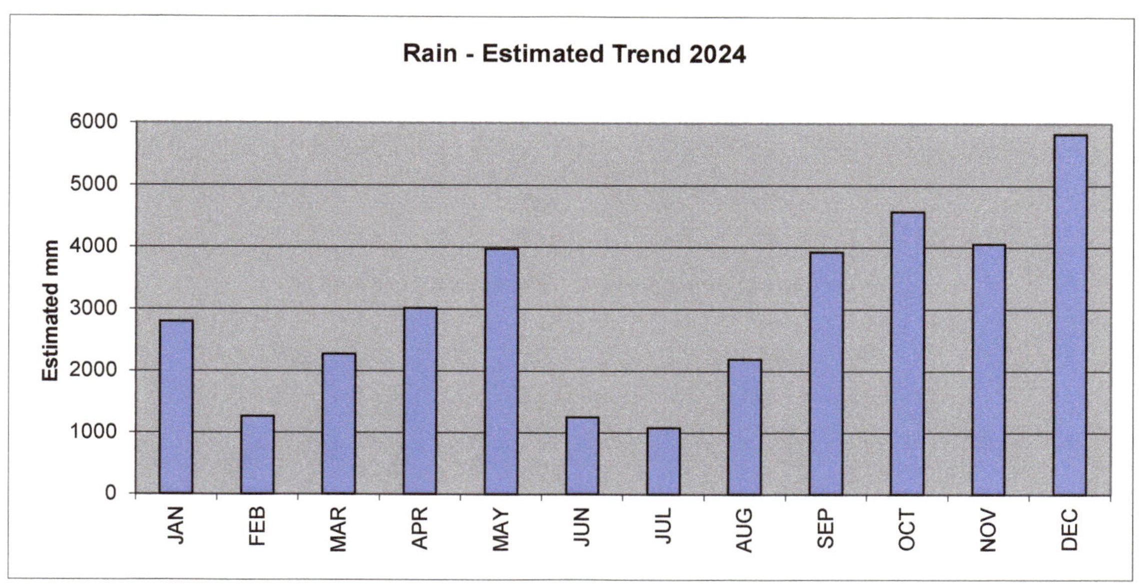

The **most precipitation** for the whole country in 2024 may be **December followed by October and November and February, June, and July** likely to be the driest. The largest amount of precipitation in one region/month may be in December in the western and southern counties. The driest period for the year may be in July in the northern and eastern counties.

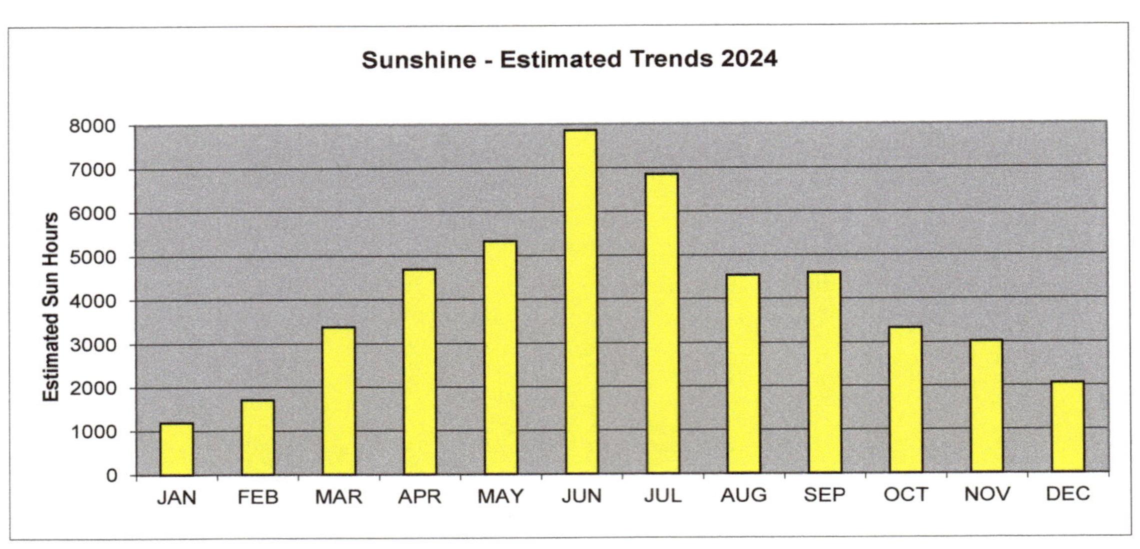
Sunshine - Estimated Trends 2024
Estimated Sun Hours
8000
7000
6000
5000
4000
3000
2000
1000
0
JAN
FEB
MAR
APR
MAY
JUN
JUL
AUG
SEP
OCT
NOV
DEC

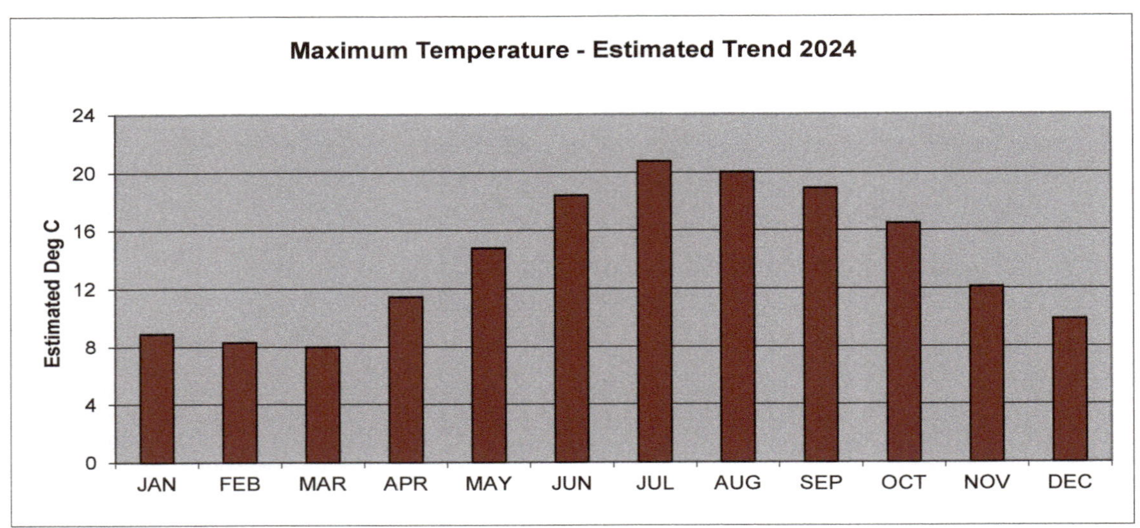
Maximum Temperature - Estimated Trend 2024
Estimated Deg C
24
20
16
12
8
4
0
JAN
FEB
MAR
APR
MAY
JUN
JUL
AUG
SEP
OCT
NOV
DEC

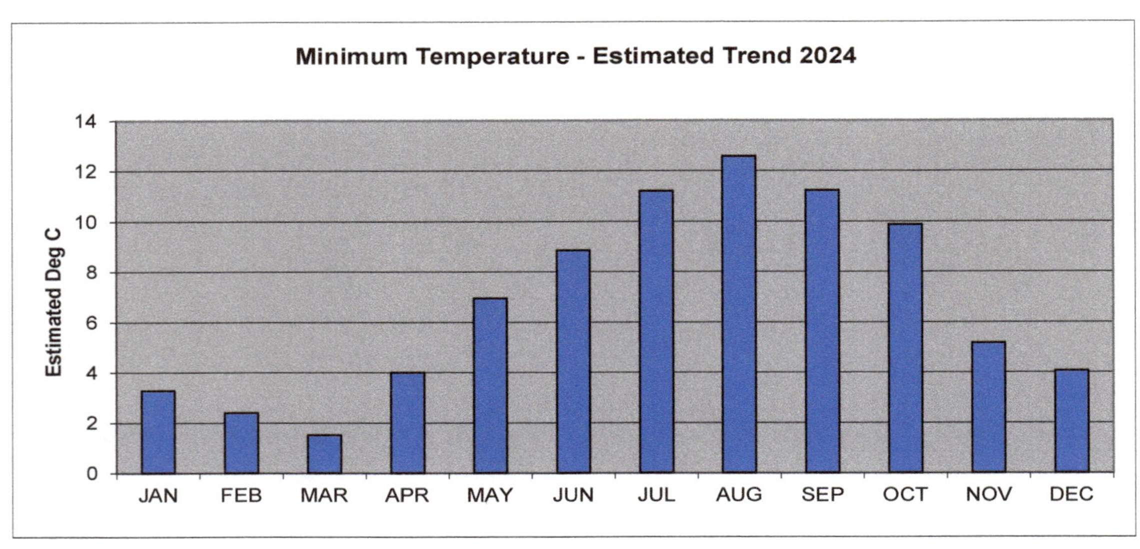
Minimum Temperature - Estimated Trend 2024
Estimated Deg C
14
12
10
8
6
4
2
0
JAN
FEB
MAR
APR
MAY
JUN
JUL
AUG
SEP
OCT
NOV
DEC

SEASONAL OVERVIEWS

WINTER

Many districts will experience an unusually drier than normal winter, with many districts only receiving around 50-60% of their usual rainfall, slightly below average sunshine overall, and warmer temperatures everywhere. The windiest period will be in the fourth week of January.

RAINFALL: The number of rain days (as measured at 1mm or more recorded each day) will be below average almost everywhere especially in the north and east, with the country as a whole averaging around 45-55 rain days compared to the norm for a typical winter of between 50-65 rain days. Only the Belmullet region is likely to see average rainfall, and the Kilkenny district wetter than average by around 25% more than is typical for the season. Overall, the Ulster Province can expect around 35-40% less rainfall than normal for the season, Connaught around 20% drier, Munster around 15-20% drier, and the Leinster Province around 40-45% drier than normal. Each month during winter will see below average rainfall with the west and the south receiving the most rainfall. The wettest spells in January will be over the second half of the month, and in February in the last week of the month, while the period over the first half of February is likely to be almost completely rain free.

SUNSHINE: By the end of winter, expected sunshine hours will be around 10-15% below normal for this season for the country as a whole, with the north west of Ulster and the northwest of the country only expected to receive around half their usual amount of sun hours. In contrast, the southwest of Ireland can anticipate up to 25% more sunshine than a typical winter. Overall, the Ulster and Connaught provinces can expect around 20—25% more cloud than normal for the season, while the Leinster Province will be around 5-10% cloudier, and only the Munster Province can anticipate above average sunshine, but still within 10% of the norm. After a sunnier than average December in 2023, February should bring a similar amount of sunshine hours as in December, but this will still be well below the February norm, while January will be considerably duller than average overall. The sunniest spells in January will be over the second half of the first week, at the end of the second week, and early in the fourth week. The days either side of the New Moon in the second week will see the most sunshine hours in February, followed by a couple of days at the end of the third week.

TEMPERATURE:
Mean air temperatures will be generally around 1–1.5°C warmer than normal for the season overall for both maximas and minimas; a trend reflected throughout all the provinces as a whole and a trend consistent throughout all the winter months. Some parts of Ireland will see much warmer maximas than normal by up to 1.5-2°C more, particularly in the Sligo to Drumsna areas, in the Newport-Munster area, as well as around Markree Castle, and minimum temperatures in the northwest and southeast also warmer by a similar variant. Only isolated pockets around the Tralee area can anticipate maximum temperatures below average but still within 1°C of the norm, and minimum temperature averages in a line from the Galway district through to Dublin slightly cooler but still within half a degree of the norm.

Maximum Temperatures: The warmest spell in December 2023 will be at the start of the fourth week, and the coolest midway through the second week. The warmest spell in January will be in the last few days of the month, and the coolest spell will be around the turn of the first week into the start of the second week, and midway through the third week. The warmest spell in February will be around the Full Moon midway through the fourth week and the coolest period around the New Moon midway through the second week. Overall, the warmest spell of winter will be in February, and the coldest spell will be in January.

Minimum Temperatures: The mildest spell in December 2023 will be midway through the fourth week, and the coolest spell will be near the end of the first week and in the first few days of December 2023.

The mildest spell in January will be in the first few days of the month and again in the last few days of January, and the coldest in the first couple of days of the second week and again towards the end of the third week. The mildest spell in February will be midway through the third week and the coolest over the first half of the second week.

Overall, the mildest spell of winter will be early in the fourth week of December 2023 and the coldest will be over the first week of December 2023. There are likely to be around 45 days with frosts observed during the winter season.

SPRING

Spring will be wetter throughout the country this year, with many districts likely to see up to 25% more than normal. Despite this, sunshine hours will be close to the average, as well as the temperatures. The windiest spells during spring will be in the fourth week of March and in the last few days of May.

RAINFALL: The number of rain days (as measured at 1mm or more recorded each day) will be around 50-60 days for the country as a whole, compared to the average range for spring of between 35-40 wet days. Overall, the Ulster Province can expect up to 30% more rain, the central provinces of Leinster and Connaught around 20-25% wetter than normal, and the Munster Province wetter by around 25-30% more. Despite this, the districts of Cork, Newport, Kilkenny, and the Ardtarmon areas can all only expect to be slightly wetter but still within 10% of the norm for the season. The wettest spell in March will be over the third week and into the start of the fourth week, between the First Quarter and clearing by the Full Moon. April will see some rain activity on most days, except for a brief spell of a couple of days mid-month. May will see some spotty shower activity over the first week, then another wet front bringing frequent showers from the end of the second week and through the first half of the third week, and finally, the last week of the month from the Full Moon onwards will see widespread rain activity on most days.

SUNSHINE: Overall, sunshine hours by the end of spring will be close to the seasonal average, and only slightly cloudier but still within 5% of the norm. Almost all areas throughout Ireland can expect below average sunshine, by around 5-10% below the average, with a few exceptions – these being in the districts around Armagh, Dumsna, Kilkenny, Shannon, and Newport, all which can expect to be slightly sunnier than average, but also still within 10% of the seasonal norm. The sunniest periods during March will be between the New Moon and First Quarter over the second week and into the start of the third week, while the last ten days of March will be particularly dull. The first week of April will be very dull while the second and third weeks April will be the sunniest period during this month, with many districts recording 8 hours of sunshine or more a day. May can expect the sunniest period to be either side of the New Moon at the turn of the first week and early second week, and again around the turn of the third week into the start of the fourth week. Much of May will see frequent daily changes from one day being dull and gloomy, to good sunshine the next day, then back to cloudy for the following day. The last week of May however, will be very dull.

TEMPERATURE:
In contrast to winter, the spring season will see fairly average to slightly below average temperatures overall for both maximum and minimum seasonal averages, but only by around 0.5°C of the norm for the country as a whole. The only areas expected to see above average maximas by the end of this season is in the Sligo to Ardtarmon area, where maximas may exceed 1-2°C above the norm, and minimas slightly warmer but still within half a degree of the norm. Minimum temperatures averages will also be above normal in the southwest of the country by between 1-2°C more than the seasonal average.

Maximum Temperatures: The warmest spell in March will be midway through third week and over the first half of the fourth week and the coolest spell will be after the New Moon midway through the second week. The warmest spell in April will be in the last ten days of the month, as well as briefly towards the end of the first week, and the coolest spell will be in the first few days of April. The warmest spell in May will be at the end of the third week and the coolest spell midway through the second week. Overall, the warmest spell of spring will be in May and the coldest spell will be in March.

Minimum Temperatures: The mildest spell in March will be midway through the fourth week and the coolest spell will be over the second week. The mildest spell in April will be towards the end of the first week, and the coolest spells will be at the very start of April as well as early in the third week. The mildest spell in May will be over the fourth week and the coolest spell midway through the first week. Overall, the mildest spell of spring will be in May and the coldest in March. March is likely to see 23 days with frosts, April 12 days and May 4 days with frosts.

SUMMER

This summer will be unusually drier than normal, with all districts only receiving around 50-60% of their usual rainfall, sunshine hours will be much higher than normal by up to 30-50% more than is typical for summer, and temperatures will be well above normal almost everywhere. The windiest period of summer will be during the fourth week of June and in the first week of July.

RAINFALL: The number of rain days (as measured at 1mm or more recorded each day) will be around 35-45 days across the country as a whole, compared to the summer average of between 30-40 days. Overall, the Ulster and Leinster Provinces can only expect around 60-65% of their usual amount of rain for the season, while the Munster and Connaught Provinces are will receive around half their usual rainfall by the end of the season. The second and third weeks of June will be dry throughout the country, as well as over much of the first and fourth weeks of July, and to a lesser extent over the fourth week of August. All other weeks will see some rain activity here and there in numerous districts. The highest amounts of rainfall will be during the first week of June, midway through the third week of July, and early in the second week and last week of August.

SUNSHINE: After a dull winter, and mediocre spring sunshine hours, summer will bring well above average sunshine for the season as a whole, compared to the norm. Overall, the Ulster and Connaught provinces can both expect around 25-30% more sunshine hours than an average summer, the Leinster Province up to 35% sunnier, and the Munster Province up to 45% sunnier with the country as a whole, averaging around 34% more sunshine than is typical for the season. In June, the sunniest periods will be the second and third weeks, while the last few days will be the cloudiest for the month, particularly in the north and west. In July, it will be the sunniest over the last ten days, while very dull and gloomy midway through the second week and again midway through the third week. August will see the sunniest period over the first couple of weeks of the month, while the cloudiest periods will be in the first half of the third week and again midway through the fourth week. All up, the longest sunniest period of summer will be in June, while the brightest sunshine hours will occur midway through the fourth week of July when 15 hours of sunshine per day may be commonly widespread for a couple of days.

TEMPERATURE:
Warmer than average maximum and minimum temperatures can be expected throughout most of the country by the end of this summer season. Maximum temperatures will averaging between 1.5-2°C above the seasonal norm almost everywhere. Minimum temperature averages however, will have more of a mix of above average and below average temperature averages, with the country as a whole likely to be around 0.5°C above the norm. Many areas in the Ulster Province and in the west of the country can anticipate maximum temperatures averaging up to 2-2.5°C above their seasonal norm. In contrast, a number of districts in the west of the country can anticipate overnight minimum temperatures of around 1-1.5°C below the seasonal norm.

Maximum Temperatures: The warmest spell in June will be over the third week and the coolest in the first few days.

The warmest spell in July will be in the last week of the month and the coolest in the first couple of days. The warmest spell in August will be midway through the first week and early in the third week, and the coolest spell will be midway through the fourth week. Overall, the warmest spell of summer will be in July, and the coolest in June.

Minimum Temperatures: The mildest spell in June will be at the end of the third week into the start of the fourth week, and again in the last few days of the month, and the coolest days will be in the first half of the first week and briefly midway through the second week. The mildest spell in July will be early in the third week and again in the last few days of the month, and the coolest midway through the fourth week. The mildest spell in August will be around the turn of the first week into the second week, and in the first couple of days of the third week, and the coolest spell will be midway through the third week, and again midway through the fourth week.

Overall, the mildest spell of summer will be at the end of July into the first week of August, and the coolest in June. Only three days with frosts can be expected over summer, and these in early June and then no further frosts are expected until November.

AUTUMN

Autumn will be wetter throughout most the country this year, with many districts likely to see around 20-30% more than normal. It will be a much sunnier than normal autumn, with many districts likely to see up to 30-40% more sunshine hours than is typical for this season. Subsequently, due to the unusually sunny autumn, temperatures will also be above normal everywhere. The windiest periods during autumn will be at the end of the first week and during the third week in October, and in the last week of November.

RAINFALL: The number of rain days (as measured at 1mm or more recorded each day), will be around 60-70 days across the country as a whole, compared to the autumn average of between 35-55 wet days. Overall, the Connaught and Leinster Provinces can expect around 15-20% more rainfall than is typical for the season, while Ulster and Munster Provinces will be between 20-30% wetter than normal. The driest spells in September will be either side of the Full Moon midway through the third week, over the fourth week in October, and the second week of November. The wettest periods in autumn will be over the second week of September and in the last couple of days of the month and into the start of October, again midway early in the third week of October, and midway through the first week and in the last week of November.

SUNSHINE: Overall, the autumn season will also see above average sunshine following a similar trend to summer, with the country as a whole expecting around 33% more sunshine hours than normal for this season. The provinces of Ulster, Connaught, and Leinster can all expect around 30-35% more sunshine hours than normal, while further south, the Munster Province can expect up to 40% more sunshine hours on average. In this breakdown, the far northwest can expect up to 50% more sunshine than normal, and the Newport to Tralee districts up to 60% more sunshine than is normal for this season. The sunniest periods for September will be just before the Full Moon midway through the third week and again midway through the fourth week, while very dull in regular waves, about every 5-6 days, as well as at the very end of the month. October will be very dull and gloomy over the last seven days of the month, while sunniest in the second half of the second week and again at the turn of the third week into the first few days of the fourth week. November will be the sunniest in the second half of the second week, while very cloudy at the end of the first week into the first few days of the second week, and again throughout the fourth week to the end of November. The sunniest period overall this autumn will be at the Full Moon in September, midway through the month.

TEMPERATURE:
Autumn temperatures will follow a similar trend to summer, with both maximum and minimum temperature averaging around 2-3°C above the norm by the end of the season. Overall temperatures will be particularly warmer than normal in the west of the country by up to 3°C or more for both maximum and minimum temperature averages, and particularly persistently warmer in the north of the country.

Maximum Temperatures: The warmest spell in September will be midway through the second week and over the third week and the coolest spell will be in the last few days of the month. The warmest spell in October will be over the first week as well as early in the fourth week, and the coolest spell will be in the last 5 days of the month. The warmest spell in November will be over the first week and early into the second week and the coolest spell will be midway through the fourth week. Overall, the warmest spell of autumn will be in September and the coldest in November.

Minimum Temperatures: The mildest spell in September will be at the end of the third week and into the start of the fourth week and the coolest spell will be at the Full Moon midway through the third week, and again midway through the fourth week. The mildest spell in October will be during the first week, and the coolest at the end of both the second week and the third week. The mildest spell in November will be at the end of the first week, and again at the Third Quarter early in the fourth week, and the coolest period will be over the second week and again in the last few days of the month.

Overall, the mildest spell of autumn will be in September and the coldest in November. No frosts are expected during autumn until the middle of November in the build-up period to the Full Moon and again in the last few days of the month.

2024 OVERALL WEATHER SUMMARY

Winds: The strongest winds of the year will be from mid to late December 2024.

Rain: The wettest day of the year likely to occur midway through the first week of November.

Dry Spells: The longest dry spells of the year will be over the first half of February, and in the middle of June during the second and third weeks of the month. The last ten days of July will also be mostly dry.

Sun: The sunniest period will be in June over the second and third weeks, while the dullest period will be the first few days of January, the second week of February, and in the few days immediately prior to Christmas Day.

Maximum Temperatures: The hottest spell in the year will be in the second half of July and the coolest spell will be

The coolest period will be around the turn of the first week and start of the second week of January, also midway through the third week of January, midway through the second weeks of February and March with similar low maximum temperatures returning at the New Moon in December in the last few days of the year.

Minimum Temperatures: The mildest spell in the year will be at the end of July and in the middle of August, and the coolest period will be just after the New Moon in March, over the second half of the second week. Frosts are likely on at least 91 days during the year.

Month	Dry spells
January	Despite some rain in most districts every day, the driest spell will be over the second half of the first week after the Third Quarter, then again on the First Quarter in the third week.
February	Over the first half of the month especially during the Third Quarter to the New Moon, and from the New Moon to the First Quarter, except for some spotty showers when the Moon Crosses the Equator around the 12th.
March	The first few days of the month just before the Third Quarter, and again after the Full Moon from around the 27th onwards to the end of the month.
April	Despite some rain almost every day, it should be dry at Southern Declination on the 1st, and again at Northern Declination around the 13th-14th.
May	On the Third Quarter on the 1st, then again around the New Moon at the end of the first week and into the first couple of days of the second week, and finally around the 19th-20th, shortly after Apogee.
June	An almost continuous dry spell is expected between the New Moon and almost up to the Full Moon throughout the second and third weeks, with only some light showers expected at the end of the third week.
July	On the New Moon during the first week, and again almost from the Full Moon to the end of the month, from around the 22nd until the 31st with only the odd exception of some spotty showers mostly in the Ulster Province around the 30th.
August	Just after the First Quarter around the end of the second week, then in the period between the Full Moon and Third Quarter, for a couple of days early in the start of the fourth week.
September	For a couple of days either side of the Full Moon during the third week, and then briefly for a day just after the Third Quarter midway through the fourth week.
October	For a couple of days either side of the Third Quarter midway through the fourth week.
November	Between the First Quarter and Full Moon from around the 10th-15th.
December	Widespread rain activity can be expected almost daily throughout the country during this month, with the only exception being a mostly dry spell in the last three to four days of the month, either side of the New Moon.

NEXT THREE YEARS

Comparing estimated rain over the next three years averaged for the whole country; 2026 should have most rain and 2025 the least, with 2024 in the middle position.

REGIONAL EXPECTATIONS BY REGIONS:

West: Wettest 2026, driest 2025, and middle 2024

South: Wettest 2026, driest 2025, and middle 2024

East: Wettest 2026, driest 2025, and middle 2024

North: Wettest 2026, driest 2025, and middle 2024

Central: Wettest 2026, driest 2025, and middle 2024

2025 may be the best year for sunshine amounts, with 2026 the cloudiest, and 2024 in the middle position.

For maximum temperatures, 2025 may be the warmest overall, with 2024 next and then 2026 the mildest.

For minimum temperatures, 2026 may be the coldest overall, 2025 the mildest, and 2024 in the middle.

JANUARY

PHASES OF THE MOON

1st	Apogee
4th	Third Quarter
4th	Crossing Equator
10th	Southern Declination
11th	New Moon
13th	Perigee #8
17th	Crossing Equator
18th	First Quarter
23rd	Northern Declination
25th	Full Moon
29th	Apogee
31st	Crossing Equator

1	2	3	4	5	6	7	8	9	10	11	12	13	14	15	16	17	18	19	20	21	22	23	24	25	26	27	28	29	30	31
Mon	Tue	Wed	Thur	Fri	Sat	Sun	Mon	Tue	Wed	Thur	Fri	Sat	Sun	Mon	Tue	Wed	Thur	Fri	Sat	Sun	Mon	Tue	Wed	Thur	Fri	Sat	Sun	Mon	Tue	Wed
A			3Q XhS						V	N		P8				XhN	1Q					^		F				A		XhS

MONTHLY SUMMARY

1st A weak frontal system brings some falls of rain and strong blustery winds possibly to gale force for a time, mostly in the west and north.

1st-3rd Mild overnight minimum temperatures throughout the country.

4th-11th Cold spell. Overnight sub-zero temperatures widespread, but milder in the south.

4th-5th Chance of widespread fog.

7th-11th A cold southeasterly flow over the country brings widespread frosts and some wintry showers, especially with spells of sleet or snow in many northern and eastern districts. Frosts are likely to be severe in inland districts.

7th Changeable weather with some good sunny spells and the chance of hail showers.

8th-9th A spell of widespread heavy rain, and possible widespread snow.

9th A warmer front crosses the country in an easterly flow bringing milder conditions throughout, but also bringing some heavy showers in many districts.

Over the next four weeks, conditions will be drier and milder than is typical for this time of year, with lighter than normal winds, and generally very dull except in the last week of this outlook period. Rainfall will be below normal almost everywhere, especially in the east and south, where less than half of their usual rainfall is likely arrive, and most rainfall can be expected midway through this outlook period. Temperatures are likely to be above normal almost everywhere with many places averaging around one degree above the norm for this time of year with the wet spell bringing widespread thick cloud mid-way through this outlook period resulting in a particularly milder than normal spell. Severe frosts however, can be expected towards the end of this outlook period. Although widespread cloud will prevail through much of the next four weeks, the fourth week will be unusually sunny for this time of year, bringing sunshine totals to above their normal almost everywhere, with almost half of the recorded sunshine hours for this outlook period expected to be recorded in this final week. During this outlook period, rain is expected on 10 to 14 days in the west and southwest, and 6 to 9 days in eastern districts, but as few as 3 wet days in the Dublin area. Thunderstorms are likely on 2 days, hail on 4 days, snow on 2 days, fog on 4 days, and frosts observed in the south on around 12 days and up to 22 frost days in inland districts.

12th-19th An anticyclonic system to the south of Ireland extends a mild but wet south to southwest flow over the country, with a series of fronts bringing rain especially to the western coastal fringes.

14th-15th Widespread fog likely.

16th-17th Wet fronts bring milder temperatures and rain to many parts, especially to the western coastal districts but less in the east as the fronts weaken as they cross the country. A cooler change brings widespread misty patches, fogs, and frosts during this period.

17th Wintry showers prevail, some of hail and chance of snow.

18th-20th Very cold spell across the country. Widespread scattered frosts.

19th Chance of snow.

20th-31st Unsettled conditions prevail throughout under a strong southwest flow. Rain or showers almost daily with the heaviest falls in western parts.

20th-23rd Rain prevails over most of the country, with some heavy falls at times along the western coastal fringes. Rainfall may be very heavy for a time in the Maam Valley region.

21st Winds possibly to gale force around the coasts.

22nd Wintry showers, some of hail likely.

23rd Chance of gale force winds in coastal parts, particularly in the northwest.

24th-25th Warmer spell in the north.

26th-27th Chance of scattered thunderstorms and hail showers.

28th Fog likely in many districts.

29th Unusually mild spell in the southeast of Ireland.

30th-31st Chance of gale force winds along the coasts and scattered thunderstorms with hail showers.

IRELAND RAINFALL ESTIMATES

January 2024

d=mainly dry, n= nonrecordable, l=light shrs, s=significant shrs, r=rain, h=hvy falls

		ULSTER PROVINCE					CONNAUGHT PROVINCE						LEINSTER PROVINCE				MUNSTER PROVINCE						
		Alderg	Hillsbo	Armag	Loughl	Malin H	Sligo	Ardtarı	Belmul	Drums	Galway	Roscoı	Markre	Edende	Kilkenı	Dublin	Shannc	Newpo	Tralee	Killarnε	Cork	All	Moon
1st	Mon	s	l	n	l	l	s	l	l	l	l	l	l	l	d	l	s	l	l	l	d	29	A
2nd	Tue	l	d	l	l	s	l	l	l	n	r	s	l	d	d	d	d	s	d	d	d	32	
3rd	Wed	n	l	d	d	l	n	d	l	s	l	s	n	n	l	d	l	s	d	d	d	28	XhS
4th	Thur	d	d	n	d	d	d	d	d	l	d	l	d	d	d	d	d	l	d	d	d	2	3Q
5th	Fri	d	d	d	d	d	d	d	d	d	d	d	d	d	d	d	d	l	d	d	d	2	
6th	Sat	d	n	d	d	d	d	d	d	d	d	d	d	n	d	l	d	d	d	d	d	2	
7th	Sun	d	d	d	d	d	d	r	d	d	d	d	d	d	d	l	d	s	d	d	n	18	
8th	Mon	d	d	d	l	d	s	r	l	s	s	l	s	l	l	l	d	h	s	h	n	85	
9th	Tue	r	h	d	r	s	r	r	s	r	r	h	r	r	s	l	s	r	h	s	r	221	
10th	Wed	s	l	r	r	r	h	s	h	r	s	s	r	l	l	s	s	s	s	r	s	163	V
11th	Thu	l	n	s	r	r	s	r	s	r	r	r	r	s	l	n	r	r	r	r	h	178	N
12th	Fri	s	d	s	n	d	n	l	l	l	d	l	l	d	d	d	l	s	l	l	d	17	
13th	Sat	l	l	d	l	s	l	l	l	l	s	l	r	l	d	d	l	n	s	l	l	42	P8
14th	Sun	n	d	l	d	n	l	n	d	d	d	n	l	n	d	n	l	d	l	d	n	7	
15th	Mon	d	d	d	d	d	n	n	l	d	l	d	l	d	d	d	d	d	n	d	l	5	
16th	Tue	d	d	d	l	d	s	s	h	s	r	s	d	d	d	d	s	r	h	r	s	114	XhN
17th	Wed	d	d	n	d	l	l	d	s	l	l	n	l	d	d	d	l	d	r	d	l	30	
18th	Thur	l	l	d	d	d	d	d	d	d	d	d	l	d	d	d	d	d	d	d	d	3	1Q
19th	Fri	d	d	d	d	d	d	d	d	d	d	d	n	d	d	l	d	d	d	d	l	4	
20th	Sat	l	s	d	s	l	s	l	s	s	r	s	s	s	s	l	l	s	h	s	s	112	
21st	Sun	l	s	l	r	s	r	r	r	s	s	s	s	s	s	s	r	r	h	h	r	169	
22nd	Mon	l	d	s	d	s	l	d	l	l	l	n	l	d	n	d	l	l	l	l	d	21	
23rd	Tue	l	s	d	s	s	r	s	r	r	r	h	s	s	l	d	h	r	h	h	s	244	^
24th	Wed	s	s	r	s	l	l	l	r	n	d	l	l	l	l	n	s	s	l	s	s	61	
25th	Thu	n	d	l	l	d	l	l	d	l	l	l	l	l	s	d	n	l	s	h	l	61	F
26th	Fri	l	l	d	s	l	s	s	r	s	s	l	s	l	s	l	s	s	r	r	s	101	
27th	Sat	l	l	l	l	l	s	l	r	s	s	s	l	l	l	l	l	s	s	l	l	70	
28th	Sun	l	l	l	s	s	l	l	n	l	l	l	l	l	d	n	l	l	l	d	d	31	
29th	Mon	l	d	l	l	r	l	l	s	n	s	n	l	d	d	d	n	r	l	d	d	48	A
30th	Tue	s	s	d	r	r	r	s	s	s	s	s	r	l	l	d	l	s	s	s	l	112	XhS
31st	Wed	l	s	s	l	s	l	l	l	l	d	l	l	d	d	l	l	l	l	l	l	36	
	Estimate:	52	53	49	100	106	117	108	155	107	127	101	112	45	47	26	87	168	188	216	83	2046	
	Average:	80	91	75	125	120	131	133	134	105	117	130	131	93	80	63	102	167	175	127	131	2309	
	Trend:	drr	drr	drr	drr	drr	drr	drr	wtr	av	av	drr	drr	drr	drr	drr	drr	av	wtr	wtr	drr	drr	

IRELAND SUNSHINE ESTIMATES
January 2024

F=fine (8-12 hours of sunshine), pc= partly cloudy (4-7 hours), c=cloudy (1-3 hours), o=overcast (0 hours)

		ULSTER PROVINCE					CONNAUGHT PROVINCE						LEINSTER PROVINCE				MUNSTER PROVINCE						
		Alderg	Hillsbo	Armag	Loughl	Malin I	Sligo	Ardtarı	Belmul	Drums	Galway	Roscoı	Markre	Edende	Kilkenı	Dublin	Shannc	Newpo	Tralee	Killarne	Cork	All	Moon
1st	Mon	o	o	o	o	o	o	o	o	o	o	o	o	o	o	o	o	o	o	o	o	7	A
2nd	Tue	o	o	o	o	o	o	o	o	o	o	o	o	o	o	o	o	o	o	o	o	3	
3rd	Wed	o	o	o	o	o	o	o	o	o	o	o	o	o	o	o	o	o	o	o	o	7	XhS
4th	Thur	o	o	o	o	o	o	o	o	o	c	o	o	o	c	o	c	c	o	o	c	17	3Q
5th	Fri	c	c	c	c	o	o	c	pc	o	pc	o	c	c	c	pc	pc	pc	c	c	c	52	
6th	Sat	o	o	o	o	o	o	o	pc	o	c	o	o	o	o	o	o	o	c	c	o	17	
7th	Sun	c	c	c	o	o	pc	c	pc	o	pc	o	o	o	o	o	c	c	pc	pc	c	44	
8th	Mon	c	c	c	o	o	o	c	o	o	o	o	c	c	c	pc	o	c	o	o	o	23	
9th	Tue	o	o	o	o	o	o	o	o	o	o	o	o	o	o	o	o	o	o	o	o	3	
10th	Wed	o	c	c	o	o	o	c	o	o	c	o	c	c	c	c	c	c	o	o	c	28	V
11th	Thu	o	o	o	o	o	o	o	o	o	o	o	o	o	c	o	o	c	o	o	c	15	N
12th	Fri	o	c	c	o	o	o	c	o	o	c	o	c	c	c	c	c	c	c	c	c	23	
13th	Sat	o	o	o	o	o	o	o	o	o	o	o	o	o	o	o	o	o	o	o	o	3	P8
14th	Sun	c	pc	pc	c	o	c	pc	o	o	c	o	c	c	pc	pc	c	pc	pc	pc	pc	59	
15th	Mon	c	c	c	c	o	o	c	o	o	o	c	c	pc	c	pc	o	c	o	o	o	31	
16th	Tue	o	o	o	o	o	o	o	o	o	o	o	o	o	o	o	o	o	o	o	o	3	XhN
17th	Wed	o	o	o	o	o	o	o	o	o	o	o	o	o	o	o	o	o	o	o	o	2	
18th	Thur	o	c	c	o	o	c	c	o	o	o	o	o	o	o	o	c	c	o	o	c	12	1Q
19th	Fri	o	c	c	o	o	o	c	o	o	o	c	c	c	c	c	o	c	pc	pc	c	33	
20th	Sat	o	o	o	o	o	o	o	o	o	o	o	o	o	o	o	o	o	o	o	o	3	
21st	Sun	o	o	o	o	o	o	o	o	o	o	o	o	o	o	o	o	o	o	o	o	3	
22nd	Mon	c	pc	pc	c	o	pc	pc	c	c	pc	c	c	pc	pc	pc	pc	pc	pc	pc	pc	85	
23rd	Tue	o	o	o	o	o	o	o	o	o	o	o	o	o	o	o	o	o	o	o	o	3	^
24th	Wed	o	o	o	o	o	o	o	o	o	o	o	o	o	o	o	c	o	o	o	o	6	
25th	Thu	c	pc	pc	c	o	pc	pc	c	o	c	o	c	pc	pc	pc	pc	pc	c	c	pc	74	F
26th	Fri	o	o	o	o	o	o	o	o	o	c	o	o	o	o	o	c	c	c	c	o	9	
27th	Sat	o	o	o	o	o	o	o	o	o	o	o	o	o	o	o	o	o	o	o	o	2	
28th	Sun	o	o	o	o	o	o	o	o	o	o	c	o	o	c	o	c	c	o	o	c	13	
29th	Mon	o	o	o	o	o	o	o	o	c	o	c	o	o	o	o	o	o	o	o	o	9	A
30th	Tue	o	o	o	o	o	o	o	o	o	o	o	o	o	o	o	o	o	o	o	o	4	XhS
31st	Wed	c	pc	pc	c	o	c	pc	c	o	c	o	c	pc	pc	pc	c	c	c	c	pc	64	
Estimate hours:		25	38	38	19	13	27	38	31	15	36	19	33	41	44	48	40	42	32	32	41	654	
Average hours:		46	46	44	41	37	47	47	46	43	51	53	54	51	53	56	49	43	44	44	53	946	
Trend:		less	av	av	less	less	less	av	less	less	less	less	less	av	av	av	av	av	less	less	less	less	

JANUARY **1** MONDAY

High pressure drifts to the south of Ireland bringing a mostly dry day with light winds and scattered fogs in the far south while low pressure to the north of the country extends a band of drizzle and the occasional outbreak of rain down through the northwest and into the southeast later in the day. Very cloudy everywhere, with southwest winds prevailing, fresher to the north, lighter to the south and decreasing with a southerly change later in the day.

Belfast: Overcast, mild, light showers, moderate breezes, chance of fog patches
Dublin: Overcast, isolated showers, light breezes, chance of misty patches
Cork: Dull and gloomy air, dry, mild, light breezes
Galway: Widespread cloud, mild day cold night, light showers, chance of fog

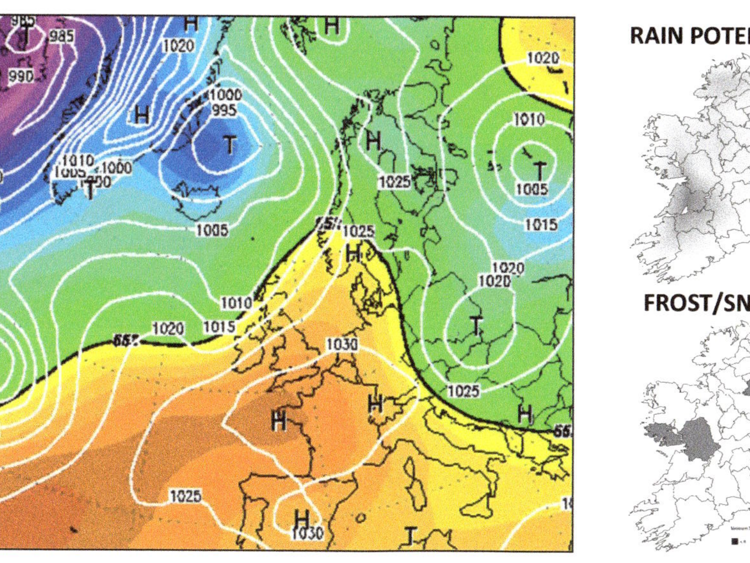

RAIN POTENTIAL

FROST/SNOW

JANUARY **2** TUESDAY

Low pressure to the north of Ireland drifts further southwest of the country directing a rain band briefly across northern and western parts while everywhere else remains mostly dry and dull throughout. Pockets of fog likely in eastern parts, while breezier south to southwest winds with occasional gusts in the west. Some showers may return to the western and southern coasts later in the day.

Belfast: Overcast, mostly dry, misty fog patches, threats of rain
Dublin: Dull and gloomy, dry, odd threats of rain, light breezes, possible overnight fog
Cork: Overcast, dry, cooler, gentle breezes, chance of light drizzle at times
Galway: Overcast, rain, moderate breezes with occasional gusts

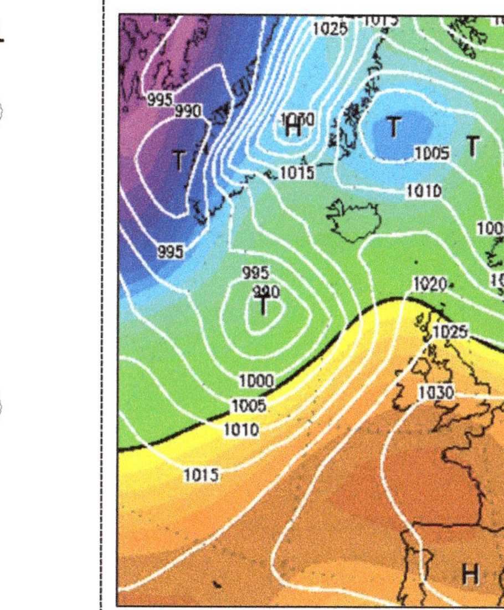

RAIN POTENTIAL

FROST/SNOW

JANUARY **3** WEDNESDAY

Light to moderate southwest winds continue to prevail throughout most of the country directed from low pressure to the north and high pressure to the south. Widespread cloud everywhere, with chance of scattered fogs, dense in some places. Cloud likely to turn to drizzle patches at times. Chance of frost pockets mostly in inland parts.

Belfast: Overcast and mild, isolated drizzly showers, light breezes
Dublin: Mostly cloudy, dry, mild, light to moderate breezes
Cork: Light breezes, widespread cloud, dry, cool, chance of shallow fog for a time
Galway: Overcast, isolated showers, milder, light breezes, chance of fog overnight

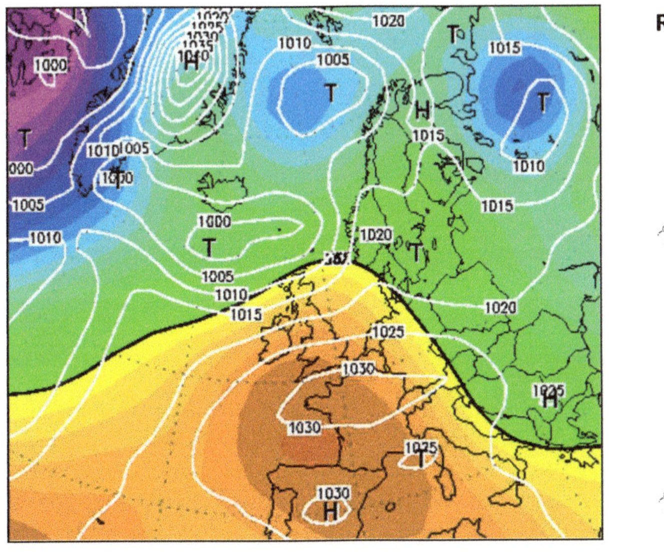

RAIN POTENTIAL

FROST/SNOW

JANUARY **4** THURSDAY

High pressure prevails directing a mostly calm to light variable flow over the country. Mostly dry with scattered fog patches almost everywhere. Some sunny spells likely mostly in the south and west. Chance of some brief showers possible later in the day along the eastern coastal fringes, and possible snow flurries in the far north.

Belfast: Overcast, mild, threats of rain, chance of fog and snow flurries
Dublin: Calm air, overcast, cooler, dry, fog likely
Cork: Cloudy, colder, dry, some overnight mists and fog patches
Galway: Cloudy, dry, gentle breezes, fog and mist patches likely

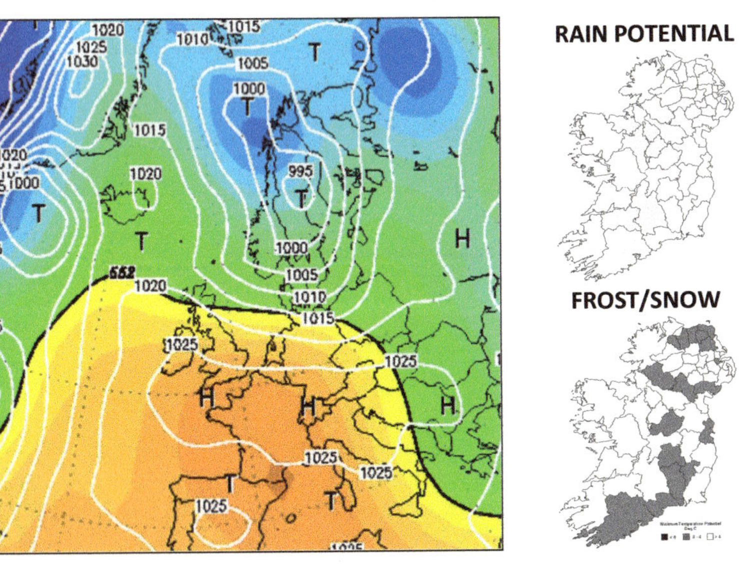

RAIN POTENTIAL

FROST/SNOW

JANUARY **5** FRIDAY

A large anticyclonic system sits over the country extending a mostly dry, calm, cool, and pleasant air throughout. Frosts and fogs likely in many districts, with widespread cloud prevailing in the central north and some sunny spells elsewhere.

Belfast: Cloudy, cooler, calm air, fog likely, chance of overnight snow flurries
Dublin: Pleasant serene air, cold, some sunny spells, fog likely
Cork: Cloudy and calm air, mostly dry with scattered fog and misty patches
Galway: Occasional sunny spells, dry, pleasant cool and calm air, fog likely

JANUARY **6** SATURDAY

A large high continues to linger over the country directing a mostly dry and light to variable flow throughout. Scattered frost and fog patches, mostly in the midlands. Occasional sunny spells in western parts with increased cloud elsewhere, with particularly dull conditions in the east and north, where some drizzle patches may be observed.

Belfast: Cold, dull and gloomy, fog likely, possible drizzle outbreaks
Dublin: Overcast, cooler, mostly dry, calm air, passing misty drizzle patches likely
Cork: Calm air, cold, widespread cloud, dry, odd threat of rain, overnight mist likely
Galway: Cloudy, dry, colder with frosts and fog patches likely, calm and serene air

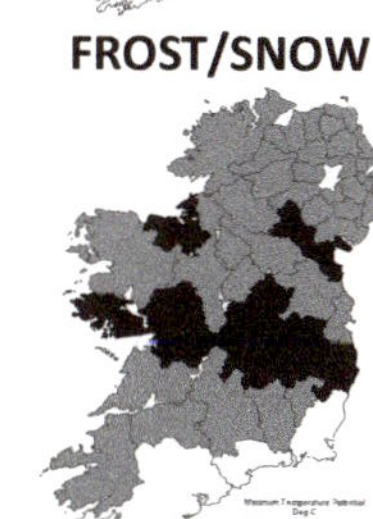

JANUARY **7** SUNDAY

The high pressure system drifts northwards off the country with low pressure moving onto the west and south resulting in a light to moderate east to southeast flow. After a mostly dry start with scattered frosts and fogs in many districts, especially in the midlands and north. Mostly dry and cloudy to the north, mostly overcast conditions with outbreaks of drizzle prevail in the east, and the chance of some wintry showers of hail or light snow may be observed.

Belfast: Cloudy, cold, frosts, light air, dry
Dublin: Overcast, cooler, isolated wintry showers, possible overnight snow or hail
Cork: Cloudy, cold, isolated drizzle and misty patches, calm air
Galway: Partly cloudy, dry, frost and fog patches, serene air

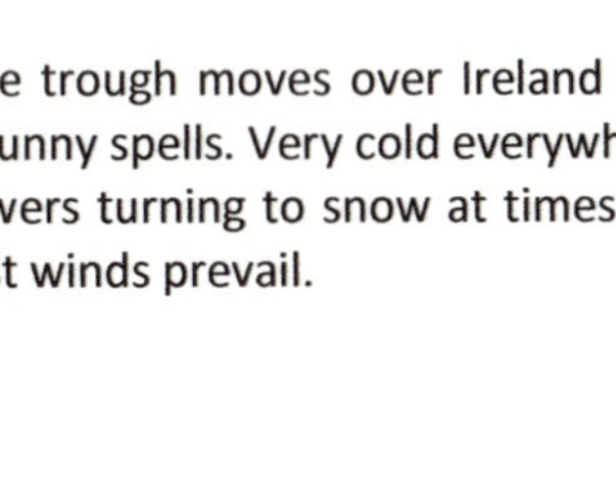

RAIN POTENTIAL

FROST/SNOW

JANUARY **8** MONDAY

A low pressure trough moves over Ireland from the west, bringing a mix of wintry showers and sunny spells. Very cold everywhere, with severe overnight frosts. Chance of wintry showers turning to snow at times particularly in the west, east and south. Light southeast winds prevail.

Belfast: Cloudy and cold, severe frosts, gentle breezes, increasing threats of rain
Dublin: Cold and frosty, isolated showers, possible light snow, some sunny spells
Cork: Overcast, cold, isolated wintry showers, calm air, chance of snow
Galway: Overcast, light breezes, cold with frosts, wintry showers, snow likely

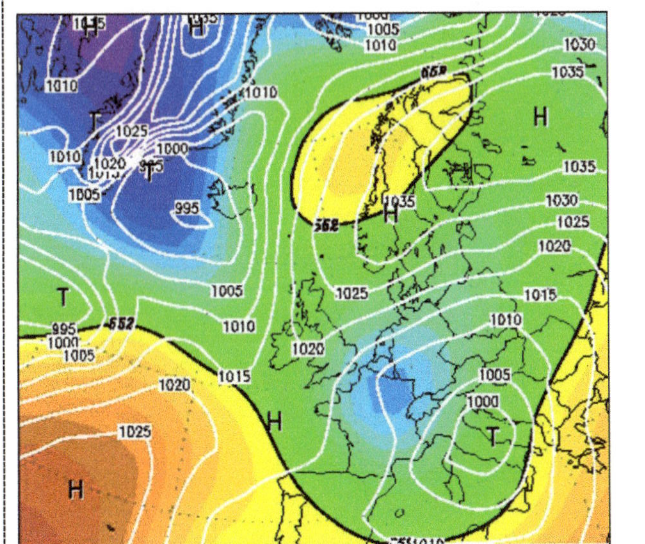

RAIN POTENTIAL

FROST/SNOW

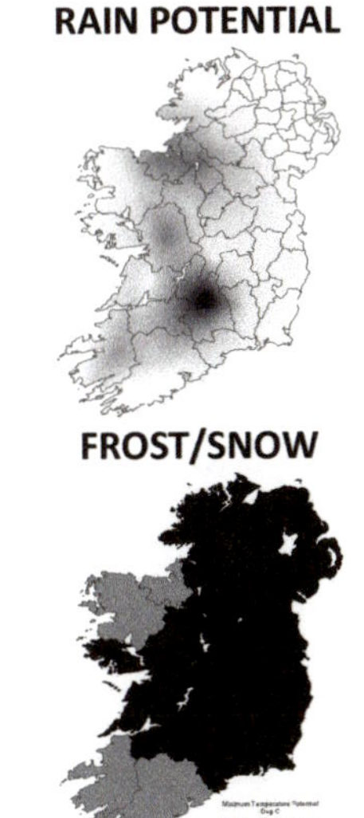

JANUARY **9** TUESDAY

Low pressure to the west of Ireland widespread cloud and strengthening southerly winds over the country, strong and gusty at times, especially in the west and possibly reaching near gales for a time. Wintry drizzle and showers in most districts, with chance of light overnight snowfalls.

Belfast: Overcast, strengthening winds, cold, heavy wintry rain, frost and snow likely
Dublin: Overcast, drizzly showers, frosts, chance of overnight snow, freshening winds
Cork: Overcast, milder, strengthening winds, wintry rain, chance of snow
Galway: Overcast, warmer, blustery at times, rain, snow possible

RAIN POTENTIAL

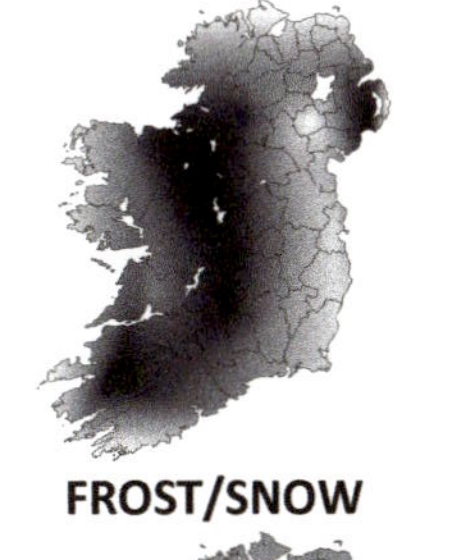

FROST/SNOW

JANUARY **10** WEDNESDAY

A depression to the northwest of Ireland extends blustery southwest winds over the country, widespread cloud, and bands of wintry showers through most districts. Showers likely to be heavy for a time in western parts, before spreading into the east. Milder temperatures.

Belfast: Cloudy, milder, brief showers and misty patches, windy at times
Dublin: Cloudy, scattered wintry showers, milder, strong winds, chance of gales
Cork: Mostly cloudy, occasional showers, unsettled winds, possible gales for a time
Galway: Cloudy, scattered showers, strong winds, possible gales

RAIN POTENTIAL

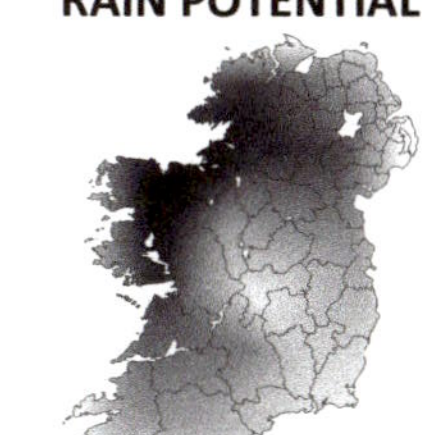

FROST/SNOW

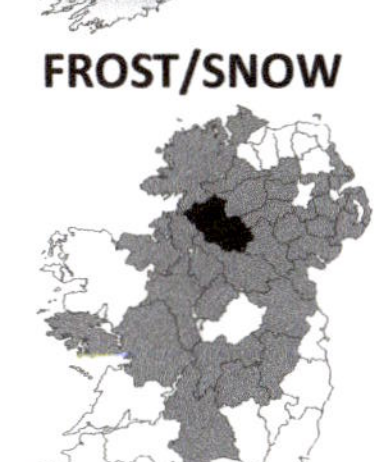

JANUARY **11** THURSDAY

Depression to the north of Ireland strengthens with southwest winds freshening to gusty at times, especially in the west and south, where winds may reach gale force for a time. A stormy air is likely to prevail with widespread cloud and scattered wintry showers throughout.

Belfast: Unsettled air, possible gales, overcast, spotty showers, chance of overnight fog
Dublin: Overcast, strong blustery winds to gales at times, drizzle patches, cooler
Cork: Stormy air, gales, cloudy, cold, heavy rain
Galway: Stormy air, overcast, rain, winds to gale force at times

RAIN POTENTIAL

FROST/SNOW

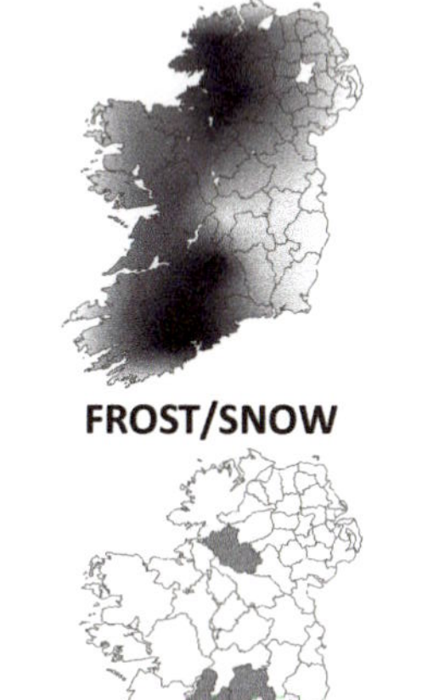

JANUARY **12** FRIDAY

A depression situated over Ireland weakens releasing a stormy air with heavy outbreaks of rain in the west and north of the country, easing later to showers and drizzly spells. Strong and unsettled blustery northwest winds prevail throughout, possibly up to gale force for a time especially in the northwest. Mostly dry in the east and south, and cold with the chance of some pockets of frost.

Belfast: Cloudy, dry with threats of rain, strong blustery winds, chance of gale gusts
Dublin: Cloudy, cold, mostly dry, odd threat of rain, strong winds at times
Cork: Cloudy, unsettled and cold blustery air, mostly dry, threats of rain
Galway: Cloudy, dry, cool, blustery with chance of gale gusts easing later

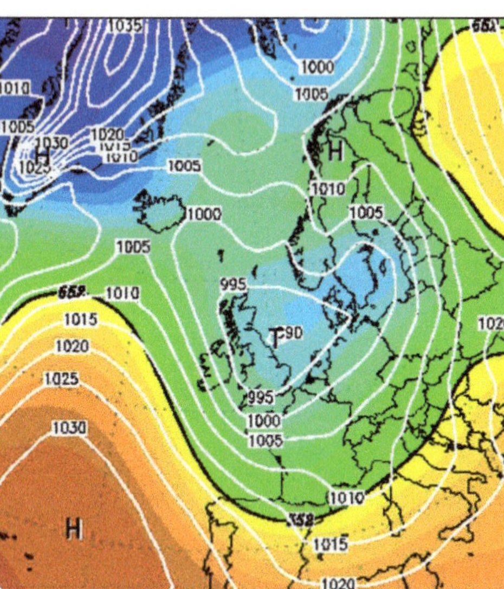

RAIN POTENTIAL

FROST/SNOW

JANUARY **13** SATURDAY

High pressure ridges over the country directing a light to moderate south to southwest flow. Widespread cloud prevails with patches of fog, drizzle and rain mostly in the western districts, while mostly dry with mists and fogs elsewhere at first with drizzly wintry showers settling in later in the day. Very cold overnight with pockets of frost likely.

Belfast: Overcast, mild day cold night, light breezes, odd drizzly shower, chance of fog
Dublin: Overcast, mostly dry, threats of rain, winds easing to light
Cork: Overcast, milder, light drizzle, mostly calm air, fog and mist likely
Galway: Overcast, mild day cold night, scattered showers and drizzles, chance of fog

RAIN POTENTIAL

FROST/SNOW

JANUARY **14** SUNDAY

The high pressure ridge slips southwards to be followed by low pressure moving onto the country from the north. Drizzly showers likely to clear the south of the country at first, allowing for a mostly dry day with occasional sunny spells and patches of fog in many other districts. Winds generally light in a variable to northwest flow, but fresher around the northern coasts.

Belfast: Mostly dry, mild, some sunny spells, overnight fog patches likely
Dublin: Sunny spells, mostly dry, frosts and fog patches, chance of isolated showers
Cork: Partly cloudy, some drizzle patches, gentle breezes
Galway: Cloudy, mostly dry, moderate breezes to mostly calm air, fog patches likely

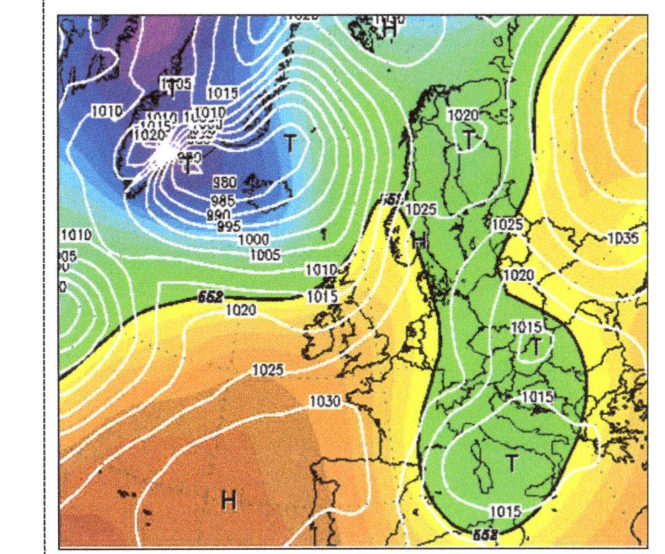

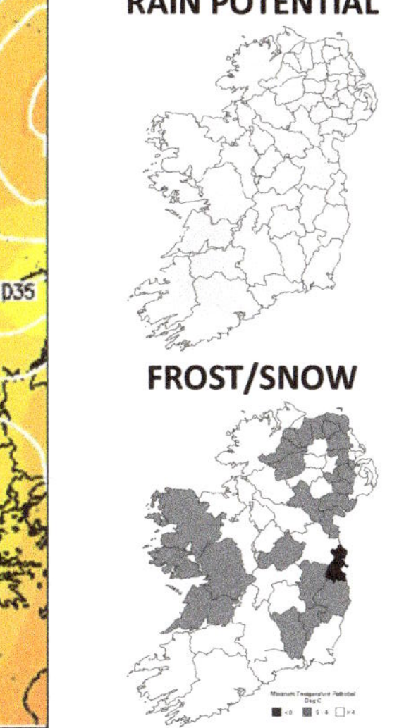

JANUARY 15 MONDAY

Low pressure sits to the northwest of Ireland while high pressure lingers to the southeast, with a light to moderate southeasterly flow prevailing. Cold and mostly dry with frost and fog patches in the north and east, while mostly overcast with patches of drizzle and fog in the west and south.

Belfast: Cloudy, dry, frost and fog patches, light southerly flow
Dublin: Cold and frosty, dry, some sunny spells, mostly calm air, fog possible
Cork: Overcast, cooler, isolated showers, calm air, fog likely
Galway: Overcast, light drizzle patches, overnight fog possible, moderate breezes

RAIN POTENTIAL

FROST/SNOW

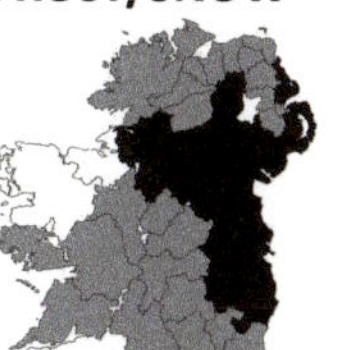

JANUARY 16 TUESDAY

A low pressure trough moves ups over the country in a moderate to fresh southeasterly low. Dull and gloomy everywhere with outbreaks of drizzle in the southwest at first spreading across the country later in the day but remaining mostly dry with threats of rain in the north and east. A band of wintry rain likely to turn to sleet in the higher reaches of the west and south of the country, with rain becoming heavy for a time in the southwest.

Belfast: Overcast, cold, frosts, mostly dry, passing threats of rain, moderate breezes
Dublin: Overcast, cold, mostly dry, overnight fog possible, breezier later in the day
Cork: Overcast, cooler, rain, moderate southerly breezes
Galway: Overcast, rain and persistent drizzles, cooler, moderate breezes

RAIN POTENTIAL

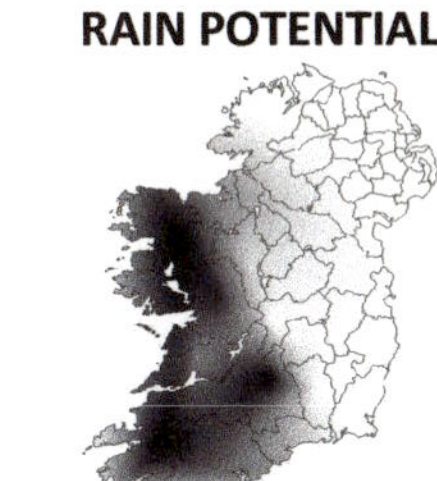

FROST/SNOW

JANUARY **17** WEDNESDAY

Weak low pressure prevails over the country bringing widespread cloud and light southerly winds throughout. Some outbreaks of drizzly rain in the west and south likely, while mostly dry elsewhere with some sunnier spells in the east. Chance of overnight patches of fog in many districts.

Belfast: Overcast, dry with passing threats of rain, fluctuating breezes, milder overnight
Dublin: Cold, mostly dry, widespread cloud, gloomy, light breezes, drizzle patches likely
Cork: Overcast, cold, passing wintry drizzle patches, moderate breezes
Galway: Colder, overcast, drizzle patches, light shifting breezes, misty patches likely

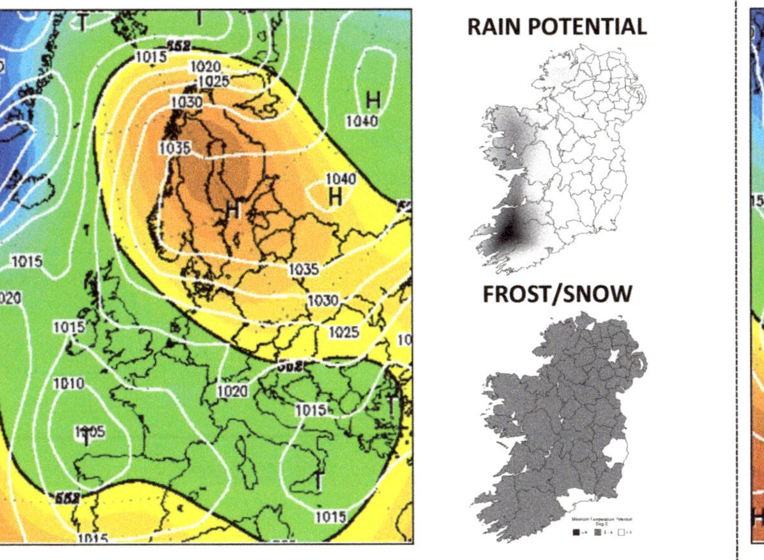

RAIN POTENTIAL

FROST/SNOW

JANUARY **18** THURSDAY

High pressure to the south and low pressure to the northwest with a variable to light southeast flow prevailing. Very cold with scattered fogs and severe frosts likely in many districts, particularly in the midlands. Some outbreaks of wintry showers possible, especially in the west. Chance of light snow in the north.

Belfast: Cloudy, cold, light showers, light breezes, chance of snow flurries
Dublin: Cold, dull and gloomy air, mostly dry, light shifting breezes, threats of rain
Cork: Cloudy, cold, mostly dry, gentle shifting breezes, morning misty patches likely
Galway: Mostly dry, widespread cloud, cold, calm air, frosts and freezing fog patches

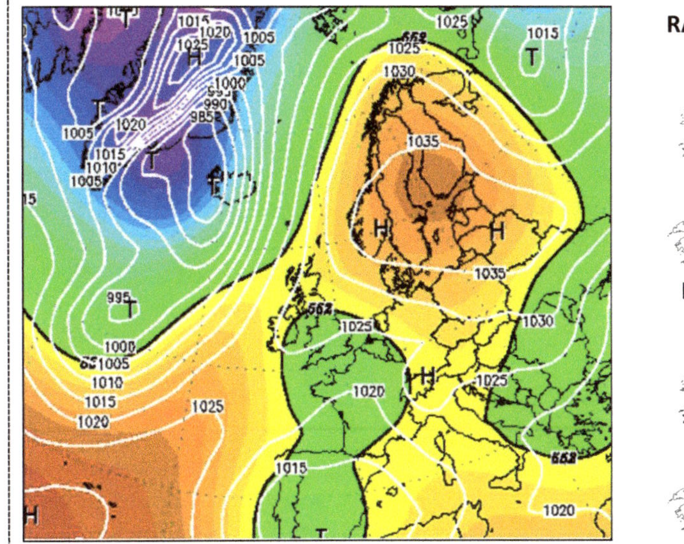

JANUARY **19** FRIDAY

An intense depression to the northwest of Ireland slowly drifts southward towards the country, while a high pressure continues to ridge over the southeast. Very cold and cloudy with light variable winds everywhere, freshening later in the day as the North Atlantic depression moves into northern parts later in the day. Mostly dry with pockets of freezing fog and increasing threats of rain and chance of overnight wintry showers, particularly in the east.

Belfast: Cloudy and cold, dry, light breezes
Dublin: Cold, cloudy, frosts, wintry showers, chance of light snow overnight
Cork: Cloudy and cool, light showers, mostly calm air, fog possible
Galway: Overcast, cold, calm air, dry, threats of rain, frosts and freezing fog likely

RAIN POTENTIAL

FROST/SNOW

JANUARY **20** SATURDAY

Depression slowly drifts onto the north of the country bringing freshening south to southwest winds throughout, possibly to gale force at times. Stormy air and very cold in most parts, with scattered wintry showers and heavy rain at times in some western parts. Chance of snow.

Belfast: Overcast, freshening breezes, cold, frosts, scattered wintry showers
Dublin: Overcast, isolated wintry showers, frosts, winds freshening later in the day
Cork: Calm, overcast, milder day cold night, light rainy spells
Galway: Milder day, gloomy air, wintry rain, frosts, light breezes

RAIN POTENTIAL

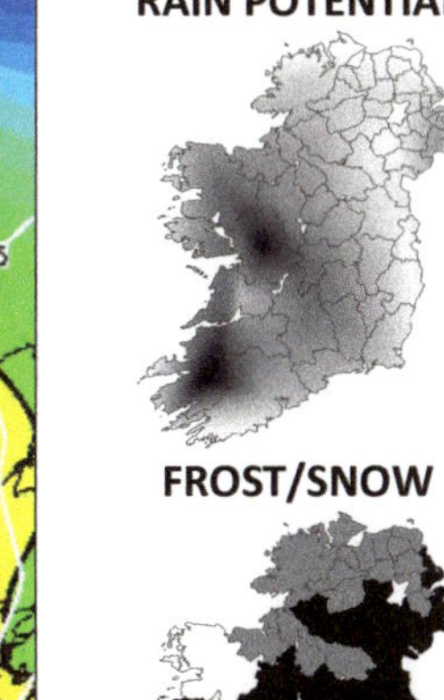

FROST/SNOW

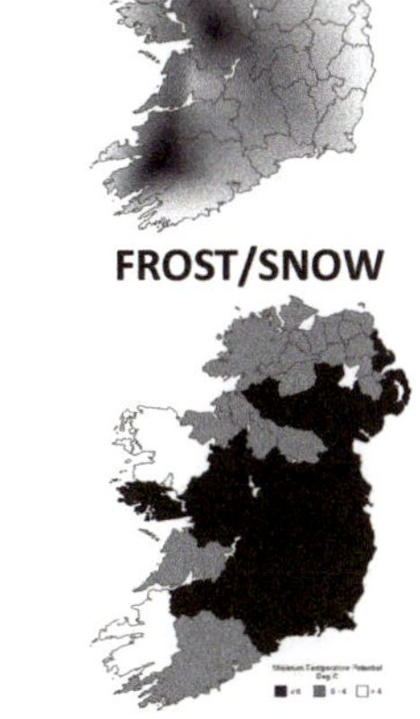

JANUARY **21** SUNDAY

A large depression centred north of Ireland extends a stormy air over the country. Dull and gloomy air with strong south to southwest winds, up to gale force at times, and widespread rain prevailing. Rain likely to be heavy at times in southern districts. Milder than of recent.

Belfast: Stormy air, overcast, milder, showery spells
Dublin: Overcast, milder, scattered showers, strong and unsettled winds, gales possible
Cork: Stormy air, overcast, rain, milder, winds likely to gale force at times
Galway: Blustery and stormy air, overcast, rain, milder

RAIN POTENTIAL

FROST/SNOW

JANUARY **22** MONDAY

Depression intensifies increasing stormy conditions throughout the country. Strong and unsettled winds everywhere. Wintry showers, some of hail, and possible snow, particularly in the west and north, while clearer spells develop particularly in the east, where some sunny breakthroughs can be expected.

Belfast: Unsettled strong winds, possible gales, milder, some sunny spells, mostly dry
Dublin: Stormy winds, mostly dry, colder, occasional sunny spells
Cork: Partly cloudy, cold, dry, strong winds to storm force likely
Galway: Stormy winds, isolated showers, overnight snow likely

RAIN POTENTIAL

FROST/SNOW

JANUARY 23 TUESDAY

Stormy conditions associated with the intense depression to the northwest of Ireland returns widespread dull and gloomy skies over the country, with squally wintry showers, stubborn drizzle spells and strong blustery winds generally of a southerly flow everywhere, with winds up to gale force at times. Rain may be heavy for a time in the west, while precipitation more isolated in the east.

Belfast: Dull and gloomy, stormy air, squally wintry showers
Dublin: Overcast, dull and mil air, strong and gusty winds, mostly dry, rain threatening
Cork: Overcast, blustery, squally showers
Galway: Overcast, stormy, wintry rain, snow and gale gusts possible

RAIN POTENTIAL

FROST/SNOW

JANUARY 24 WEDNESDAY

Unsettled and blustery southerly winds prevail as the depression to the northwest of Ireland continues to dominate the country. Stormy air throughout. A band of heavy rain crosses the northwestern districts and move down through the country in a southeasterly tract, clearing the southeast later in the day. Winds may reach gale force for a time.

Belfast: Milder, overcast, light showers, strong winds and possible gale gusts
Dublin: Overcast, mild, isolated showers, strong winds, stormy air at times
Cork: Unsettled strong winds with gusts, overcast, squalls
Galway: Overcast, dry with passing threats of rain, blustery, gale gusts likely

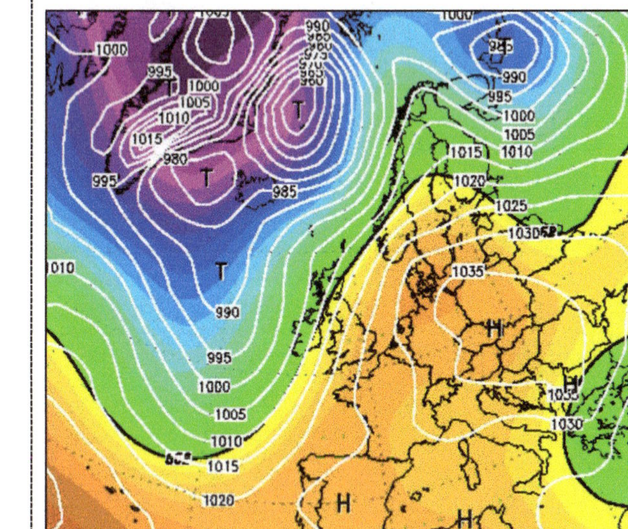

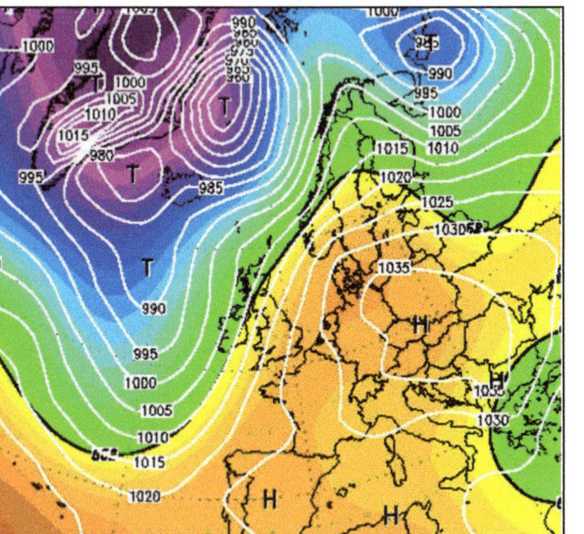

RAIN POTENTIAL

FROST/SNOW

JANUARY **25** THURSDAY

A depression to the west of Ireland directs a lighter generally westerly flow over the country, backing southerly and strengthening later in the day. Some light showers mostly in the south and west, while mostly dry with occasional sunny spells elsewhere. Possible patches of fog overnight in the north and frosts in the west. Cloud increasing later in the day.

Belfast: Dry, some sunny spells, chance of overnight fog, winds freshening later
Dublin: Partly cloudy, dry, cooler, winds easing to light for a time
Cork: Changeable, mix of brief sunny spells and spotty showers, colder
Galway: Mostly cloudy, colder, spotty showers, lighter breezes, frost pockets

RAIN POTENTIAL

FROST/SNOW

JANUARY **26** FRIDAY

North Atlantic depression moves back down over Ireland bringing strong southerly winds and a stormy air up through the country. Overcast with drizzly showers almost everywhere, with just the odd sunny breakthrough, mostly in the west. Winds may reach gale force for a time.

Belfast: Overcast, stormy, strong blustery winds, possible gales, drizzly showers, cooler
Dublin: Overcast and stormy air, squally light rain, chance of gusty gales at times
Cork: Overcast, cold, windy with possible gale gusts, scattered showers
Galway: Cloudy, milder day, scattered showers, stormy air with strong gusty winds

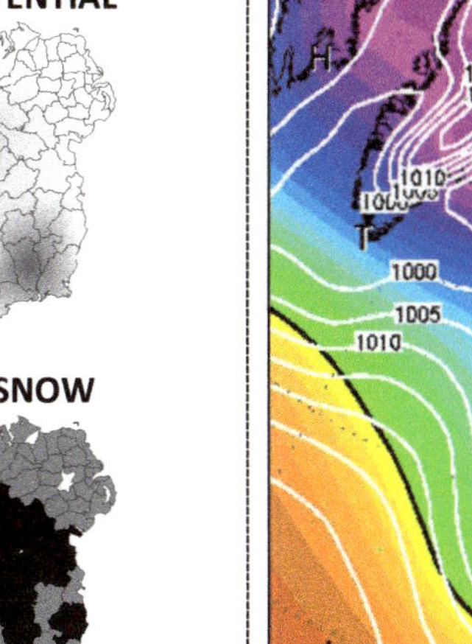

RAIN POTENTIAL

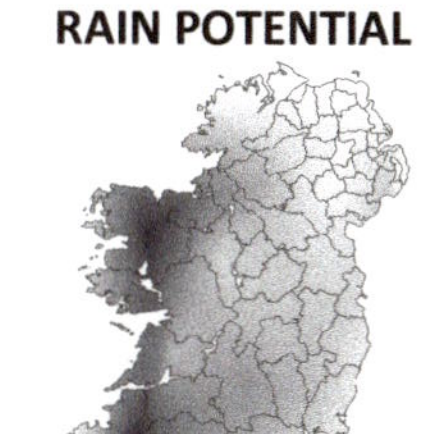

FROST/SNOW

JANUARY 27 SATURDAY

Depression weakens as it lingers over Ireland with variable winds easing to moderate. Dull and gloomy throughout with the odd spotty showers here and there. Chance of some patches of fog overnight mostly in the north, while wintry showers including light snow possible in the west, turning to heavier rain as a front spreads eastward for a time.

Belfast: Overcast, light showers, cool, winds easing to light, chance of fog patches
Dublin: Overcast, cool, isolated showers, winds easing to moderate
Cork: Widespread cloud, spells of light misty rain, cold overnight, winds easing
Galway: Overcast, windy, colder, wintry showers, snow likely

JANUARY 28 SUNDAY

Low pressure continues to prevail over Ireland bring moderate northwesterly winds, freshening and backing westerly later in the day. Overcast throughout with patches of drizzle and fog here and there. Chance of some short sunny spells mostly in the east.

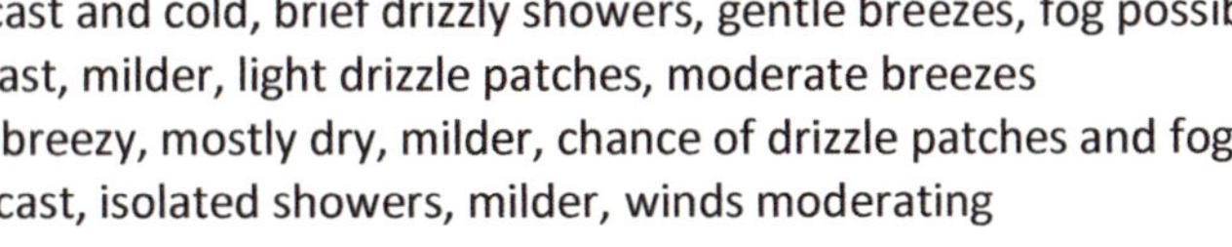
Belfast: Overcast and cold, brief drizzly showers, gentle breezes, fog possible
Dublin: Overcast, milder, light drizzle patches, moderate breezes
Cork: Cloudy, breezy, mostly dry, milder, chance of drizzle patches and fog
Galway: Overcast, isolated showers, milder, winds moderating

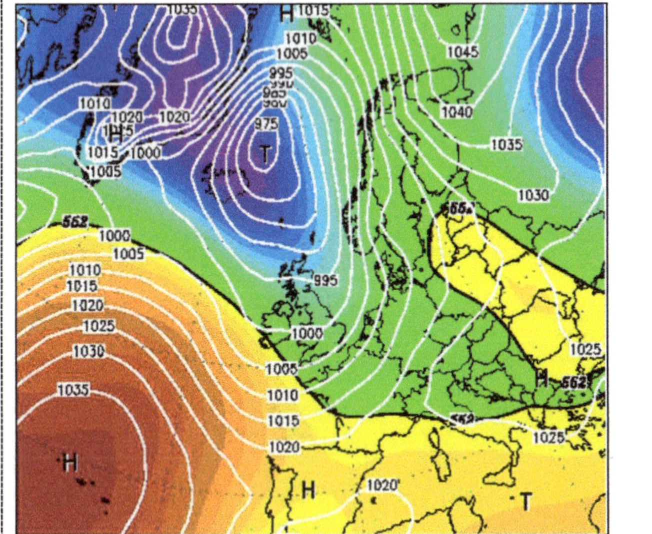

JANUARY 29 MONDAY

Low pressure directs a fresh to moderate westerly flow across the country. Widespread cloud, dull and gloomy throughout with chance of overnight mist or fog patches, particularly in the north and east. Mostly dry at first with isolated coastal showers developing, and rain in the north and west later in the day.

Belfast: Widespread cloud, dry, milder, moderate breezes, chance of fog patches
Dublin: Dry, dull and gloomy air, milder, fresh breezes at times, chance of misty patches
Cork: Mostly overcast, dry, warmer, moderate breezes
Galway: Overcast, scattered showers, strong blustery winds at times

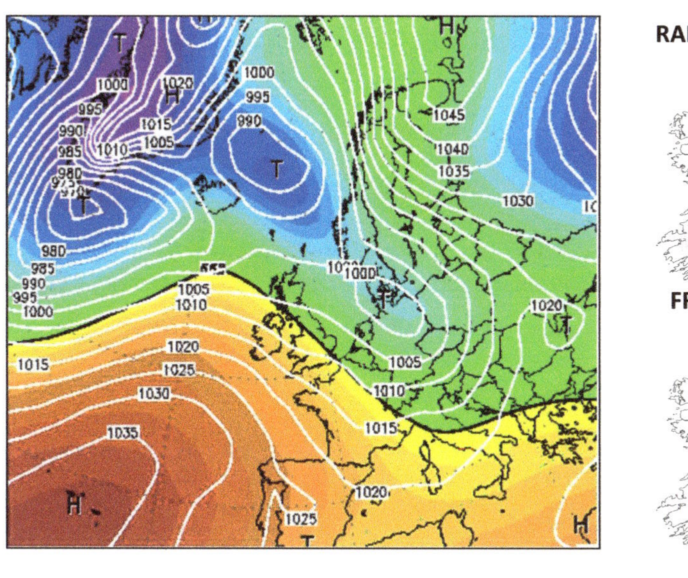

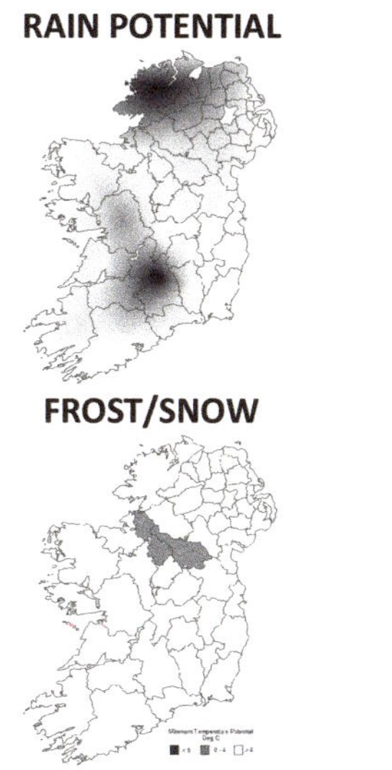

JANUARY 30 TUESDAY

Depression northwest of Ireland brings further widespread cloud, scattered showers and strengthening southwest to westerly winds across the country. Winds likely to be very gusty at times, possible to gale force later in the day and rain and drizzle likely to be persistent in the northwest.

Belfast: Stormy air, scattered showers and drizzles, strengthening winds with gusts
Dublin: Mild, dry, dull and gloomy air with passing threats of rain, freshening winds
Cork: Overcast, windy spells, light showers
Galway: Overcast, blustery with winds strengthening later in the day, showers

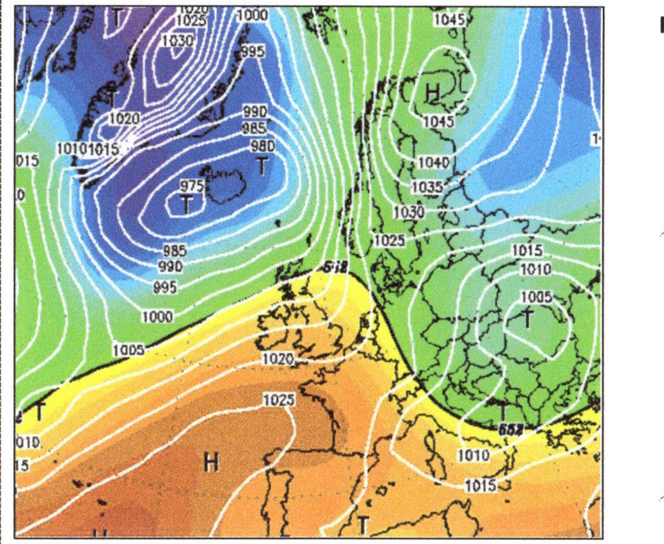

JANUARY 31 WEDNESDAY

Stormy air returns as the depression to the north of Ireland strengthens. An unsettled day with a mix of showers and brief sunny spells. Some showers may be wintry at times, with possible snow overnight, particularly in the west. Southwesterly winds prevail, possibly up to gale force with occasional gusts at times.

Belfast: Mix of sunny spells and showers, blustery, possible gales, mild day cold night
Dublin: Stormy air, mix of sunny spells and light rain, cold, freezing fog overnight
Cork: Strong unsettled winds, some sunny spells, isolated showers, colder
Galway: Stormy air, cloudy, colder, mostly dry, chance of overnight snow

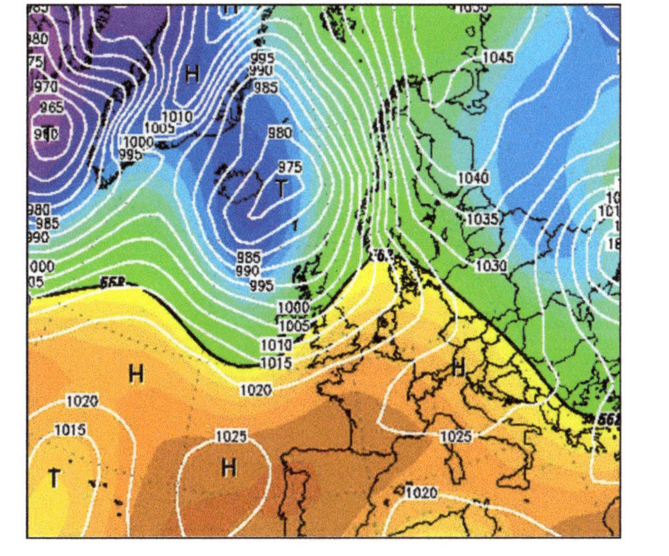

RAIN POTENTIAL

FROST/SNOW

FEBRUARY

PHASES OF THE MOON

2nd	Third Quarter
6th	Southern Declination
9th	New Moon
10th	Perigee #4
13th	Crossing Equator
16th	First Quarter
19th	Northern Declination
24th	Full Moon
25th	Apogee
27th	Crossing Equator

1	2	3	4	5	6	7	8	9	10	11	12	13	14	15	16	17	18	19	20	21	22	23	24	25	26	27	28	29
Thur	Fri	Sat	Sun	Mon	Tue	Wed	Thur	Fri	Sat	Sun	Mon	Tue	Wed	Thur	Fri	Sat	Sun	Mon	Tue	Wed	Thur	Fri	Sat	Sun	Mon	Tue	Wed	Thur
	3Q				V			N	P4			XhN			1Q			^					F	A		XhS		

MONTHLY SUMMARY

1st-11th An anticyclonic system to the south of the country extends high pressure over Ireland for a few days followed by another to the northeast. Mostly dry and sunny with light winds prevailing. Widespread frosts develop towards the end of this spell.
8th-11th Widespread frosts likely. Chance of freezing fog patches and severe ground frosts mostly in inland parts.
9th-10th Very sunny spell in many parts of the country, particularly in southern and western coastal fringes.
10th Fog likely. Very cold with severe frosts in many districts.

Over the next four weeks a couple of high pressure systems mostly situated to the southeast of Ireland will bring, will result in below average rainfall throughout the country. The first high pressure system will cover much of the first half of this coming outlook period, directing a mild south to southwest airflow over the country, then another high pressure flow, this time to the northwest, in the second half, with cooler north to northeast winds prevailing. A short low pressure period with a deep depression will prevail midway through the outlook period between the two high pressure dominance. While rainfall amounts will be below average throughout the country over the next four weeks, it will be particularly dry in the south, where a number of places may only see around a quarter of rainfall amounts to what is typical for this time of year. Sunshine hours will be above average in the northwest, while below normal in the east, with the best sunshine hours likely to be enjoyed during the second half of this outlook period. Temperatures will reflect the high pressure dominance during this period, with mean air temperatures slightly above normal in most parts, and up to a degree or more above the average in the north and northwest. Due to the prevalence of high pressure during this outlook period, winds will be generally lighter than is typical for this time of year. During this outlook period there are likely to be around 11 and 13 wet days in some western and southwestern districts, and around 6 to 10 wet days elsewhere. There may be up to 3-4 days with thunderstorms, hail on 9 days, snow briefly in the last week of this outlook period, and gales on 1 day. Frosts days will be slightly more than norm for this time of year at around 14 to 19 ground frosts in inland parts, while low than normal elsewhere with some coastal areas possibly not seeing any frosts. Fogs are likely to be only seen on around 7 days.

11th-12th Very cold with severe frosts in many districts.

12th-21st High pressure to the southeast of the country brings a mostly settled period of dry conditions with frequent cloud as it slowly crosses the country. Winds generally a light south to southwest flow, with slightly warmer than normal temperatures and few frosts.
13th-15th Scattered patches of fog likely.
18th Passage of a cold front brings a brief spell of spotty light rain here and there.
22nd-25th A mixed of heavy rain at times as well as fog over the country. Very heavy rain likely in the Maam Valley early in this spell.
20th-21st Widespread frosts, with very low overnight temperatures prevailing, especially in eastern counties.
22nd-28th Low pressure to the northwest of Ireland directs a series of fronts over the country with spells of heavy rain now and then, clearing later to showers and some thunderstorms at times.
23rd-26th Warmer than normal spell.
24th-25th Rain eases to showers. Chance of thunderstorms.
25th Very blustery with chance of gale winds in the northwest.
25th-27th Milder spell with strong winds prevailing, occasionally to gale force around the western coasts.
25th-28th Chance of thunderstorms and hail showers.
27th A deep depression is likely to pass near the north coasts.
28th-3rd March Bright and sunny spell over much of the country.
29th Widespread frosts.
29th-10th March High pressure lingers to the northwest of the country extending a cool north to northeast flow over the country. Mostly dry spell with colder temperatures than normal and widespread ground frosts throughout.

IRELAND RAINFALL ESTIMATES

February 2024

d=mainly dry, n= nonrecordable, l=light shrs, s=significant shrs, r=rain, h=hvy falls

		ULSTER PROVINCE					CONNAUGHT PROVINCE						LEINSTER PROVINCE				MUNSTER PROVINCE						
		Alderg	Hillsbo	Armag	Loughl	Malin H	Sligo	Ardtarl	Belmul	Drums	Galway	Roscoi	Markre	Edende	Kilkenn	Dublin	Shannc	Newpo	Tralee	Killarne	Cork	All	Moon
1st	Thur	l	d	l	d	d	d	d	l	d	d	d	d	d	d	d	d	d	l	d	d	4	
2nd	Fri	d	d	d	d	d	d	d	d	d	d	d	d	d	d	d	d	d	d	d	d	0	3Q
3rd	Sat	d	d	d	d	d	d	d	d	d	d	d	d	d	d	d	d	d	d	d	d	0	
4th	Sun	d	d	d	d	d	d	d	d	d	d	d	d	d	d	d	d	d	d	d	d	0	
5th	Mon	d	d	d	d	d	d	d	d	d	d	d	d	d	d	d	d	d	d	d	d	0	
6th	Tue	d	d	d	d	d	d	d	d	d	d	d	d	d	d	d	d	d	d	d	d	0	V
7th	Wed	d	d	d	d	d	d	d	d	d	d	d	n	d	d	l	d	d	d	d	d	1	
8th	Thu	d	d	d	d	d	d	d	d	d	d	d	d	d	d	d	d	d	d	d	d	1	
9th	Fri	d	d	d	d	d	d	d	d	d	d	d	d	d	d	d	d	d	d	d	d	0	N
10th	Sat	d	d	d	d	d	d	d	d	d	d	d	n	d	d	d	d	d	d	d	d	0	P4
11th	Sun	d	d	d	d	d	d	d	d	d	d	d	n	d	d	d	d	d	d	d	d	0	
12th	Mon	d	d	d	d	d	d	l	d	l	n	l	d	n	d	d	d	l	l	n	d	7	XhN
13th	Tue	d	d	d	d	d	d	d	d	d	d	d	d	d	d	d	d	d	d	d	d	1	
14th	Wed	d	d	d	d	d	d	d	d	d	d	d	d	d	d	d	d	d	d	d	d	0	
15th	Thur	d	d	d	d	d	d	d	d	d	d	d	d	d	d	d	d	d	d	d	d	0	
16th	Fri	d	d	d	d	d	d	d	d	d	d	d	d	d	d	d	d	d	d	d	d	0	1Q
17th	Sat	d	d	d	d	d	d	d	d	d	d	d	d	d	d	d	d	d	d	d	d	0	
18th	Sun	l	l	d	s	s	s	s	s	s	s	s	s	l	l	s	l	r	l	s	l	82	
19th	Mon	l	d	l	d	n	l	l	l	l	d	l	d	d	d	d	n	l	l	n	d	12	^
20th	Tue	d	d	d	d	d	d	d	l	d	d	d	l	d	d	d	d	d	d	d	d	2	
21st	Wed	d	d	d	d	d	n	d	d	d	d	n	s	d	d	d	d	l	d	l	d	4	
22nd	Thu	l	s	d	r	l	r	r	s	s	r	s	r	s	s	l	s	h	l	s	l	151	
23rd	Fri	l	l	s	l	s	l	l	r	l	l	l	s	l	d	l	l	s	l	s	d	59	
24th	Sat	l	s	l	s	l	s	s	r	s	l	l	s	s	s	s	l	r	h	r	s	121	F
25th	Sun	s	s	s	r	l	r	r	h	r	r	s	s	s	r	r	s	r	r	h	r	204	A
26th	Mon	s	l	s	s	s	s	s	s	s	s	s	r	l	l	l	s	s	s	s	l	122	XhS
27th	Tue	l	l	l	s	s	l	s	r	s	l	s	r	l	d	d	s	s	s	r	l	94	
28th	Wed	l	d	l	d	d	d	l	s	l	l	n	l	d	d	d	l	l	r	l	l	37	
29th	Wed	d	d	d	d	d	d	d	l	d	d	d	n	d	l	d	l	d	l	d	d	6	
Estimate:		26	27	25	47	36	45	48	80	44	50	40	56	29	29	26	38	90	78	70	24	908	
Average:		58	68	54	91	87	92	94	97	78	88	98	92	70	57	49	76	127	141	103	98	1717	
Trend:		drr	drr	drr	drr	drr	drr	drr	drr	drr	drr	drr	drr	drr	drr	drr	drr	drr	drr	drr	drr	drr	

		ULSTER PROVINCE					CONNAUGHT PROVINCE						LEINSTER PROVINCE				MUNSTER PROVINCE						
		Alderg	Hillsbo	Armag	Loughl	Malin H	Sligo	Ardtarı	Belmul	Drums	Galway	Roscoı	Markre	Edende	Kilkenı	Dublin	Shannc	Newpo	Tralee	Killarne	Cork	All	Moon
1st	Thur	c	c	c	c	o	c	c	c	o	o	o	c	c	pc	pc	o	c	c	c	pc	51	
2nd	Fri	o	o	o	o	c	o	o	o	o	o	o	o	o	o	o	c	o	o	o	o	7	3Q
3rd	Sat	o	o	o	o	o	o	o	o	o	o	c	o	o	o	o	o	o	o	o	o	5	
4th	Sun	o	o	o	o	o	c	o	o	o	o	o	o	o	o	o	o	o	o	o	o	6	
5th	Mon	o	c	c	o	o	o	c	pc	c	c	c	c	c	c	c	c	c	pc	pc	o	41	
6th	Tue	c	c	c	c	c	o	c	pc	o	c	c	c	c	o	c	o	o	o	o	o	29	V
7th	Wed	o	o	o	c	c	o	o	o	o	o	o	o	o	o	o	o	o	pc	pc	o	22	
8th	Thu	c	pc	pc	c	c	pc	pc	pc	o	pc	o	c	c	pc	pc	pc	pc	pc	pc	pc	91	
9th	Fri	pc	F	F	c	c	F	F	F	o	pc	o	pc	pc	F	pc	pc	pc	pc	pc	F	121	N
10th	Sat	c	c	c	o	o	o	c	F	o	pc	o	c	c	c	c	pc	pc	pc	pc	pc	65	P4
11th	Sun	c	c	c	c	c	c	c	pc	o	c	o	c	c	o	c	o	o	o	o	o	26	
12th	Mon	c	c	c	o	o	o	c	o	o	o	o	c	c	c	c	o	o	c	c	o	22	XhN
13th	Tue	o	o	o	o	o	o	o	o	o	c	o	o	o	o	o	c	c	c	c	o	13	
14th	Wed	o	o	o	o	o	o	o	o	o	o	o	o	o	o	o	o	o	o	o	o	5	
15th	Thur	o	o	o	o	o	o	o	o	o	o	o	o	o	o	o	o	o	o	o	o	2	
16th	Fri	o	o	o	c	c	o	o	o	c	o	c	o	o	o	o	o	o	o	o	o	8	1Q
17th	Sat	o	o	o	c	c	o	o	o	c	o	c	o	o	o	o	o	o	o	o	o	7	
18th	Sun	o	o	o	o	c	o	o	o	c	o	c	o	o	o	o	o	o	o	o	o	7	
19th	Mon	c	pc	pc	c	o	c	pc	c	o	c	o	c	pc	pc	pc	pc	pc	pc	pc	c	69	^
20th	Tue	c	c	c	c	o	c	c	pc	o	pc	o	c	c	c	c	pc	pc	c	c	pc	59	
21st	Wed	c	c	c	c	c	o	c	o	o	o	c	c	c	c	c	c	c	o	o	o	23	
22nd	Thu	o	o	o	c	c	o	o	c	c	o	c	c	o	o	o	o	o	o	o	o	14	
23rd	Fri	o	o	o	o	c	o	o	o	c	o	c	o	o	o	o	o	o	o	o	o	8	
24th	Sat	o	o	o	o	o	o	o	o	o	o	c	c	o	o	o	o	o	o	o	o	12	F
25th	Sun	c	c	c	c	c	c	c	c	c	c	c	c	c	pc	pc	c	c	c	c	c	55	A
26th	Mon	c	pc	pc	c	c	pc	pc	pc	c	pc	c	c	pc	pc	pc	pc	pc	c	c	pc	81	XhS
27th	Tue	c	c	c	c	c	o	c	o	c	o	c	c	c	c	c	c	c	o	o	c	25	
28th	Wed	c	pc	pc	c	c	pc	pc	c	c	c	c	c	pc	pc	pc	o	c	c	c	c	64	
29th	Thur	c	c	c	c	c	pc	c	pc	c	pc	c	c	o	c	o	c	c	pc	pc	c	60	
Estimate hours:		43	53	53	38	33	46	53	69	29	62	31	45	52	56	60	54	55	56	56	52	997	
Average hours:		64	64	62	63	62	62	62	67	60	64	64	65	62	64	70	66	61	62	60	64	1268	
Trend:		less	less	av	less	less	less	av	av	less	av	less	less	av	av	av	less	av	av	av	less	less	

FEBRUARY 1 THURSDAY

A large high pressure system sits over the country bringing a mostly dry and light west to southwest flow throughout. Some sunny spells in the south and east, with increased cloud in the west and north. Chance of some drizzle patches mostly in the west, and fog pockets in the east and west. Frosts likely mostly in inland parts.

Belfast: Mostly cloudy and dry, cool, moderate breezes
Dublin: Partly cloudy, cold, dry, light breezes, chance of frosts in sheltered pockets
Cork: Scattered sunny spells, dry, mild day cold night, calm air, fog possible
Galway: Mostly overcast and dry, calm, overnight frost and shallow fog patches likely

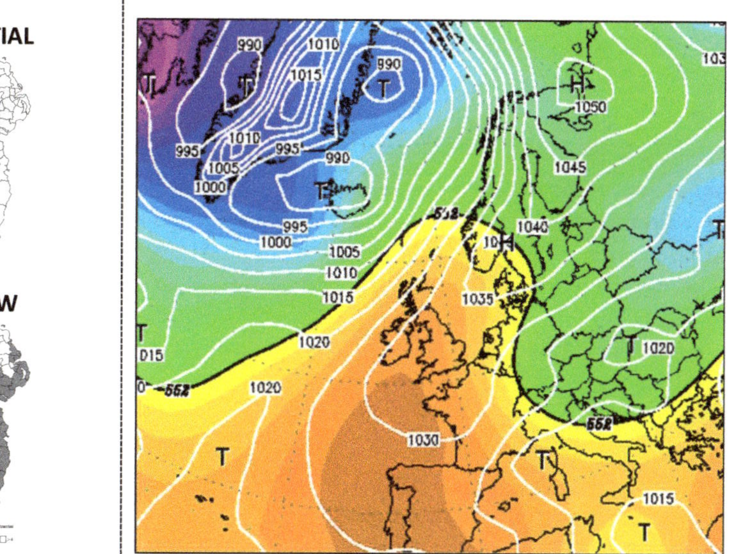

RAIN POTENTIAL

FROST/SNOW

FEBRUARY 2 FRIDAY

A large anticyclonic system continues to linger over Ireland bringing a mostly dry and overcast day throughout with light to moderate southerly breezes prevailing, fresher at times in the north and west. Chance of some drizzly spells mostly around the western coasts.

Belfast: Overcast, dry, passing threats of rain, mild day cold night, moderate breezes
Dublin: Overcast, dry, frosts, calm and serene air
Cork: Dry, overcast, dull and gloomy air, light to mostly calm air
Galway: Mostly overcast and dry, light breezes

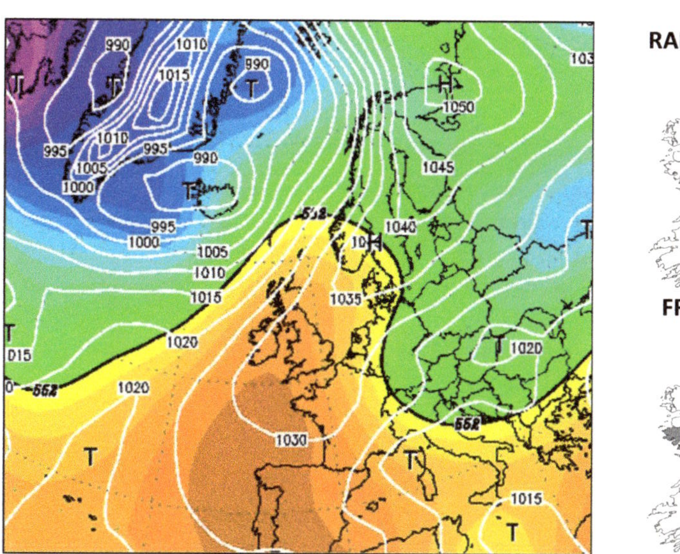

RAIN POTENTIAL

FROST/SNOW

FEBRUARY 3 SATURDAY

High pressure continues to prevail over Ireland bringing further widespread cloud and mostly dry conditions with winds generally light south-easterlies, but stronger at times in the northwest. Pockets of overnight frosts likely in some eastern districts, and possible light showers for a brief time in the far northwest. Milder than of recent, and some brief sunny breakthroughs likely in the Midlands.

Belfast: Overcast, cool, dry, fresh breezes at times
Dublin: Dull and gloomy air, dry, cold, light breezes
Cork: Overcast, dry with passing threats of rain, light breezes
Galway: Overcast, dry, moderate breezes, milder

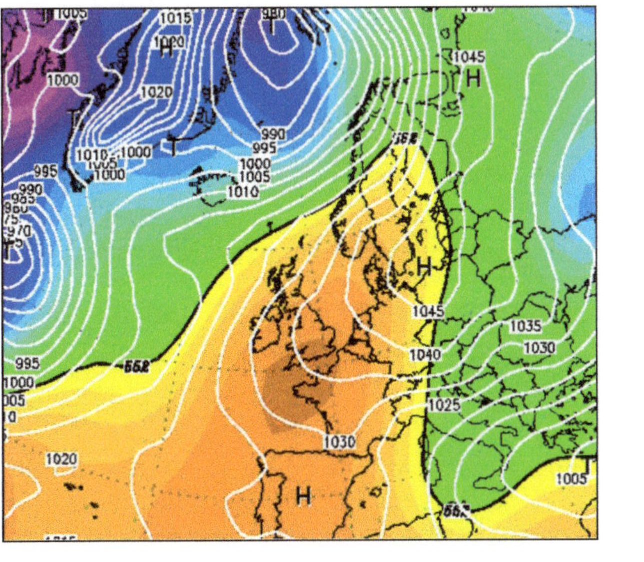

RAIN POTENTIAL

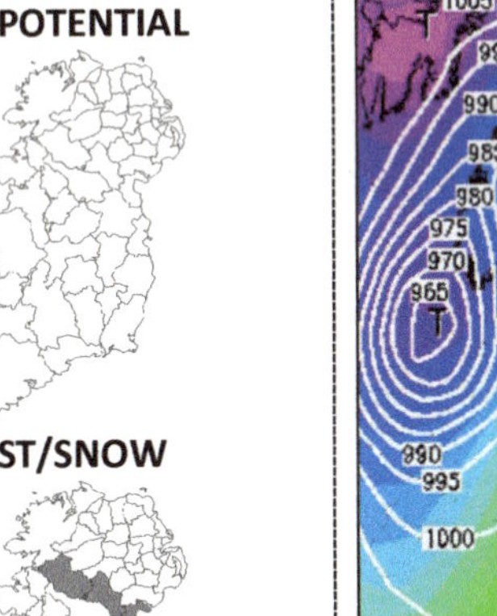

FROST/SNOW

FEBRUARY 4 SUNDAY

High pressure drifts off Ireland allowing a low pressure trough to slowly move onto the country in a light southeast backing northeasterly flow. Dull and gloomy everywhere, with some light misty drizzles likely in the far northern districts, while mostly dry elsewhere.

Belfast: Overcast, light breezes, cold, dr, odd threat of rain, chance of misty patches
Dublin: Overcast, dry, gloomy, cold, light shifting breezes
Cork: Overcast, cooler, dry, threats of rain, light breezes
Galway: Overcast, mostly dry with threats of rain, moderate breezes

RAIN POTENTIAL

FROST/SNOW

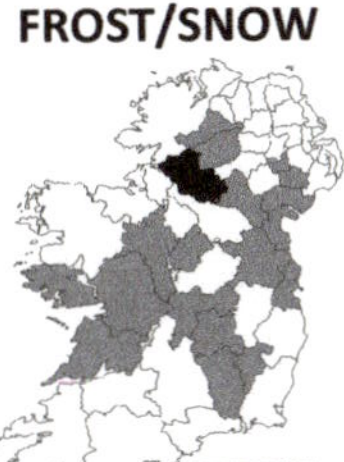

FEBRUARY 5 MONDAY

High pressure drifts back over Ireland in a gentle easterly flow, bringing a dry and mostly calm and serene air throughout the country, with pockets of frosts and mists likely mostly in the west, north, and inland parts. Widespread cloud prevails although some sunny spells are likely to break through, particularly in the western half of the country.

Belfast: Mostly calm air, cloudy, cool, dry, mists likely
Dublin: Cloudy and mostly calm air, slightly warmer, dry
Cork: Mostly cloudy, dry, cool, light breezes to mostly calm air
Galway: Dry, mostly calm, brief sunny spells, misty patches and frost pockets likely

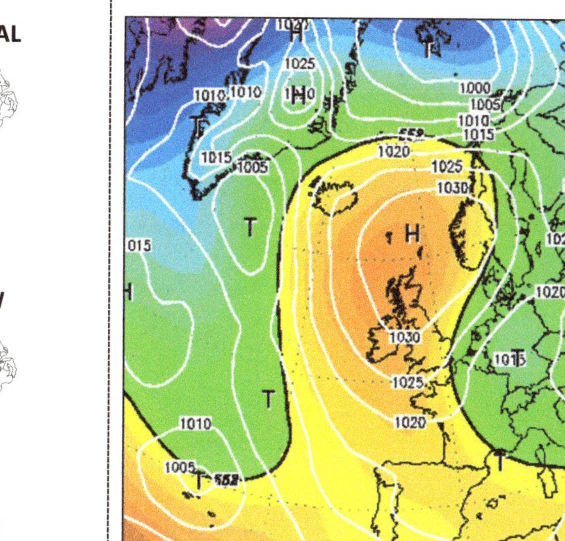

RAIN POTENTIAL

FROST/SNOW

FEBRUARY 6 TUESDAY

High pressure continues to dominate the country. Mostly cloudy and dry throughout with some pockets of frost here and there and the odd sunny breakthrough. Winds generally light and in an east to southeast flow. Some freshening breezes around the southern coasts.

Belfast: Calm, cloudy, cold, dry
Dublin: Cloudy, dry, light shifting breezes
Cork: Overcast, cold and dry, some threats of rain, windy at times
Galway: Partly cloudy, dry, cold, light breezes

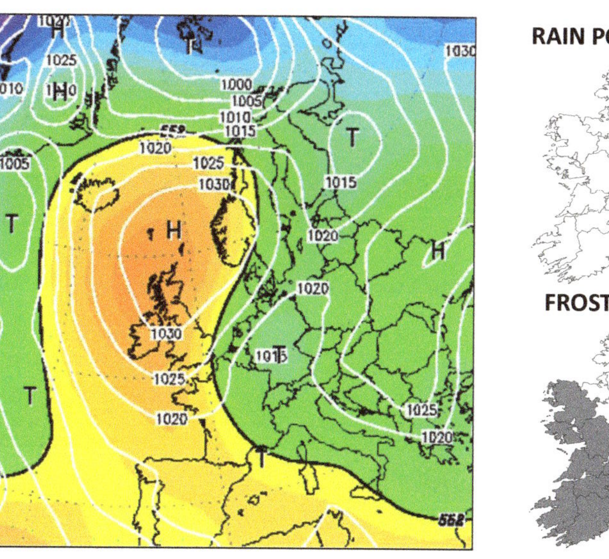

RAIN POTENTIAL

FROST/SNOW

FEBRUARY 7 WEDNESDAY

A large anticyclonic system continues to linger over Ireland in a light northerly flow throughout, fresher around the exposed coasts at times. Widespread cloud prevails and mostly dry except for the chance of some localized showers along the eastern coasts, and sunnier spells in the far southwest. Scattered frosts, some severe in the northwest and midlands region likely.

Belfast: Dull and gloomy air, cold with frosts, dry, mostly calm
Dublin: Calm and quiet air, dull, gloomy, drizzle patches, cold
Cork: Mostly overcast, cold and dry, chance of drizzle patches, winds easing to light
Galway: Cold, overcast and gloomy, mostly dry, frost patches possible, light breezes

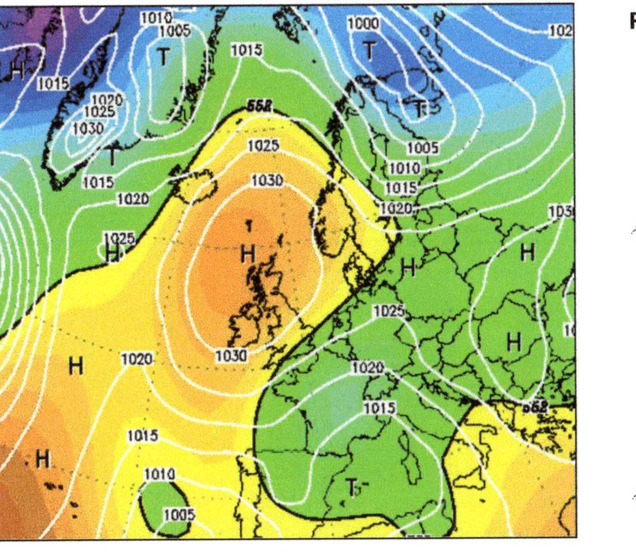

RAIN POTENTIAL

FROST/SNOW

FEBRUARY 8 THURSDAY

Light northeasterly breezes prevail under the influence of a continuing anticyclonic system. Very cold overnight with pockets of frost in sheltered parts. Mostly dry throughout except for the chance of some isolated showers towards the east. Good sunny spells almost everywhere, except in the far north, where cloudier conditions will prevail.

Belfast: Calm, dry, cold, serene air, chance of pockets of frosts
Dublin: Occasional sunny spells, dry, frosts, gentle breezes
Cork: Dry, cold, occasional sunny spells, gentle breezes
Galway: Dry and cold, frosts, occasional sunny spells, gentle breezes

RAIN POTENTIAL

FROST/SNOW

FEBRUARY **9** FRIDAY

Anticyclonic system remains stalled over Ireland bringing light variable breezes, mostly sunny and dry conditions throughout. Very cold overnight with widespread frosts and scattered fog patches likely, with fog slow to clear in many districts, particularly in central parts.

Belfast: Sunny and dry, cold, severe frosts, calm air with fog patches likely
Dublin: Sunny, dry, cold, frosts, calm and serene atmosphere
Cork: Sunny, dry, cool, light westerly breezes at times
Galway: Mostly sunny, cool and dry, overnight frosts, light winds

RAIN POTENTIAL

FROST/SNOW

FEBRUARY **10** SATURDAY

Anticyclonic. Mostly dry throughout in a mostly calm to very light easterly air. Very cold with severe frosts in many districts and freezing fogs which may be slow to clear in some places and may form some light snow flurries in northern parts. Mostly sunny in the west, with increased scattered cloud to the east, and mostly cloudy in the northern and southern districts.

Belfast: Cloudy, calm, very cold, severe frosts, freezing fogs, chance of snow flurries
Dublin: Cloudy, dry, scattered frosts, serene atmosphere
Cork: Partly cloudy, dry and cold, chance of overnight frost and fog pockets
Galway: Mostly sunny, dry, severe frosts, calm air, mist and fog patches overnight

RAIN POTENTIAL

FROST/SNOW

FEBRUARY 11 SUNDAY

High pressure remains stalled over Ireland bringing a calm and serene air throughout the country. Very cold overnight with frost and fog patches in many districts. Frost may be very severe in central parts, and fogs may be slow to clear. Mostly dry everywhere with a mix of some sunny pockets and thick cloud elsewhere. Chance of freezing fogs in the north that may form some snow flurries.

Belfast: Calm, cloudy, very cold, severe frosts, freezing fog likely, possible snow flurries
Dublin: Cloudy, colder with severe frosts in places, dry, calm air
Cork: Overcast, dry, cold, calm and gloomy air
Galway: Cloudy, dry, cold, severe frosts, calm air, misty patches likely

FEBRUARY 12 MONDAY

Anticyclonic, mostly dry, cloudy, and calm throughout with light south to southeast winds prevailing. Scattered frosts and fog patches likely, with severe frosts likely in inland parts. Chance of some foggy patches in the some southern parts turning to drizzle for a time.

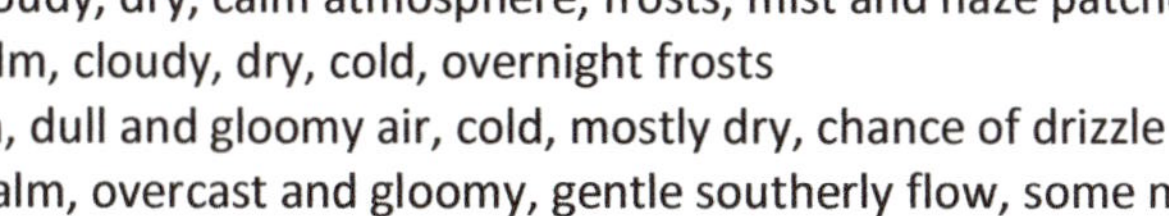

Belfast: Cloudy, dry, calm atmosphere, frosts, mist and haze patches likely
Dublin: Calm, cloudy, dry, cold, overnight frosts
Cork: Calm, dull and gloomy air, cold, mostly dry, chance of drizzle outbreaks
Galway: Calm, overcast and gloomy, gentle southerly flow, some misty drizzles

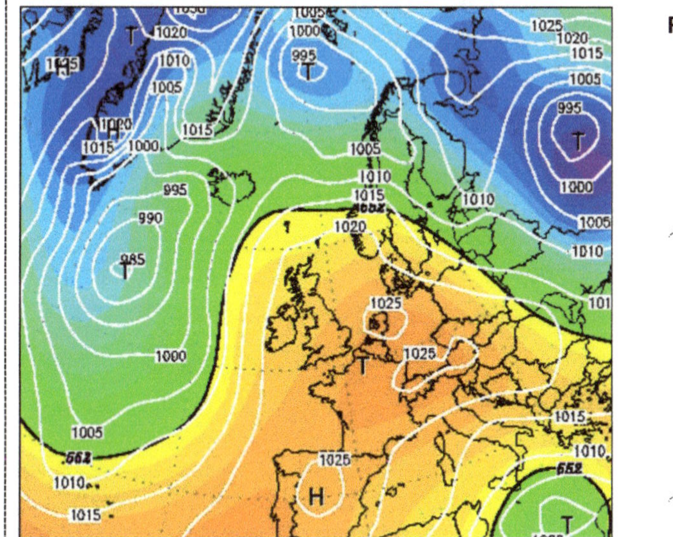

FEBRUARY **13** TUESDAY

High pressure remains stalled over Ireland with low pressure to the west directing a light southerly flow at times. Widespread cloud throughout with patches of rain or drizzle in some western and northern parts, while mostly dry with mists and fog elsewhere. Some pockets of frosts likely mostly in central parts and chance of misty drizzles turning to light snow in the north.

Belfast: Overcast, light southerly air, mostly dry, fog likely, chance of snow flurries
Dublin: Overcast, cold with scattered frosts and fog patches, light breezes, mostly dry
Cork: Overcast, mostly dry with passing threats of rain, moderate southerly flow
Galway: Cloudy, milder, moderately southerly flow, dry

RAIN POTENTIAL

FROST/SNOW

FEBRUARY **14** WEDNESDAY

A large anticyclonic system continues to linger over Ireland bringing mostly dry conditions very light variable breezes throughout. Dull and gloomy with widespread cloud everywhere, with scattered fog in many districts and chance of some frost in eastern parts. Chance of fog patches in the north turning to light snow at times.

Belfast: Dull and gloomy, calm, dry, threats of rain, fog likely, chance of snow flurries
Dublin: Overcast, shallow overnight fog patches likely, dry, calm air
Cork: Overcast, dry, calm, fog likely
Galway: Overcast, dry, calm, mists likely

RAIN POTENTIAL

FROST/SNOW

FEBRUARY **15** THURSDAY

Widespread cloud continues to prevail throughout Ireland as the large high pressure system remains stalled over the country. Winds generally light and variable, with the occasional moderately southwesterly change mostly around the northwest coasts. Dry with fog in most parts, except for the chance of some freezing fog patches turning to snow flurries possible in the north and threats of rain in the west.

Belfast: Overcast, calm, mostly dry, frosts, freezing fog patches, snow flurries possible
Dublin: Cold, calm, dry, widespread cloud, mist patches likely
Cork: Calm and serene air, overcast, dry, cold, chance of misty patches
Galway: Overcast, milder, dry, passing threats of rain, calm atmosphere

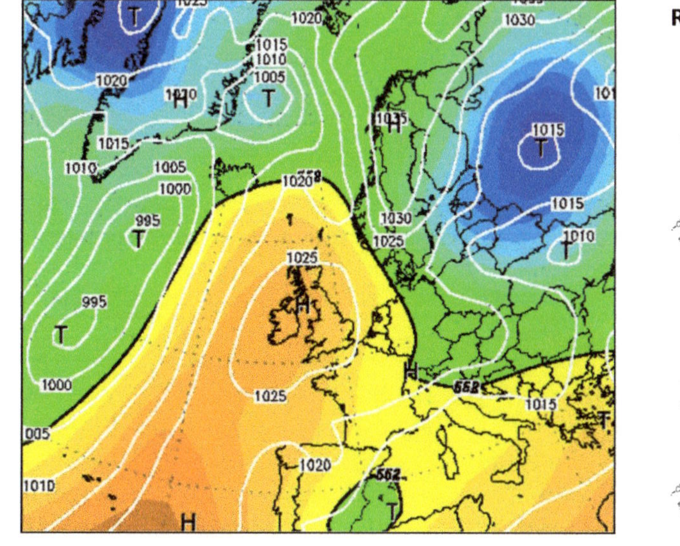

RAIN POTENTIAL

FROST/SNOW

FEBRUARY **16** FRIDAY

High pressure remains stalled over Ireland as it strengthens. Very dull and gloomy everywhere, with scattered frosts and fog patches and light variable breezes to mostly calm air, except for some moderate southwesterly breezes mostly around the northwestern coastal fringes. Mostly dry throughout with passing threats of rain.

Belfast: Calm, very dull, milder day, frosts and fog patches, dry with odd threat of rain
Dublin: Overcast and gloomy, cold, dry, calm air
Cork: Overcast and dry, passing threats of rain, cool, calm
Galway: Cool, dry, dull and gloomy, calm and misty air

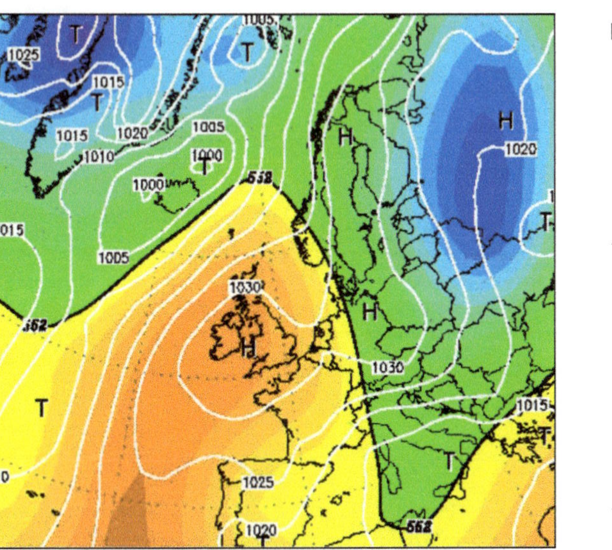

RAIN POTENTIAL

FROST/SNOW

FEBRUARY **17** SATURDAY

The large anticyclonic system that has dominated of recent begins to slowly drift southwards in a gentle to moderate southwest flow, fresher with some gusts developing around the northwest coasts. Overcast conditions prevail everywhere with increasing threats of rain, some misty patches and the occasional outbreak of light rain or drizzle spells.

Belfast: Overcast, dry, increasing threats of rain, freshening breezes, possible mists
Dublin: Overcast, dry with increasing threats of rain, milder, moderate breezes
Cork: Overcast, dry, milder, increasing threats of rain, light winds freshening later
Galway: Overcast, dry with increasing threats of rain, milder, moderate breezes

FEBRUARY **18** SUNDAY

The high pressure ridge moves to the south of Ireland allowing low pressure troughs to move into the north in a strengthening westerly flow. Overcast and mostly dry with scattered misty patches at first before a stormy cold front crosses the country ahead of an unsettled northwesterly flow bringing scattered wintry showers and possible gale gusts in the north at first, moving down into the southern districts later in the day.

Belfast: Overcast, milder, blustery winds, drizzly showers
Dublin: Overcast, occasional showers, freshening winds with occasional gusts
Cork: Overcast, drizzle outbreaks, fluctuating winds, strong at times
Galway: Unsettled air, overcast, scattered showers, milder, blustery at times

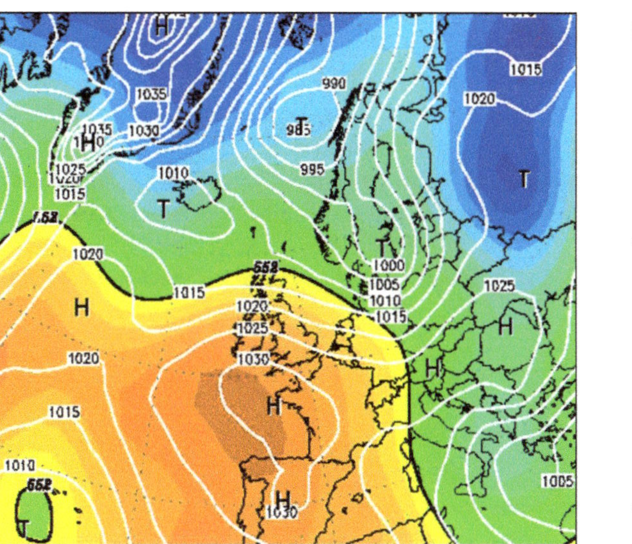

FEBRUARY 19 MONDAY

A high pressure system ridges to the southwest while a depression drifts to the east of Ireland with a fresh north to northwesterly airflow prevailing, reaching possible gale gusts at times. Light showers clearing to mostly dry with some sunny spells throughout, except for the chance of some lingering showers mostly in the northwest of the country. Chance of overnight frosts developing.

Belfast: Dry, some sunny spells, mild day cold night, unsettled air, gale gusts at times
Dublin: Partly cloudy and dry, blustery at times, cooler, overnight frosts possible
Cork: Mostly dry, windy at times, brief sunny outbreaks, chance of light drizzle
Galway: Dry and colder, brief sunny spells, fluctuating winds

RAIN POTENTIAL

FROST/SNOW

FEBRUARY 20 TUESDAY

High pressure ridges back up over the country again, directing a light and cold northerly air throughout the country, with severe frosts in many districts overnight. Mostly dry and cloudy except for the chance of some coastal showers in the west. Some sunny outbreaks likely, more frequently in the east.

Belfast: Cloudy, dry, mild day cold night, possible frost pockets, cold northerly winds
Dublin: Cloudy, colder, dry, frosts, light breezes
Cork: Partly cloudy, dry, cool, chance of overnight frosts, light northerly flow
Galway: Partly cloudy, dry, cold with chance of overnight frost pockets

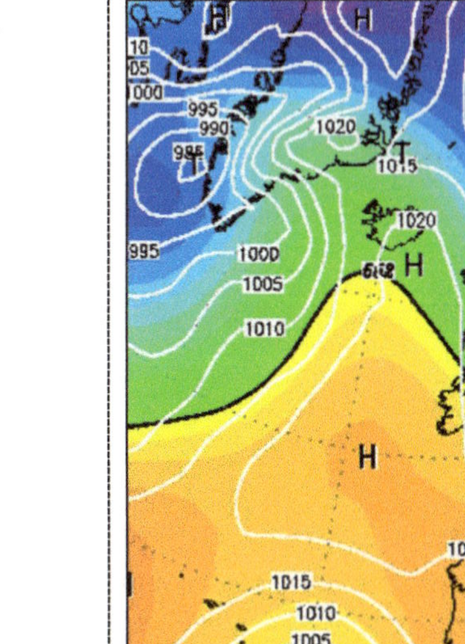

RAIN POTENTIAL

FROST/SNOW

FEBRUARY 21 WEDNESDAY

Anticyclonic. Very cold and dry with severe frosts in many districts. Widespread cloud prevails, particularly in the west and south, with some sunny breakthroughs in the east and north. Winds generally variable and light, backing south to southeasterly later in the day, fresher at times around the western and northern coastal fringes.

Belfast: Cloudy, colder, dry, light breezes, severe frosts likely
Dublin: Cloudy and dry, cold, frosts, light shifting breezes
Cork: Overcast, cooler, dry with increasing threats of rain, shifting breezes
Galway: Mostly overcast and dry, increasing threats of rain, light breezes, frosts

RAIN POTENTIAL

FROST/SNOW

FEBRUARY 22 THURSDAY

A depression to the northwest of Ireland pushes the high to the east of the country. Mostly overcast throughout with wintry light rain and drizzle patches starting at first in the west and slowly spreading eastward, although chance of some brief sunny breakthroughs possible mostly in the north. Winds generally light southerlies at first, turning southwesterly and fresher, particularly around the coasts.

Belfast: Overcast, frosts, scattered wintry showers, windy with gusts at times
Dublin: Overcast, drizzle patches, slightly warmer, light breezes
Cork: Overcast, patches of light rain, milder, fluctuating breezes
Galway: Dull and gloomy, wintry rain and drizzle, freshening winds later in the day

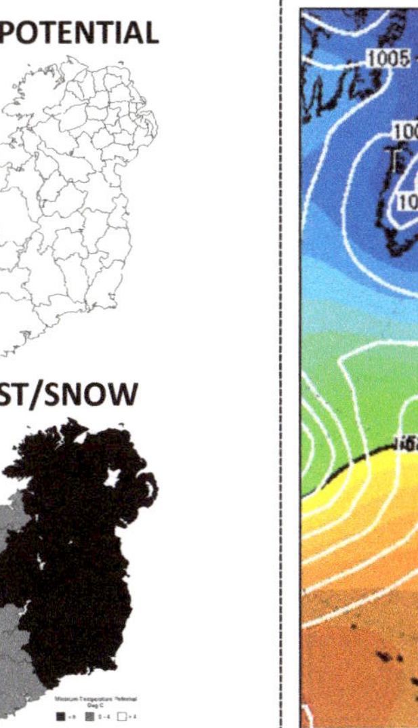

RAIN POTENTIAL

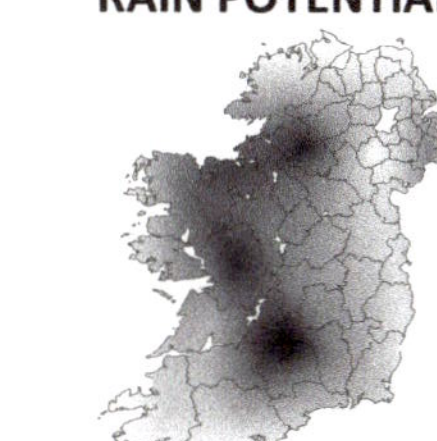

FROST/SNOW

FEBRUARY 23 FRIDAY

An intense depression west of Ireland directs a wet front in a breezy southwest airflow over the country. Scattered outbreaks of rain throughout, with some heavy falls here and there. Rain contracts later to the north, with showers turning to drizzles continuing to the south.

Belfast: Overcast, milder, passing drizzle outbreaks, blustery
Dublin: Milder, heavy cloud, light showers, freshening winds with occasional gusts
Cork: Unsettled and blustery air, milder, mostly dry, chance of isolated showers
Galway: Overcast, milder, outbreaks of light rain, windy with occasional gusts

FEBRUARY 24 SATURDAY

A stormy wet front associated with a depression to the west of Ireland crosses the country, spreading outbreaks of rain through most districts, sometimes heavy in places. Dull and gloomy throughout with showers becoming squally at times. Winds generally light to moderate in a southerly flow, but fresh and gusty at times with the passage of rain bands, and particularly blustery around the coastal fringes.

Belfast: Overcast, passing showers, milder, breezy spells, chance of overnight fog
Dublin: Overcast, milder, breezy at times, scattered showers
Cork: Overcast, windy, squally showers
Galway: Overcast, stormy air, light showers, gusty winds at times

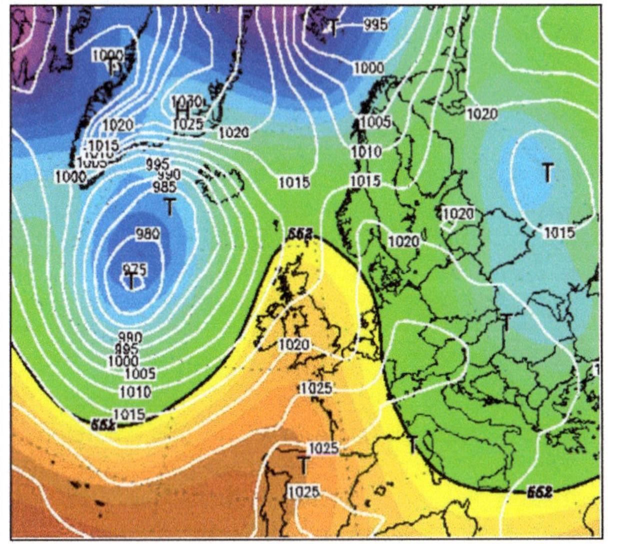

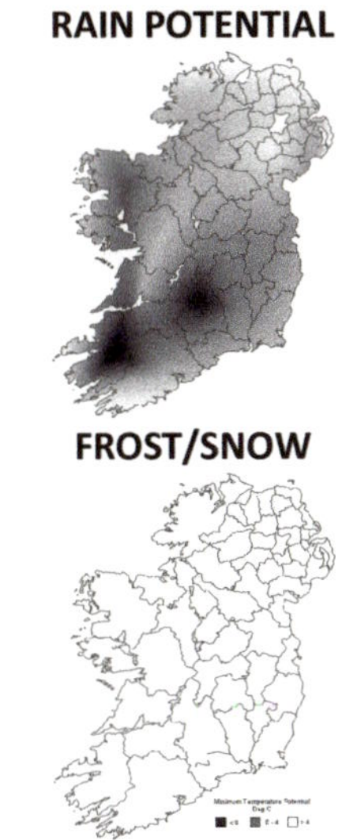

FEBRUARY **25** SUNDAY

Low pressure system directs a strong south to southwest flow across the country, possibly up to gale force at times. An active frontal system associated with the depression crosses the country bringing a mix of sunny spells and scattered showers, some of them heavy particularly around the eastern coasts. Wintry showers likely to turn to persistent rain for a time later in the day, particularly in the east and south.

Belfast: Mix of showers and some sunny spells, strengthening winds later in the day
Dublin: Changeable, rain, some sunny spells, strong winds with occasional gusts
Cork: Changeable, rain, some sunny breakthroughs, cold, unsettled winds
Galway: Unsettled air, cloudy, rain

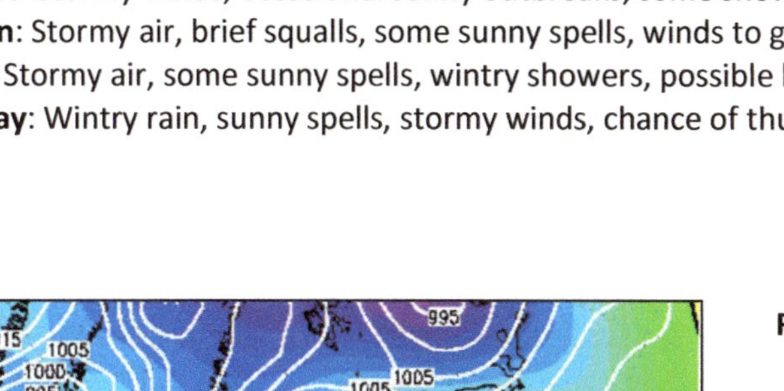

FEBRUARY **26** MONDAY

A depression centered to the immediate northwest of Ireland extends a stormy air throughout the country. Strong and unsettled southwesterly winds prevail, with gust possibly up to gale force at times. A mix of some sunny spells and wintry showers everywhere, including the chance of thunderstorms, hail showers and possible snow in the higher reaches.

Belfast: Stormy winds, occasional sunny outbreaks, some showers, mild day cold night
Dublin: Stormy air, brief squalls, some sunny spells, winds to gale force at times likely
Cork: Stormy air, some sunny spells, wintry showers, possible hail and gale winds
Galway: Wintry rain, sunny spells, stormy winds, chance of thunderstorms and snow

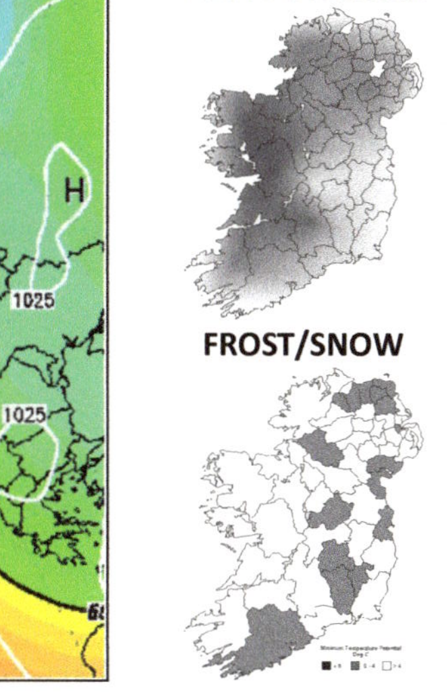

FEBRUARY **27** TUESDAY

Stormy conditions continue as the depression to the north of Ireland directs a strong and blustery southwest flow throughout, with winds up to gale force at times likely. Mixed conditions with thunderstorms and wintry scattered showers everywhere and the occasional sunny spell. Rain activity heaviest in the west, where some precipitation of hail or snow may be observed.

Belfast: Stormy air, cloudy, squally drizzle patches, cooler
Dublin: Stormy, cloudy, mostly dry, threats of rain, some drizzle likely, possible gales
Cork: Cloudy, wintry squally showers, cold, stormy air with gales likely
Galway: Stormy, overcast, squally showers, gale winds and thunderstorms possible

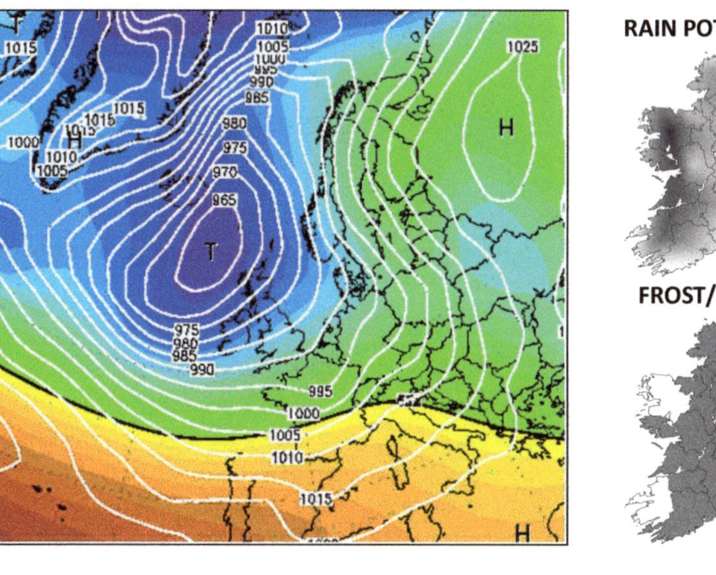

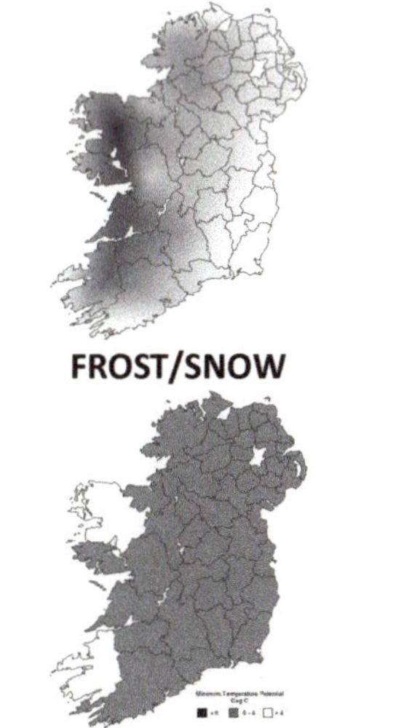

FEBRUARY **28** WEDNESDAY

Low pressure prevails over the country. Very cold with widespread overnight frosts, some severe in inland parts. Showers likely to become more persistent turning to rain as they spread in a northeasterly tract, with heavy rain at times likely in the southwest, and lighter rain elsewhere. Some sunny spells later in the day. Winds generally in a moderate westerly flow. Chance of light snow in some western parts.

Belfast: Cooler, windy, dry, some sunny spells
Dublin: Dry, some sunny spells, winds easing but blustery for a time, overnight frosts
Cork: Cloudy, blustery, brief showers, chance of gale gusts
Galway: Storm, cloudy, brief wintry showers, chance of thunderstorms and light snow

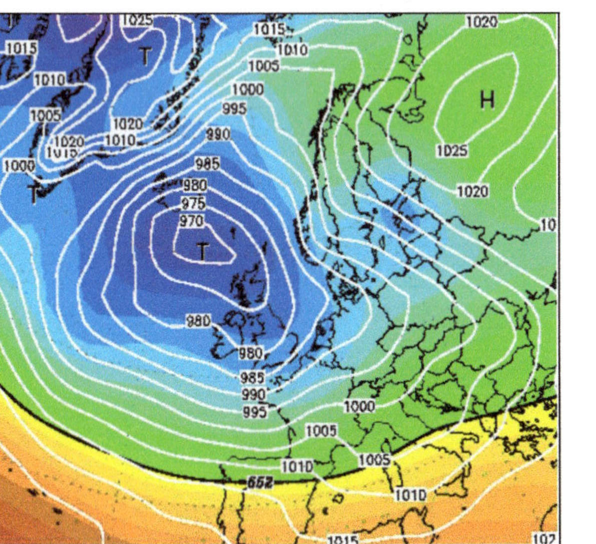

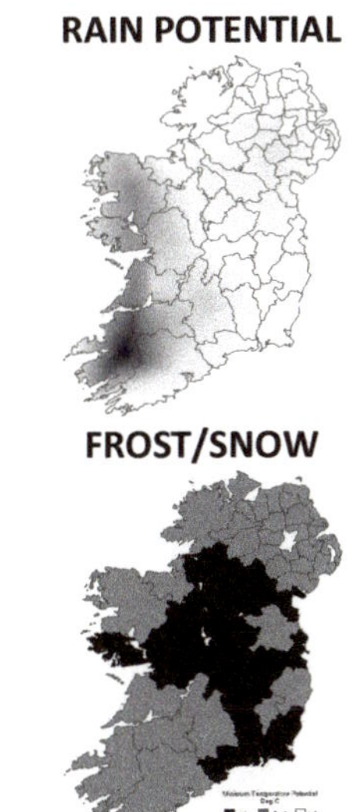

FEBRUARY 29 THURSDAY

Low pressure trough forms over the country with the depression centre to the northwest of Ireland and another forming to the southeast. Very cold with light variable winds everywhere, and severe and widespread overnight frosts. Mostly dry with some sunny spells during the day. Chance of overnight wintry showers, mostly around the coasts, and some possibly turning to snow flurries. Freezing fogs likely to develop, particularly in the north and east.

Belfast: Dry, calm air, sunny spells, overnight frosts, chance of snow flurries and fog
Dublin: Overcast, colder, frosts, light breezes to mostly calm air, fog possible
Cork: Cloudy, calmer air, dry, cool
Galway: Dry, occasional sunny breakthroughs, winds easing to light, frosts overnight

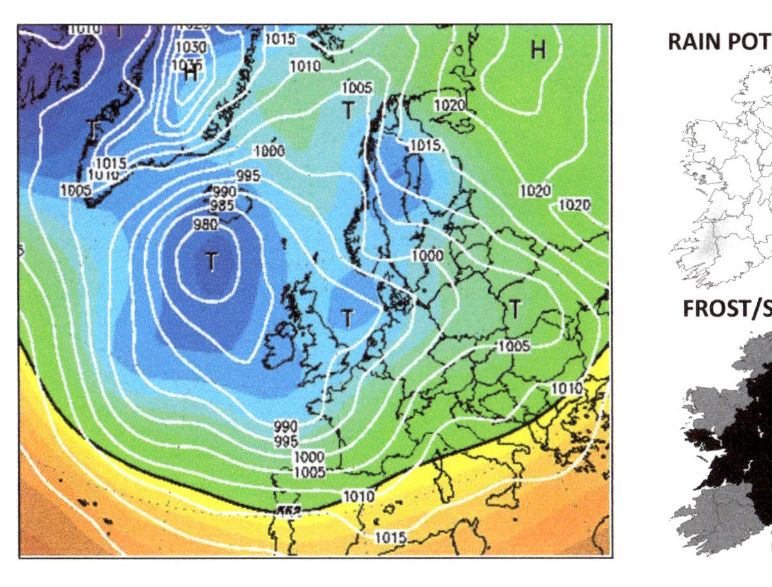

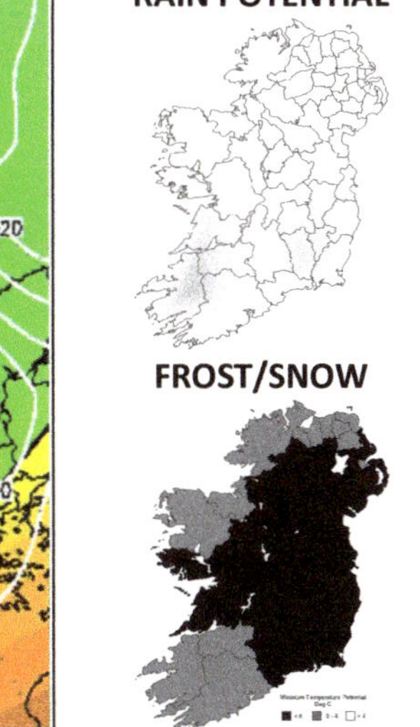

MARCH

PHASES OF THE MOON

3rd	Third Quarter
5th	Southern Declination
10th	New Moon
10th	Perigee #1
12th	Crossing Equator
17th	First Quarter
17th	Northern Declination
23rd	Apogee
25th	Full Moon
25th	Crossing Equator

1	2	3	4	5	6	7	8	9	10	11	12	13	14	15	16
Fri	Sat	Sun	Mon	Tue	Wed	Thur	Fri	Sat	Sun	Mon	Tue	Wed	Thur	Fri	Sat
		3Q		V					N P1		XhN				

17	18	19	20	21	22	23	24	25	26	27	28	29	30	31
Sun	Mon	Tue	Wed	Thur	Fri	Sat	Sun	Mon	Tue	Wed	Thur	Fri	Sat	Sun
1Q ^						A		F XhS						

MONTHLY SUMMARY

29th-10th March High pressure lingers to the northwest of the country extending a cool north to northeast flow over the country. Mostly dry spell with colder temperatures than normal and widespread ground frosts throughout.

1st Widespread frosts with very severe frosts likely in the southeast.

1st-3rd Bright and sunny spell over much of the country but duller in the east.

1st-6th Chance of hail showers.

3rd-5th Wintry showers, including those of snow, prevail mostly in the east but spread to more widespread later.

9th-10th The passage of a cold front brings colder temperatures, with wintry showers in many districts, including hail showers and snow.

10th Severe frosts may be widespread.

Over the next four weeks air pressure systems will be lower than normal for much of this period, except for a higher pressure spell midway through as a large anticyclone crosses to the north of the country. Winds will fairly typical for this time of year with the strongest winds likely during the second week of this outlook period. Most rain will be seen at the start and end of the next four weeks, with mostly dry conditions in the middle. As such, rainfall amounts are likely to be above normal throughout much of the country, and up to twice as much as normal in parts of the east and northeast, mostly due to a couple of significant wet weather periods, but a little below in some parts of the northwest.

Sunshine hours will be higher than normal at the start of the next four weeks, before widespread mostly cloudy conditions settle in for the remainder, resulting in below average sunshine hours for this time of year. Overall, sunshine hours will be average to slightly above average in the north, while below normal elsewhere.

It will also be much cooler than is normal for this time of year, especially in the first week ahead when light snowfalls may be widespread, and again in the middle of this outlook period, when severe frosts will occur and temperatures regularly below -10°C in inland parts and -5°C elsewhere. During this outlook period there is likely to be around 15-20 wet days (when 1mm or more rainfall is measured), 4 days when thunderstorms may be heard, 12 days with hail showers, 9 days with snowfalls and widespread fogs on 7 days.

11th-16th A low pressure system situated in the North Sea directs a very cold northerly flow over the country. Widespread showers, wintry at first, and including hail showers and snow, will prevail, heavier along the northern and western coasts. Showers ease as the winds decrease, allowing for widespread severe frosts and freezing fog patches and long sunny spells to become established, particularly in the south. Sunniest in the southeast.

12th-14th Very cold spell with widespread severe frosts and fog patches particularly in inland counties.

13th-15th Chance of thunderstorms.

14th Widespread sunny spell.

16th-23rd Atlantic frontal systems cross the country bringing unsettled and very dull conditions throughout, with rain or showers each day and slightly warmer than normal temperatures.

17th Widespread fog likely for a time. Dull and gloomy everywhere.

17th-18th Spells of heavy rain. Warmer than usual in most districts except in the far north.

19th-20th Chance of hail showers and snowfalls.

19th-23rd Strong and blustery south to southwest winds prevail, occasionally up to gale force at times around the coasts.

19th Thunderstorms possible. Chance of gale force winds in the far northwest.

20th Some good sunny spells likely.

21st Very wet spell.

21st-22nd Widespread ground frosts and fog. Mostly overcast everywhere.

22nd-24th Spell of warmer temperatures.

23rd Outbreaks of heavy rain likely.

24th-2nd April A mostly dry spell as high pressure north of the country prevails. Winds generally in an east to northeast flow. Some wintry showers likely mostly around the northeast coasts. Cooler, with severe frosts developing at times.

26th Very high pressure prevails over the country.

26th-27th Wintry showers of hail and snow likely.

28th Very sunny day in many parts of the country, particularly in the northwest.

31st Severe frosts likely. Chance of snow.

IRELAND RAINFALL ESTIMATES

March 2024

d=mainly dry, n= nonrecordable, l=light shrs, s=significant shrs, r=rain, h=hvy falls

		ULSTER PROVINCE						CONNAUGHT PROVINCE					LEINSTER PROVINCE				MUNSTER PROVINCE						
		Alderg	Hillsbo	Armag	Loughl	Malin H	Sligo	Ardtarı	Belmul	Drums	Galway	Roscoı	Markre	Edende	Kilkenı	Dublin	Shannc	Newpo	Tralee	Killarnє	Cork	All	Moon
1st	Fri	d	d	d	d	d	d	d	d	d	d	d	d	d	d	d	d	d	d	d	d	1	
2nd	Sat	d	d	d	n	d	l	l	d	d	d	d	n	d	n	d	d	d	d	d	d	5	
3rd	Sun	d	d	d	d	d	d	d	n	d	d	d	d	l	l	l	d	l	d	d	d	7	3Q
4th	Mon	d	d	d	d	d	d	d	d	d	d	d	d	l	d	l	d	d	d	d	s	3	
5th	Tue	l	l	d	l	d	l	d	d	l	l	l	d	s	l	s	l	l	d	n	l	18	V
6th	Wed	l	d	l	d	d	d	d	d	d	d	l	d	d	s	l	d	d	l	l	l	16	
7th	Thu	n	d	l	d	d	d	d	d	d	d	d	d	d	n	d	d	d	d	d	d	1	
8th	Fri	d	d	d	n	d	l	d	d	d	d	d	d	d	d	d	d	d	d	d	d	1	
9th	Sat	l	l	d	s	l	l	l	n	l	n	l	d	n	d	d	d	s	d	l	d	24	
10th	Sun	l	l	l	s	l	s	s	l	s	l	l	s	d	d	l	d	s	l	l	d	45	N P1
11th	Mon	l	d	l	s	l	s	s	r	l	d	n	l	d	d	d	d	s	s	s	d	46	XhN
12th	Tue	d	d	d	d	s	l	d	l	d	d	d	l	d	d	d	d	d	d	n	d	11	
13th	Wed	d	d	d	l	s	s	l	l	l	d	l	s	d	d	l	d	l	d	d	d	27	
14th	Thur	l	d	l	l	l	s	l	l	l	l	l	s	d	d	d	d	s	l	s	d	34	
15th	Fri	d	d	d	s	s	l	l	l	d	d	l	l	d	l	d	l	l	s	l	d	29	
16th	Sat	s	r	n	s	s	s	s	s	r	r	r	s	r	r	s	l	r	s	s	l	144	
17th	Sun	s	r	s	r	l	r	r	s	r	h	r	h	s	s	s	s	h	r	r	l	195	1Q ^
18th	Mon	s	s	s	s	l	l	l	h	s	l	s	s	s	l	s	s	s	r	s	s	116	
19th	Tue	s	r	s	r	l	r	r	s	s	s	s	r	l	l	l	r	r	r	r	s	147	
20th	Wed	s	d	s	l	s	l	l	l	s	l	d	h	d	d	l	l	s	l	l	l	59	
21st	Thu	r	h	l	h	h	r	h	r	h	r	h	s	h	r	r	r	h	s	l	s	262	
22nd	Fri	s	l	r	s	r	s	s	d	l	l	l	l	l	s	l	d	l	l	r	l	77	
23rd	Sat	r	h	d	r	s	s	s	s	r	l	r	r	s	r	l	s	r	r	s	s	167	A
24th	Sun	l	l	s	l	l	l	l	s	l	d	l	l	l	d	d	d	d	d	d	d	22	
25th	Mon	n	d	l	d	d	d	d	l	d	d	d	d	n	d	s	d	d	d	d	d	7	F XhS
26th	Tue	l	l	d	d	l	d	d	d	d	l	l	d	s	s	l	d	d	d	l	l	11	
27th	Wed	n	l	d	d	d	d	d	d	d	d	d	d	l	d	l	d	d	d	d	l	6	
28th	Thur	d	d	d	d	d	d	d	d	d	d	d	d	d	d	d	d	d	d	d	d	0	
29th	Fri	d	d	d	d	d	d	d	d	d	d	d	d	d	d	d	d	d	d	d	d	0	
30th	Sat	d	d	d	d	d	d	d	d	d	d	d	d	d	d	l	d	d	d	d	d	1	
31st	Sun	d	d	d	d	d	d	d	d	d	d	d	d	d	d	l	d	d	d	d	l	2	
	Estimate:	73	91	53	109	77	100	86	72	78	61	75	112	64	47	50	50	96	77	66	47	1482	
	Average:	67	77	66	96	88	109	111	99	88	95	109	109	77	63	53	79	141	116	84	98	1824	
	Trend:	av	wtr	drr	wtr	drr	av	drr	drr	drr	drr	drr	av	drr	drr	av	drr	drr	drr	drr	drr	drr	

IRELAND SUNSHINE ESTIMATES

March 2024

F=fine (8-12 hours of sunshine), pc= partly cloudy (4-7 hours), c=cloudy (1-3 hours), o=overcast (0 hours)

		ULSTER PROVINCE					CONNAUGHT PROVINCE						LEINSTER PROVINCE				MUNSTER PROVINCE						
		Alderg	Hillsbo	Armag	Loughl	Malin H	Sligo	Ardtarı	Belmul	Drums	Galway	Roscoı	Markre	Edende	Kilkenı	Dublin	Shann	Newpo	Tralee	Killarne	Cork	All	Moon
1st	Fri	c	pc	pc	c	c	pc	pc	F	c	pc	c	c	c	pc	c	c	pc	F	F	pc	91	
2nd	Sat	pc	pc	pc	c	o	F	pc	F	o	F	o	c	pc	pc	pc	F	F	pc	pc	pc	119	
3rd	Sun	pc	pc	pc	c	c	F	pc	F	c	F	c	c	c	pc	pc	F	pc	pc	pc	F	118	3Q
4th	Mon	c	c	c	c	c	pc	c	pc	c	pc	c	c	o	c	o	pc	c	pc	pc	c	67	
5th	Tue	c	o	o	c	c	o	o	c	c	c	c	o	o	o	o	o	o	c	c	o	18	V
6th	Wed	c	c	c	c	c	pc	c	pc	c	pc	c	c	c	c	c	c	c	c	c	o	54	
7th	Thu	pc	pc	pc	c	c	pc	pc	c	c	pc	c	pc	pc	pc	pc	pc	pc	c	c	pc	90	
8th	Fri	c	c	c	c	c	c	c	c	c	c	c	c	c	c	c	o	c	c	c	o	41	
9th	Sat	c	c	c	c	c	c	c	c	c	c	c	c	c	pc	c	c	c	pc	pc	pc	64	
10th	Sun	pc	pc	pc	pc	pc	pc	pc	pc	pc	pc	pc	pc	pc	pc	F	pc	pc	pc	pc	pc	134	N P1
11th	Mon	pc	pc	pc	pc	pc	pc	pc	pc	pc	pc	pc	pc	pc	F	F	F	F	F	F	F	138	XhN
12th	Tue	pc	pc	pc	pc	pc	pc	pc	pc	pc	pc	pc	pc	pc	pc	pc	pc	pc	c	c	c	121	
13th	Wed	pc	pc	pc	pc	pc	pc	pc	pc	pc	pc	pc	pc	pc	pc	pc	F	F	pc	pc	pc	142	
14th	Thur	pc	pc	pc	pc	pc	pc	pc	pc	F	pc	pc	pc	F	F	F	F	F	F	F	F	163	
15th	Fri	pc	pc	pc	pc	pc	pc	pc	pc	pc	pc	pc	pc	pc	pc	pc	pc	pc	F	F	F	148	
16th	Sat	c	c	c	c	c	c	c	c	pc	c	c	c	pc	pc	pc	pc	pc	pc	pc	pc	76	
17th	Sun	o	o	o	o	o	o	o	o	o	o	o	o	o	o	o	o	o	o	o	o	0	1Q ^
18th	Mon	c	c	c	c	c	c	c	c	c	c	c	c	c	c	o	c	c	pc	pc	pc	43	
19th	Tue	c	c	c	c	c	c	c	c	c	c	c	c	c	pc	c	c	c	pc	pc	pc	61	
20th	Wed	F	F	F	F	F	F	F	F	F	F	F	F	F	F	F	F	F	pc	pc	pc	167	
21st	Thu	o	o	o	o	o	o	o	o	o	o	o	o	o	o	c	o	o	o	o	o	9	
22nd	Fri	o	o	o	o	o	o	o	o	o	o	o	o	o	o	o	o	o	o	o	o	1	
23rd	Sat	o	o	o	o	o	o	o	o	o	c	o	o	o	o	o	c	o	o	o	o	8	A
24th	Sun	c	c	c	c	c	c	c	c	c	pc	pc	c	c	pc	c	pc	pc	pc	pc	pc	80	
25th	Mon	o	o	o	o	o	o	o	o	o	o	o	o	o	c	o	o	o	c	c	c	16	F XhS
26th	Tue	c	c	c	c	c	c	c	c	c	c	c	c	c	o	c	c	c	o	o	o	30	
27th	Wed	o	o	o	o	o	o	o	o	o	o	o	o	o	o	o	o	o	o	o	o	0	
28th	Thur	c	c	c	c	c	pc	c	pc	c	pc	pc	c	c	o	c	c	c	o	o	o	60	
29th	Fri	o	o	o	o	o	o	o	o	c	pc	c	o	c	pc	c	F	pc	F	F	F	78	
30th	Sat	c	c	c	c	c	c	c	c	o	o	c	c	o	o	o	o	o	c	c	c	19	
31st	Sun	o	o	o	o	o	o	o	o	o	o	o	o	o	o	o	o	o	o	o	o	1	
Estimate hours:		95	105	105	90	85	115	105	118	90	122	89	90	98	112	96	127	119	133	133	129	2158	
Average hours:		93	93	89	95	96	96	97	102	93	105	109	102	106	103	112	104	92	97	95	109	1987	
Trend:		av	more	more	av	less	more	av	more	av	more	less	less	av	av	less	more	more	more	more	more	more	

MARCH **1** FRIDAY

A wide low pressure trough sits over Ireland with a depression to the south of the country and another low pressure centre to the northwest, bringing a light to moderate north to northeast flow throughout. Very cold with severe frosts almost everywhere, with some mist and fog patches here and there. Mostly dry and sunny day, although chance of some isolated wintry showers or snow flurries possible, mostly to the east.

Belfast: Partly cloudy, cold, severe overnight frosts, dry, light shifting breezes
Dublin: Cloudy, cold, mostly dry, chance of freezing fog and snow flurries
Cork: Partly cloudy, dry, cold, frost pockets likely, winds freshening later in the day
Galway: Partly cloudy, dry, cold with overnight frosts, light shifting breezes

RAIN POTENTIAL

FROST/SNOW

MARCH **2** SATURDAY

Low pressure weakens to the southeast of the country while high pressure ridges over the northwest with a light to moderate northeast flow prevailing, stronger at times around the coasts. Mostly dry with good sunny spells throughout, and cold with scattered frosts, mostly in inland parts. Chance of some isolated wintry showers around coastal fringes in the northwest and east.

Belfast: Mostly sunny, dry, frosts, cold northerly breezes
Dublin: Partly cloudy, isolated showers, cold, breezy spells, chance of snow flurries
Cork: Mostly sunny, cold and dry, windy at times
Galway: Very sunny, dry, frost pockets, light breezes freshening later in the day

RAIN POTENTIAL

FROST/SNOW

MARCH 3 SUNDAY

High pressure moves over the country directing a light to moderate north to northeast flow, stronger at times along the northern and eastern coasts, and possibly with occasional gale gusts. Very cold everywhere with severe frosts mostly in inland parts. Mostly sunny and dry, except for some isolated wintry showers mainly in the east and southeast, where some showers of hail, sleet, or snow may be observed.

Belfast: Partly cloudy, cool, dry, frosts, moderate breezes
Dublin: Partly cloudy, isolated wintry showers, fresh breezes
Cork: Dry, very sunny and cool, chance of frost pockets, cool northerly breezes
Galway: Sunny, dry, chance of overnight frost, fresh breezes

MARCH 4 MONDAY

High pressure continues to direct an east to northeast flow over the country, strong and gusty at times. A mix of sunny spells and wintry showers likely in many districts, particularly over the eastern half of the country. Cold with frosts mostly in the west, and wintry showers of hail or light snow possible to the east.

Belfast: Cloudy, mostly dry with increasing threats of rain, breezy spells, some frost
Dublin: Overcast, brief wintry showers, breezy spells, hail or light snow possible
Cork: Colder, increasing cloud, light wintry showers, windy spells, chance of snow
Galway: Mostly sunny, cold and dry, pockets of overnight frost, breezy spells

MARCH **5** TUESDAY

High pressure prevails directing a light northeast flow later freshening with a northerly change, particularly around the northern and eastern coasts. Widespread cloud throughout with outbreaks of wintry rain or drizzle in many parts, particularly in the north and east, while more isolated to the west and south. Some precipitation may fall as light snow in some eastern parts.

Belfast: Overcast, cold, brief wintry showers, moderate breezes
Dublin: Overcast, moderate breezes, occasional wintry showers
Cork: Overcast, cold, light snow, fresh winds
Galway: Overcast, shifting breezy spells, cold with frost pockets, spells of light rain

RAIN POTENTIAL

FROST/SNOW

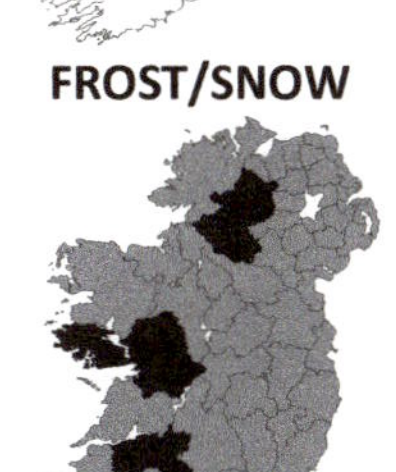

MARCH **6** WEDNESDAY

High pressure continues to linger over Ireland while low pressure sits southeast of the country with a moderate northeasterly flow prevailing. A mixed day with mostly dry conditions and occasional sunny spells in the north and central west, while the passage of a rain band brings cloudier skies and passing wintry showers elsewhere. Chance of some light snow in the south of Ireland.

Belfast: Cold, dry, brief sunny spells, moderate winds strengthening later in the day
Dublin: Cold, cloudy, brief wintry showers, moderate breezes
Cork: Overcast, breezy, cold, brief wintry showers, chance of snow
Galway: Partly cloudy, dry and cold, moderate breezes

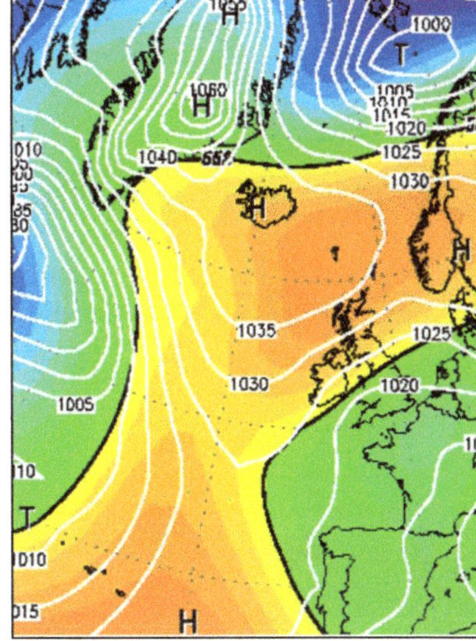

RAIN POTENTIAL

FROST/SNOW

MARCH **7** THURSDAY

A ridge of high pressure continues to linger over Ireland directing a fresh to strong east to northeast flow throughout. Cold and mostly dry with scattered overnight pockets of frosts and frequent sunny spells.

Belfast: Partly cloudy, dry, windy with occasional gusts
Dublin: Dry, occasional sunny spells, strong winds with gusts, cold
Cork: Partly cloudy, cold, dry, strong gusty winds at times
Galway: Partly cloudy, cooler, dry, strong and blustery winds

RAIN POTENTIAL

FROST/SNOW

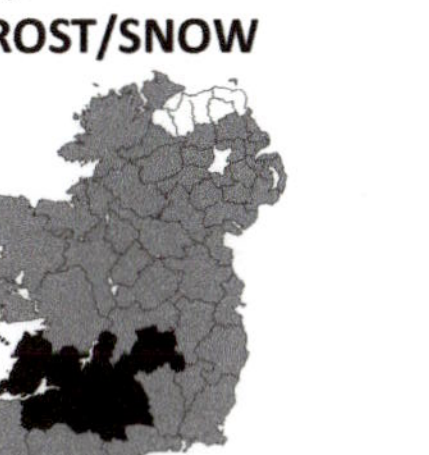

MARCH **8** FRIDAY

High pressure continues to linger to the northwest of Ireland while low pressure sits to the south. Winds generally moderate to fresh in a northeast flow, but possibly up to gale force at times around coastal fringes. Mostly dry and cloudy although some isolated showers may be observed here and there, mostly overnight.

Belfast: Cloudy, dry, cold, winds easing
Dublin: Cloudy, dry, blustery
Cork: Overcast, dry, strong winds at times, cold
Galway: Cloudy, windy, dry, milder

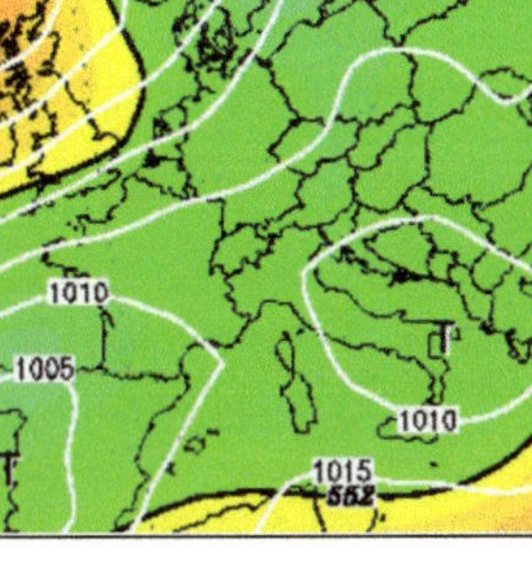

RAIN POTENTIAL

FROST/SNOW

MARCH 9 SATURDAY

High pressure builds back over Ireland bringing very cold temperatures and severe frosts in many districts overnight. Winds generally light and variable at first, before freshening with a northwest change. A mixed day with some sunny spells and passing wintry showers, including the chance of possible snow in some northern parts.

Belfast: Brief wintry showers, mostly cloudy, cold, frosts, light breezes, chance of snow
Dublin: Colder with frosts, cloudy, mostly dry with odd threat of rain, breezy at times
Cork: Partly cloudy, cold, mostly dry with odd threat of rain, shifting breezes
Galway: Cloudy, isolated wintry showers, frosts, light breezes

RAIN POTENTIAL

FROST/SNOW

MARCH 10 SUNDAY

A depression to the northwest of Ireland directs a stormy cold front over the country with wintry showers, including the likelihood of sleet and snow, into the north of the country at first and into southern parts later in the day. Precipitation likely to be more persistent over the northern half of the country while more isolated to the south. Cold with frosts and some good sunny spells in many districts. Winds generally in a north to northwest flow, strong at times in the north with possible gale gusts, while lighter to the south.

Belfast: Sunny spells, frosts, wintry showers, chance of light snow overnight, windy
Dublin: Sunny, windy, cold, frosts, light showers, chance of overnight snow flurries
Cork: Mostly sunny, dry, cold with overnight frost pockets, cold northerly flow
Galway: Cold, partly cloudy, light wintry showers, chance of snow, windy spells

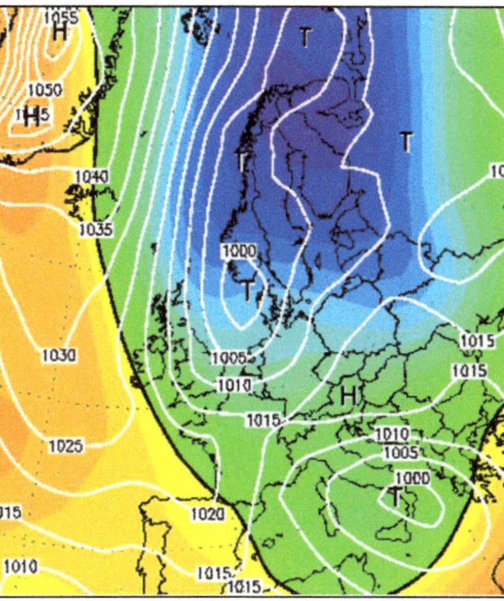

RAIN POTENTIAL

FROST/SNOW

MARCH 11 MONDAY

A depression moves to the northwest of Ireland while a degrading high pressure ridges to the west, with a light to moderate northwest flow prevailing, more blustery at times in northern parts. Very cold throughout with severe overnight frosts. Some long sunny spells in most districts, but interspersed with wintry showers; some of hail, sleet, and snow likely, particularly in the north and west.

Belfast: Partly cloudy, dry, colder, frosts, unsettled winds, chance of light snow
Dublin: Dry, sunny, cold with severe frosts, breezy spells, chance of snow flurries
Cork: Dry, cold, frosts, sunny, chance of overnight snow flurries
Galway: Partly cloudy, dry, colder northerly breezes, frosts, chance of snow flurries

RAIN POTENTIAL

FROST/SNOW

MARCH 12 TUESDAY

A large depression moves onto Ireland bringing very cold temperatures and light northwesterly breezes. Widespread severe frosts throughout, with wintry showers, possibly of snow, mostly in the northwest. Generally dry and mostly sunny elsewhere with chance of scattered mist and freezing fogs in sheltered parts.

Belfast: Mostly sunny and dry, cold with severe frosts, fluctuating winds
Dublin: Mostly sunny, dry, severe frosts, light breezes, chance of misty patches
Cork: Unsettled and changeable, cloudy, dry, very cold with severe frosts
Galway: Partly cloudy, dry, severe frosts, chance of flurries and fog patches

RAIN POTENTIAL

FROST/SNOW

MARCH 13 WEDNESDAY

A large depression continues to linger over Ireland directing a very cold and generally light north to northwesterly flow. Very cold everywhere with severe frosts in most districts. Mostly dry and sunny with scattered fogs at first before a cold front moves onto the north of Ireland and moves southward down through the country bringing snow into northern parts at first and later into central and western parts.

Belfast: Dry and cold, severe frosts, some sunny spells, chance of fog and snow flurries
Dublin: Light shifting breezes, partly cloudy, overnight snow, cold with severe frosts
Cork: Mostly sunny and dry, cold, severe frosts, light breezes
Galway: Mostly sunny, dry, cold, severe frosts, fog and snow flurries possible

RAIN POTENTIAL

FROST/SNOW

MARCH 14 THURSDAY

Very cold and frosty conditions continue under the influence of low pressure centres situated to both the south of Ireland as well as to the northeast. Mostly dry and sunny with severe frosts and pockets of freezing fog in many districts. Some overnight wintry showers likely, mostly in the form of light snow. Winds generally light in a northerly flow.

Belfast: Mostly sunny, dry, severe frosts, moderate breezes, chance of overnight snow
Dublin: Cold with severe frosts, sunny, dry, light shifting breezes
Cork: Very sunny, dry, cold with severe frosts, light shifting breezes
Galway: Sunny, dry, severe frosts, light snow likely, gentle breezes, possible fog

RAIN POTENTIAL

FROST/SNOW

MARCH **15** FRIDAY

Low pressure trough crosses the country in a light north to northwest flow, with stronger breezes in northern parts. Very cold temperatures continue to bring further widespread frosts in many districts, and scattered fogs in the north. Mostly sunny day throughout, with overnight wintry showers moving in from the west and spreading eastwards, with some falls of snow likely, mostly in the west.

Belfast: Partly cloudy, dry, frosts, occasional windy spells
Dublin: Mostly sunny and dry, cold, frosts, moderate breezes
Cork: Very sunny, milder, mostly dry, chance of brief drizzle spells, light breezes
Galway: Partly cloudy, dry, milder day, overnight frosts, breezy spells, possible snow

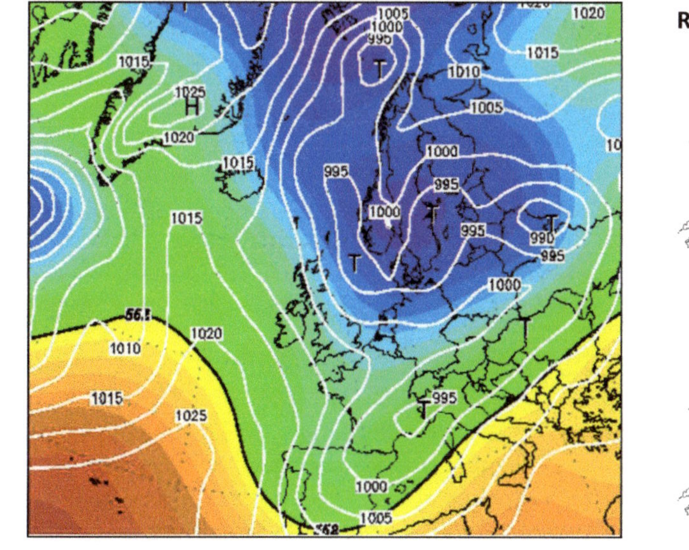

MARCH **16** SATURDAY

Low pressure moves away in a light to moderate north to northwest flow while high pressure ridges to the southwest of the country and another depression to the northwest advances with a southerly change. Cold with widespread frost in most districts, and fog patches likely in northern and southern parts. Increasing cloud with the passage of a band of rain as it crosses the country in an easterly tract, bringing widespread wintry showers throughout. Chance of precipitation falling as snow, particularly in the north and west.

Belfast: Light breezes, cloudy, wintry rain, frosts, chance of snow and fog patches
Dublin: Light breezes, cold with frosts, some sunny spells, brief showers likely
Cork: Partly cloudy, brief showers, milder, light breezes, chance of fog overnight
Galway: Cloudy, overnight frosts, rain, possible snow, building winds later in the day

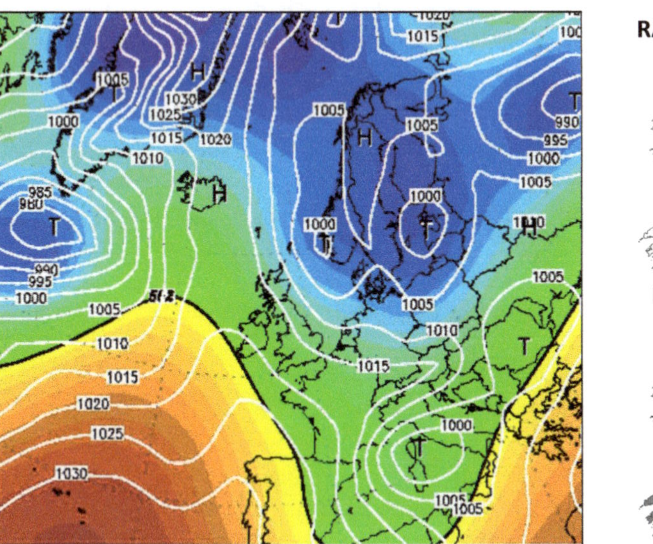

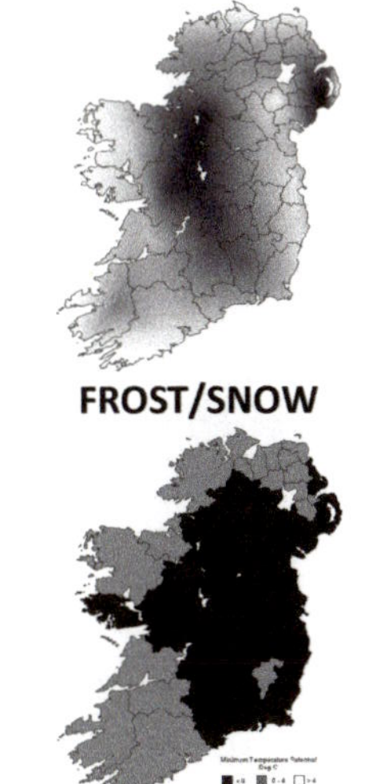

MARCH **17** SUNDAY

High pressure southwest of Ireland and low pressure to the northwest, with a moderate to fresh southwesterly flow prevailing. Widespread cloud brings a dull and gloomy day everywhere, with wintry rain in most districts. Chance of snow in the north, and storms in the south. Milder with few frosts, except in isolated parts in the midlands.

Belfast: Overcast, milder, wintry rain, snow possible, breezy at times, chance of fog
Dublin: Milder, overcast, scattered showers, windy with occasional gusts
Cork: Overcast, light showers, warmer, moderate breezes freshening later in the day
Galway: Stormy air, milder, overcast and dull, heavy rain and persistent drizzle

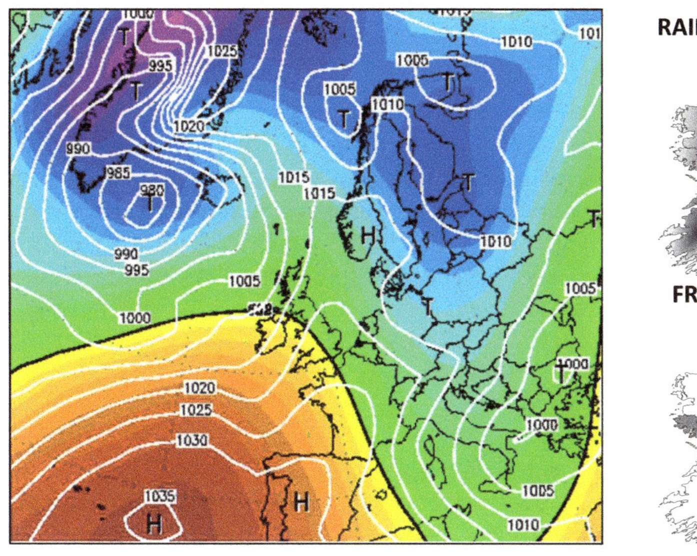

RAIN POTENTIAL

FROST/SNOW

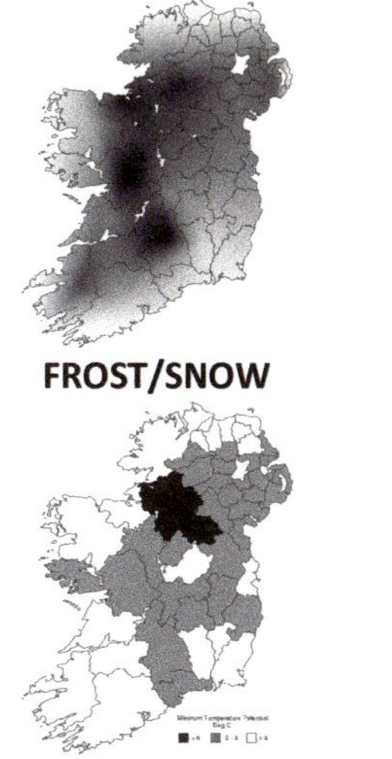

MARCH **18** MONDAY

Depression to the northwest of Ireland extends light southerly winds at first, freshening later with occasional gusts with a northwesterly change. Stormy air develops in many districts, with widespread cloud and scattered outbreaks of rain throughout. Chance of thunderstorms, especially in western parts.

Belfast: Unsettled and breezy, cloudy, occasional showers, milder
Dublin: Stormy air, overcast, squally showers, milder
Cork: Changeable, mix of showers and sunny spells, mild day cold night, blustery
Galway: Cloudy, windy, milder, light rain, chance of thunderstorms

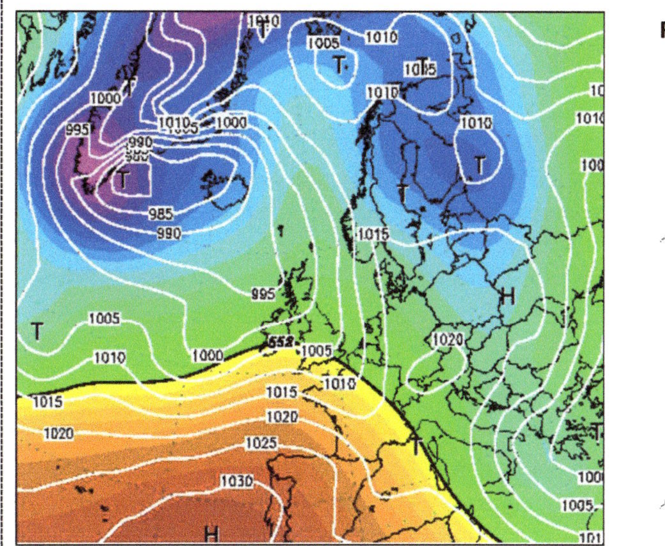

RAIN POTENTIAL

FROST/SNOW

MARCH 19 TUESDAY

Low pressure to the west of Ireland and another to the east, with a low pressure trough between them which directs an unsettled southerly flow over the country. A cold front between the two low pressure centres spreads widespread cloud and unsettled bands of rain throughout, with heavy falls at times in some western districts and the chance of wintry showers of snow or hail. Brief sunny breakthroughs likely between the showers.

Belfast: Cloudy, rain, mild day cold night, fluctuating breezes
Dublin: Cloudy, brief showers, fluctuating winds, mild day cold night
Cork: Changeable, partly cloudy, wintry showers, unsettled breezes, possible snow
Galway: Cloudy, unsettled winds, scattered wintry showers, chance of snow and hail

RAIN POTENTIAL

FROST/SNOW

MARCH 20 WEDNESDAY

A large depression covers the country bringing moderate northwest winds throughout, easing later in the day. Changeable throughout with good sunny spells and the occasional wintry showers, some of the possibly of snow or hail and accompanying gustier winds. Winds likely to reach gale force for a time in western and southern parts.

Belfast: Sunny, cooler, mostly dry, windy with occasional gusts
Dublin: Sunny, brief wintry showers, possible hail and frosts, strong gusty winds
Cork: Partly cloudy, cold, light wintry showers, strong gusty winds
Galway: Sunny, blustery winds, possible gales, squally wintry showers, possible snow

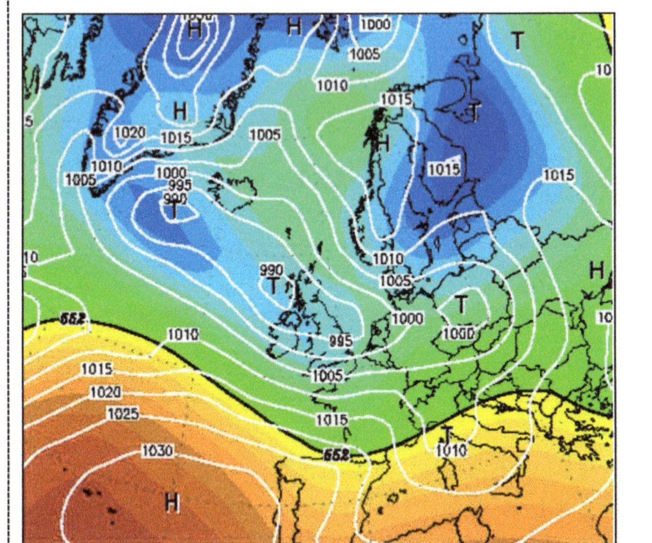

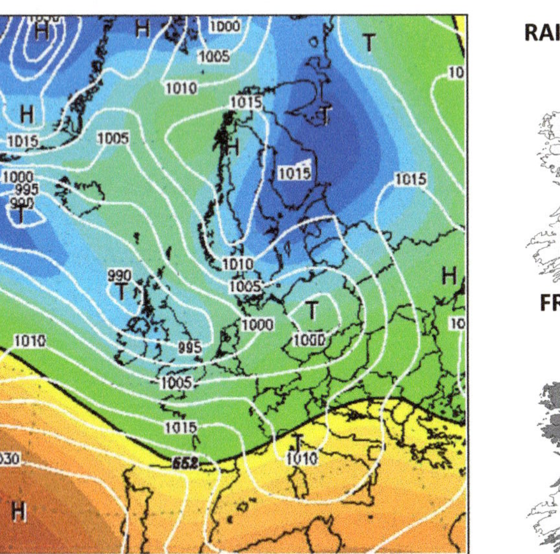

RAIN POTENTIAL

FROST/SNOW

MARCH 21 THURSDAY

Low pressure weakens over the country as the depression contracts northwards while high pressure sits to the far south of Ireland. Gentle south to southeasterly winds prevail, freshening later in the day. A wet front crosses the southern coasts and moves up over the country throughout the day, bringing a stormy air and widespread rain throughout, with heavy falls in many districts. Some precipitation likely to be wintry and possibly fall as snow, particularly in the west.

Belfast: Overcast, heavy rain with wintry overnight showers, shifting breezy spells
Dublin: Cloudy, colder, rain, light breezes later freshening
Cork: Overcast, scattered showers, fluctuating winds
Galway: Overcast, rain and drizzle, moderate breezes, chance of overnight snow

RAIN POTENTIAL

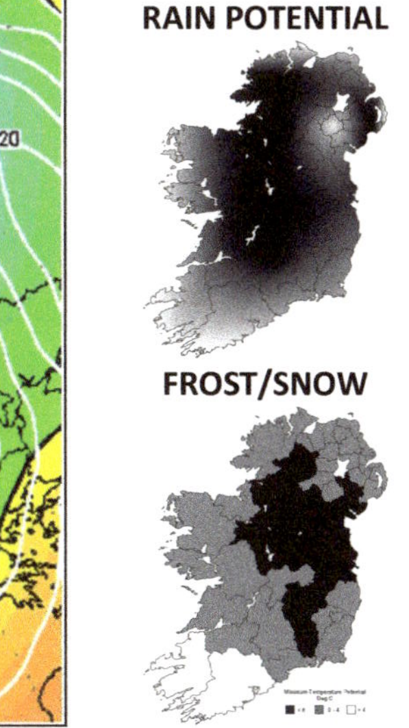

FROST/SNOW

MARCH 22 FRIDAY

Low pressure intensifies over Ireland bringing a stormy air with unsettled strong and blustery winds fluctuating southeasterly to southwesterly before easing later in the day. Winds may reach gale force for a time, especially in the north and west. Widespread cloud throughout with persistent rain in the north, and light and more drizzly rain elsewhere.

Belfast: Overcast, strong and blustery winds, chance of gales, spotty showers, colder
Dublin: Overcast, milder, isolated drizzly showers, strong winds, possible gusts
Cork: Unsettled and blustery at times, milder, overcast, light rain
Galway: Stormy air, overcast, brief drizzly showers

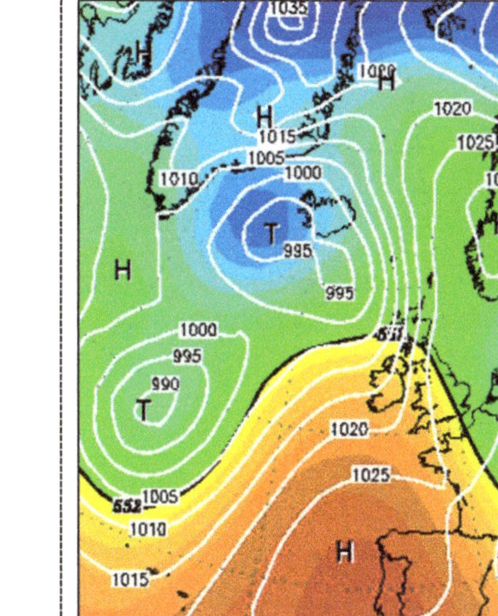

RAIN POTENTIAL

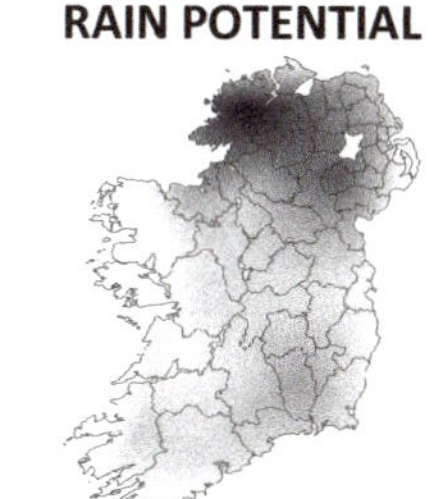

FROST/SNOW

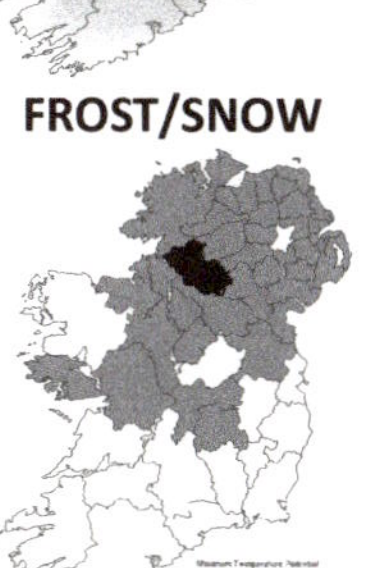

MARCH **23** SATURDAY

Stormy air throughout the country with blustery southeasterly winds prevailing, with winds up to gale force at times. Widespread cloud with persistent rain and drizzle everywhere, with outbreaks of heavy at times in many places.

Belfast: Stormy, overcast, milder, heavy rain
Dublin: Overcast, scattered showers, strong blustery winds, chance of gales at times
Cork: Overcast, rain, strong and blustery winds
Galway: Cloudy, milder, strong winds, chance of gales at times, brief squally showers

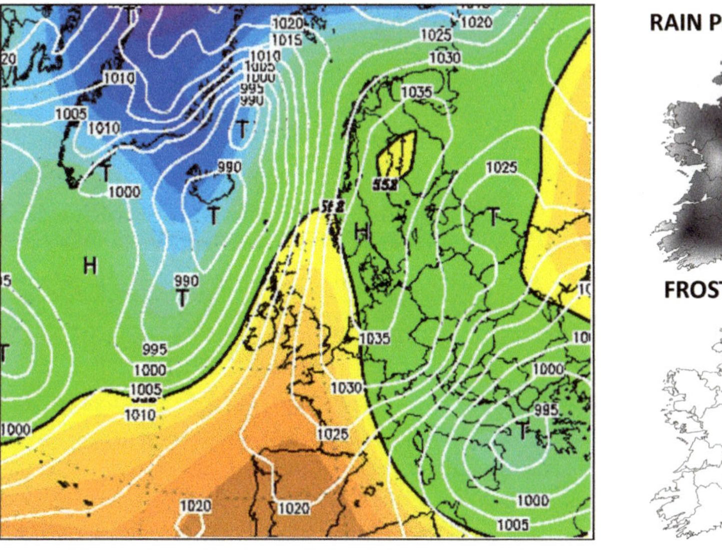

RAIN POTENTIAL

FROST/SNOW

MARCH **24** SUNDAY

A depression over Ireland weakens with west to southwest winds easing to mostly light and variable. Light showers and drizzle lingers over the northern half of the country, while clearing to mostly dry with increasing sunny spells to the south. Chance of overnight fog patches mostly in the north and west.

Belfast: Cloudy, light drizzly showers, fluctuating winds
Dublin: Cloudy, milder, mostly dry with passing threats of rain, changeable winds
Cork: Partly cloudy, milder, dry, shifting windy spells
Galway: Changeable, partly cloudy, dry, fluctuating winds

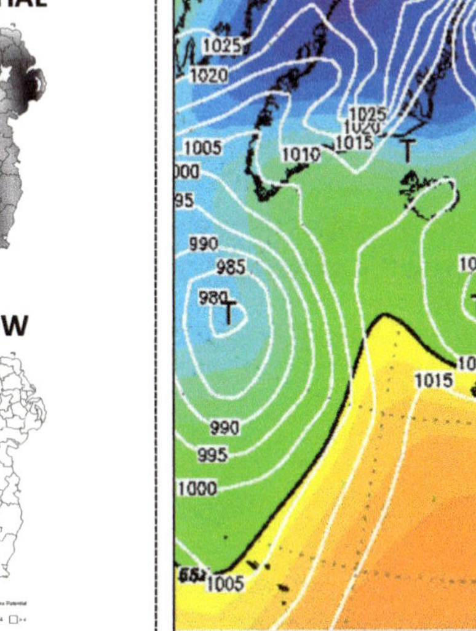

RAIN POTENTIAL

FROST/SNOW

MARCH **25** MONDAY

A high pressure system to the east of Ireland ridges over the country bringing light and variable breezes in a generally northeasterly flow, fresher around the northern and eastern coasts. Widespread cloud throughout brings a mostly dull and gloomy day with passing threats of rain everywhere, some fog patches mostly in the north and west, and the occasional light drizzly showers mostly in the east.

Belfast: Overcast, dry, threats of rain, light shifting breezes, chance of fog patches
Dublin: Overcast, light breezes to mostly calm air, occasional showery patches
Cork: Frequent cloudy spells, dry, colder, light shifting breezes
Galway: Overcast, colder, dry, light shifting breezes, chance of mist and fog patches

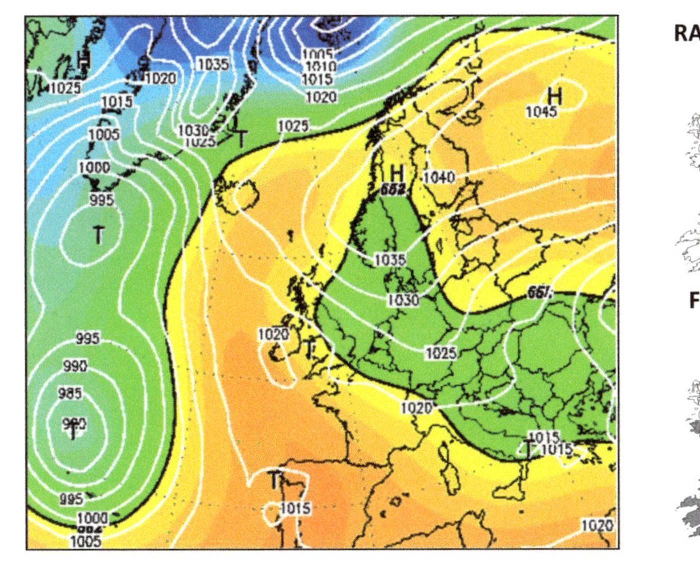

MARCH **26** TUESDAY

High pressure prevails over the country directing a fresh easterly flow with the occasional gusts at times. Mostly cloudy and very cold everywhere with light wintry showers in most districts, with some precipitation possibly falling as snow here and there.

Belfast: Cloudy, colder, light showers, breezy, occasional gusts likely, chance of snow
Dublin: Colder, cloudy, windy spells, wintry showers, snow likely
Cork: Overcast, colder, brief showers, chance of light snow, moderate breezes
Galway: Freshening breezes, cloudy, isolated showers, chance of light snow and mists

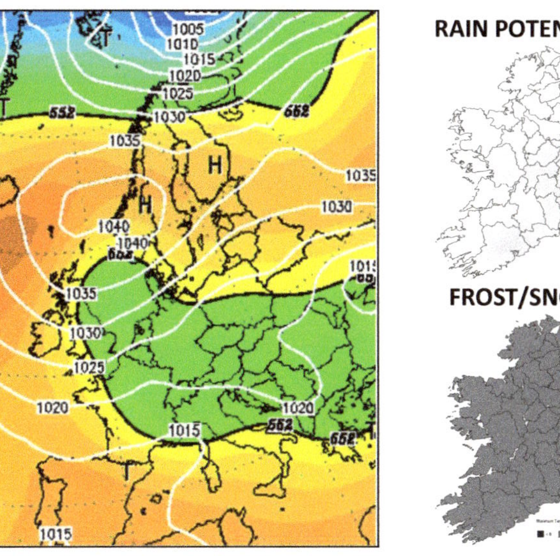

MARCH **27** WEDNESDAY

Unsettled and blustery easterly winds prevail under the influence of a high pressure system centred to the north of the country. Widespread cloud and cold everywhere. Mostly dry in the west, with isolated wintry showers elsewhere, some possibly falling as hail or snow at times.

Belfast: Stormy air, overcast, cold, isolated wintry showers, possible snow flurries
Dublin: Strong winds to gales, overcast, brief wintry showers, hail and snow likely
Cork: Stormy air, cold, mix of snow showers and light rain, gusty at times
Galway: Overcast, cold, frosts, strong gusty winds, dry with passing threats of rain

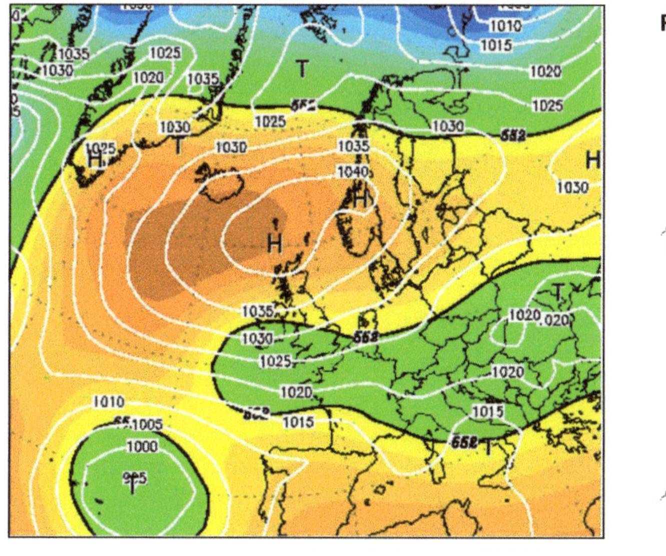

MARCH **28** THURSDAY

High pressure continues to prevail over the country directing a fresh east-to-northeast flow throughout, with gusts possibly up to gale force at times around coastal fringes. Mostly dry everywhere, with widespread cloud in the north and east, and slightly sunnier in the north and west.

Belfast: Brief sunny spells, dry, cold
Dublin: Cloudy, cold and dry, winds easing
Cork: Overcast, dry, winds easing to light, cold
Galway: Partly cloudy, milder, dry, fresh winds

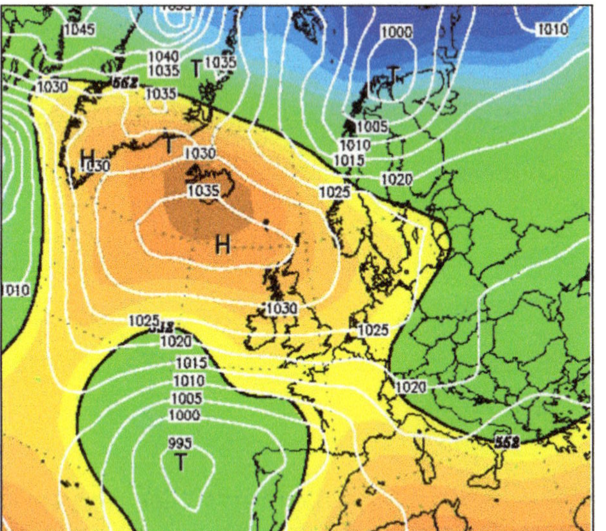

MARCH 29 FRIDAY

High pressure weakens over Ireland while a low pressure system stalls to the south of the country, generating a fresh easterly flow throughout, easing to lighter as the day progresses. Mostly dry everywhere with good sunny spells in the south and west, while cloudier in the east, and dull and gloomy in the north. Scattered frosts likely overnight, especially in the southwest.

Belfast: Overcast, dry, slightly warmer, moderate breezes
Dublin: Cloudy, milder, dry, moderating breezes
Cork: Very sunny and dry, milder, moderate breezes
Galway: Partly cloudy, colder, dry, fresh breezes

RAIN POTENTIAL

FROST/SNOW

MARCH 30 SATURDAY

High pressure continues to contract in a northwesterly tract while low pressure remains stalled south of the country. Winds generally moderate to fresh in a northeasterly flow throughout. Widespread cloud and mostly dry everywhere, except for the chance of some light drizzly showers in the east and possible snow flurries in the north. Severe frosts likely in sheltered parts of the midlands.

Belfast: Cloudy, mostly dry, threats of rain, moderate breezes, chance of snow flurries
Dublin: Moderate breezes, overcast, cold, brief drizzly showers
Cork: Cloudy, cooler, dry, moderate breezes
Galway: Overcast, dry with passing threats of rain, cold, moderate breezes

RAIN POTENTIAL

FROST/SNOW

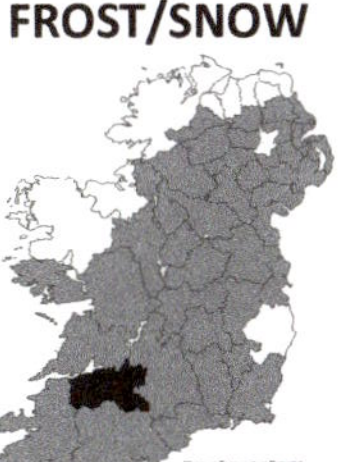

MARCH 31 SUNDAY

Low pressure trough moves onto Ireland in a moderate east to north-easterly flow. Widespread cloud prevails with overnight drizzly wintry showers likely in the east and south, some possibly of snow, while mostly dry in the north and west. Frosts likely in inland parts.

Belfast: Overcast, cold, dry with threats of rain easing, light breezes
Dublin: Overcast, cold, moderate shifting breezes, light wintry showers, snow possible
Cork: Overcast, colder, brief showery spells, light breezes at times
Galway: Overcast, cold, dry with occasional threat of rain, light breezes

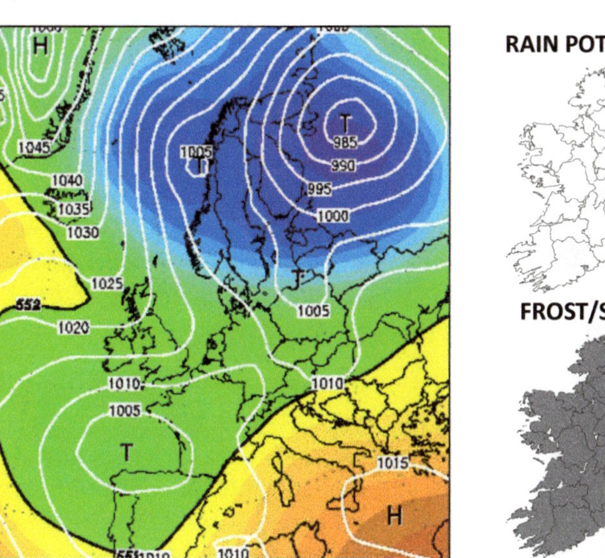

APRIL

PHASES OF THE MOON

1st	Southern Declination
2nd	Third Quarter
7th	Perigee #5
8th	New Moon
8th	Crossing Equator
13th	Northern Declination
15th	First Quarter
20th	Apogee
22nd	Crossing Equator
23rd	Full Moon
28th	Southern Declination

1 Mon	2 Tue	3 Wed	4 Thur	5 Fri	6 Sat	7 Sun	8 Mon	9 Tue	10 Wed	11 Thur	12 Fri	13 Sat	14 Sun	15 Mon	16 Tue	17 Wed	18 Thur	19 Fri	20 Sat	21 Sun	22 Mon	23 Tue	24 Wed	25 Thur	26 Fri	27 Sat	28 Sun	29 Mon	30 Tue
V	3Q					P5	N XhN					^		1Q					A		XhS	F					V		

MONTHLY SUMMARY

24th March - 2nd April A mostly dry spell as high pressure north of the country prevails. Winds generally in an east to northeast flow. Some wintry showers likely mostly around the northeast coasts. Cooler everywhere, with severe frosts developing at times.
1st Very sunny over the northern half of the country.
1st-2nd Severe frosts likely.
3rd-6th Widespread fog likely.
3rd-10th Milder spell with regular outbreaks of rain or showers, some heavy, as Atlantic frontal systems prevail, directing a strong southwest to westerly flow.
5th-6th Very wet and milder spell with the heaviest rain in the northwest.
7th Very low pressure prevails over the country.
7th-8th Chance of hail showers.
9th Warmer than usual maximum temperatures.
10th Thunderstorm bands likely to cross the northern half of the country, bringing driving showers and the chance of tornado formations.

At the start of the next four-week look ahead period, a deep depression passes close to the western coastline of Ireland with low pressure values being mostly consistent throughout the period. It is only during the third week and in the last few days of the fourth week that a brief respite of high pressure touches the country. Some rain activity is likely to occur almost daily over the coming four weeks, although amounts are likely to be generally small, and with amounts well below average in the south and southeast for this time of year. Conversely, above average rainfall can be expected in northern parts. Sunshine hours will be above average for this time of year everywhere, with the south and eastern provinces of Leinster and Munster particularly sunnier than average. The first and last weeks of this look-ahead period will both be the sunniest, with the middle weeks the dullest. After a cold start to this outlook period over the first week with some unusually low temperatures for this time of year, milder temperatures settle in towards the end of the second week. As a result, the overall expectation of temperatures by the end of this outlook period will be above normal throughout the country by around a degree warmer. Conversely, low minimum overnight temperatures will result in more ground frosts than is typical for this time of year. Generally, the first half of this outlook period will be the windiest, but despite this, winds will be below normal for this time of year in southern parts of the country. In contrast, winds will be above normal in northern parts. Calmer winds over the second half of this outlook period allow for the development of regular patches of fog. During this outlook period there will be around 7-10 days with rain in the east while up to 18-20 days of rain in the northwest. There is likely to be around 5 days when thunderstorms are observed and hail on 6 days. There will be about 16-21 days with ground frosts, mostly in inland parts, snow on 2 days, and fog on 6 days.

11th-20th Low pressure brings a cold and showery spell, with fresh to strong south to southwest flow at first backing later to west to northwesterly.
11th-13th An intense depression moves across the country bringing scattered showers, thunderstorms and hail, especially in eastern districts.
13th-20th Cold and frosty throughout the country, with frosts becoming less widespread towards the end of this week. Very sunny spell in most districts.
14th-15th Brief respite with winds easing and clearing skies, allowing for severe frosts to develop, particularly severe in inland parts, particularly in the north Midlands.
16th-19th Chance of hail showers during this period.
17th Strong gusty winds at times, especially in the northwest.
17th-19th A mixed period with long sunny spells between bands of thunderstorms and heavy rain at times crossing the country, particularly in eastern parts. Wintry showers in the west of Ireland may bring falls of sleet and snow.
20th Winds ease again. Severe frosts in many parts, including in eastern counties.
21st-10th May Frontal systems cross the country in an easterly tract, bringing milder temperatures and outbreaks of showers in all districts, heavier falls at times in the west and north.
21st-30th Warmer maximum temperatures settle in with most districts recording maximums daily temperatures between 10-14°C.
28th-30th Outbreaks of heavy rain at times in the west and north of Ireland.
29th-30th Chance of fog patches.

IRELAND RAINFALL ESTIMATES

April 2024

d=mainly dry, n= nonrecordable, l=light shrs, s=significant shrs, r=rain, h=hvy falls

		ULSTER PROVINCE					CONNAUGHT PROVINCE						LEINSTER PROVINCE				MUNSTER PROVINCE						
		Alderg	Hillsbo	Armag	Loughl	Malin H	Sligo	Ardtar	Belmul	Drums	Galway	Roscom	Markre	Edende	Kilkenn	Dublin	Shann	Newpo	Tralee	Killarne	Cork	All	Moon
1st	Mon	d	d	d	d	d	d	d	d	d	d	d	d	d	d	d	d	d	d	l	d	1	V
2nd	Tue	l	l	d	d	d	d	d	d	d	l	l	s	s	s	n	d	l	n	r	s	44	3Q
3rd	Wed	s	s	l	l	l	s	s	l	s	l	l	s	s	s	r	s	l	r	l	r	92	
4th	Thu	s	s	s	s	r	s	s	l	r	r	s	s	s	l	s	l	r	s	s	s	122	
5th	Fri	s	r	s	s	r	r	r	h	h	h	r	r	r	r	l	s	h	h	h	h	273	
6th	Sat	l	l	s	s	l	s	s	l	l	s	l	s	s	s	l	r	s	s	r	l	90	
7th	Sun	l	l	l	d	s	d	d	l	l	d	d	l	d	d	l	s	d	s	l	l	31	P5 XhN
8th	Mon	s	r	n	s	l	s	s	s	r	h	r	s	s	s	s	r	s	r	r	r	178	N
9th	Tue	r	r	s	s	s	s	s	s	l	s	s	s	l	l	l	l	s	s	s	l	105	
10th	Wed	s	s	s	l	l	l	l	l	s	r	s	d	l	l	l	s	s	s	s	l	88	
11th	Thur	s	l	r	r	l	s	s	r	r	s	r	s	s	s	s	l	r	s	s	l	136	
12th	Fri	s	l	s	l	l	l	n	l	l	d	l	s	s	n	r	d	l	l	l	d	43	
13th	Sat	l	l	l	d	n	d	d	d	l	d	l	d	l	d	n	n	d	d	d	d	10	^
14th	Sun	l	d	l	d	d	d	d	n	d	d	d	d	d	d	d	d	d	d	d	d	2	
15th	Mon	d	d	d	h	l	l	l	s	d	l	d	l	d	d	d	d	s	d	d	d	34	1Q
16th	Tue	l	l	n	n	s	l	n	l	s	l	l	l	l	d	l	l	l	l	l	s	34	
17th	Wed	s	s	l	r	s	s	s	l	s	s	r	s	s	l	l	l	s	l	s	l	99	
18th	Thu	s	s	s	s	l	l	l	l	l	n	n	l	l	l	l	l	s	l	n	l	45	
19th	Fri	l	l	l	s	l	l	l	d	l	l	l	l	l	l	l	d	l	d	d	d	21	
20th	Sat	l	l	l	s	l	r	r	s	s	r	s	s	l	l	d	d	h	d	s	d	105	A
21st	Sun	l	d	l	l	r	s	l	s	l	l	s	l	l	d	l	s	l	s	n	l	62	XhS
22nd	Mon	n	d	l	l	l	l	n	d	l	s	l	n	l	l	l	l	l	d	d	d	18	
23rd	Tue	n	l	d	l	l	l	l	l	l	d	l	n	d	d	d	l	l	l	l	n	17	
24th	Wed	d	d	d	n	d	l	n	d	l	l	s	s	d	d	d	l	l	s	l	l	16	F
25th	Thur	d	d	d	l	s	s	l	l	d	l	n	n	d	d	d	d	d	s	l	l	21	
26th	Fri	n	d	l	l	l	s	l	s	l	l	l	s	d	d	d	d	l	l	d	d	23	
27th	Sat	l	l	n	s	l	s	s	s	s	d	s	s	l	d	s	d	s	d	l	d	61	
28th	Sun	s	s	l	l	l	l	l	l	s	l	l	l	l	s	s	l	l	r	h	s	90	V
29th	Mon	l	l	l	d	l	d	n	l	l	l	d	l	d	d	d	d	l	s	l	s	19	
30th	Tue	l	d	l	d	l	d	d	l	d	d	d	d	d	l	d	l	d	r	s	r	39	
Estimate:		87	95	78	112	82	96	84	90	107	118	102	90	64	61	64	78	127	142	146	97	1917	
Average:		58	61	58	70	65	78	80	72	67	72	81	78	66	56	54	59	97	110	81	77	1439	
Trend:		w tr	w tr	w tr	w tr	w tr	w tr	av	w tr	w tr	w tr	w tr	w tr	av	av	av	w tr	w tr	w tr	w tr	w tr	w tr	

IRELAND SUNSHINE ESTIMATES

April 2024

F=fine (8-12 hours of sunshine), pc= partly cloudy (4-7 hours), c=cloudy (1-3 hours), o=overcast (0 hours)

		ULSTER PROVINCE					CONNAUGHT PROVINCE						LEINSTER PROVINCE				MUNSTER PROVINCE						
		Alderg	Hillsbo	Armag	Loughl	Malin H	Sligo	Ardtar	Belmul	Drums	Galway	Rosco	Markre	Edende	Kilkenr	Dublin	Shannc	Newpo	Tralee	Killarne	Cork	All	Moon
1st	Mon	F	F	F	F	F	F	F	F	pc	pc	pc	F	pc	pc	F	c	c	o	o	o	121	V
2nd	Tue	c	c	c	c	c	o	c	o	c	o	o	c	c	c	c	o	o	o	o	o	20	3Q
3rd	Wed	o	o	o	o	o	o	o	o	o	o	o	o	o	o	o	o	o	o	o	o	0	
4th	Thu	c	c	c	c	c	o	c	o	c	o	o	c	c	c	c	o	c	o	o	o	27	
5th	Fri	o	o	o	o	o	o	o	o	o	o	o	o	o	o	o	o	o	o	o	o	0	
6th	Sat	c	c	c	c	c	c	c	c	c	o	o	c	c	c	c	o	o	o	o	o	21	
7th	Sun	o	o	o	o	o	o	o	o	o	o	o	o	o	o	o	c	o	o	o	o	11	P5 XhN
8th	Mon	pc	pc	pc	pc	pc	pc	pc	pc	pc	pc	pc	pc	pc	c	pc	pc	c	o	o	o	85	N
9th	Tue	c	c	c	c	c	c	c	c	c	c	c	c	c	c	c	c	c	o	o	o	37	
10th	Wed	c	c	c	c	c	c	c	c	c	c	c	c	c	pc	pc	pc	pc	c	c	c	71	
11th	Thur	pc	pc	pc	pc	pc	pc	pc	pc	pc	pc	pc	pc	pc	pc	pc	pc	pc	pc	pc	pc	116	
12th	Fri	pc	pc	pc	pc	pc	pc	pc	pc	pc	F	pc	pc	pc	pc	pc	F	F	F	F	F	141	
13th	Sat	F	F	F	F	F	F	F	F	F	F	F	F	F	F	F	F	F	F	F	F	184	^
14th	Sun	F	F	F	F	F	F	F	F	F	F	F	F	F	F	F	F	F	F	F	F	226	
15th	Mon	F	F	F	F	F	F	F	F	F	F	F	F	F	F	F	F	F	F	F	F	206	1Q
16th	Tue	c	c	c	c	c	c	c	c	pc	pc	c	c	pc	pc	c	pc	pc	F	F	F	94	
17th	Wed	pc	pc	pc	pc	pc	pc	pc	pc	pc	pc	pc	pc	pc	pc	pc	pc	pc	pc	pc	pc	112	
18th	Thu	F	F	F	F	F	F	F	F	F	F	F	F	F	F	F	F	F	F	F	F	186	
19th	Fri	F	F	F	F	F	F	F	F	F	F	F	F	F	F	F	F	F	F	F	F	170	
20th	Sat	pc	pc	pc	pc	pc	c	pc	c	F	pc	pc	pc	F	F	F	F	F	F	F	F	153	A
21st	Sun	pc	pc	pc	pc	pc	pc	pc	pc	pc	pc	pc	pc	pc	pc	pc	pc	pc	pc	pc	pc	136	XhS
22nd	Mon	o	o	o	o	o	o	o	o	c	o	o	o	c	c	c	o	c	c	c	c	20	
23rd	Tue	pc	pc	pc	pc	pc	pc	pc	pc	pc	pc	pc	pc	pc	pc	F	pc	pc	pc	pc	pc	128	
24th	Wed	F	F	F	F	F	pc	F	pc	pc	F	pc	F	pc	pc	F	F	F	pc	pc	pc	155	F
25th	Thur	c	c	c	c	c	o	c	o	c	o	o	c	c	c	c	c	c	o	o	o	18	
26th	Fri	pc	pc	pc	pc	pc	pc	pc	pc	pc	pc	pc	pc	pc	pc	c	pc	pc	pc	pc	pc	95	
27th	Sat	pc	pc	pc	pc	pc	c	pc	c	pc	pc	c	pc	pc	pc	pc	pc	pc	pc	pc	pc	107	
28th	Sun	o	o	o	o	o	o	o	o	o	o	o	o	o	c	c	o	o	o	o	o	14	V
29th	Mon	F	F	F	F	F	F	F	F	pc	F	F	F	pc	pc	pc	F	F	pc	pc	pc	163	
30th	Tue	c	c	c	c	c	pc	c	pc	c	c	pc	c	c	c	c	o	o	o	o	o	54	
Estimate hours:		145	145	145	145	145	131	145	131	147	141	136	145	147	149	159	152	150	138	138	138	2871	
Average hours:		151	151	133	152	153	138	140	158	137	149	155	145	151	146	156	148	140	150	150	156	2957	
Trend:		av	av	more	av	av	av	av	less	av	av	less	av	av	av	av	av	av	less	less	less	average	

APRIL **1** MONDAY

Low pressure system to the northwest of Ireland directs a light to moderate northeasterly flow over the country, backing southeasterly later in the day. Mostly dry, cool, and sunny with overnight frosts over the northern half of the country while cloudier and milder with the chance of some isolated showers in the south.

Belfast: Sunny, dry, mostly calm and cool serene air with severe frosts
Dublin: Sunny, cold, dry, frosts, light breezes
Cork: Overcast, cold and dry with increasing threats of rain, light shifting breezes
Galway: Partly cloudy, dry, cold with frosts, gentle breezes

RAIN POTENTIAL

FROST/SNOW

APRIL **2** TUESDAY

A weak low pressure trough crosses the country with a light to moderate southeasterly flow with occasional gusts, directed from an intense depression to the southwest of Ireland, with winds backing easterly later in the day. Widespread cloud throughout with outbreaks of light rain in most districts except for mostly dry along the western coastal fringes. Overnight frosts likely over the northern half of the country.

Belfast: Cloudy, pockets of frost, spells of light rain, moderate breezes
Dublin: Mostly cloudy, isolated showers, cold with frost, gentle shifting breezes
Cork: Overcast, cold, rain, freshening breezes later in the day
Galway: Overcast, light drizzly showers, cold, frosts, winds freshening later in the day

RAIN POTENTIAL

FROST/SNOW

APRIL **3** WEDNESDAY

A very large depression to the southwest of Ireland moves up over the country in a fresh to blustery southeast flow. Stormy air with widespread cloud and scattered showers throughout. Winds may reach gale force for a time in the south and west.

Belfast: Overcast, milder, blustery winds, scattered showers
Dublin: Unsettled blustery winds, overcast, rain, stormy air, milder
Cork: Stormy air, overcast, rain, milder, winds to gale force possible at times
Galway: Overcast, milder, strong winds to gale force at times, light squally showers

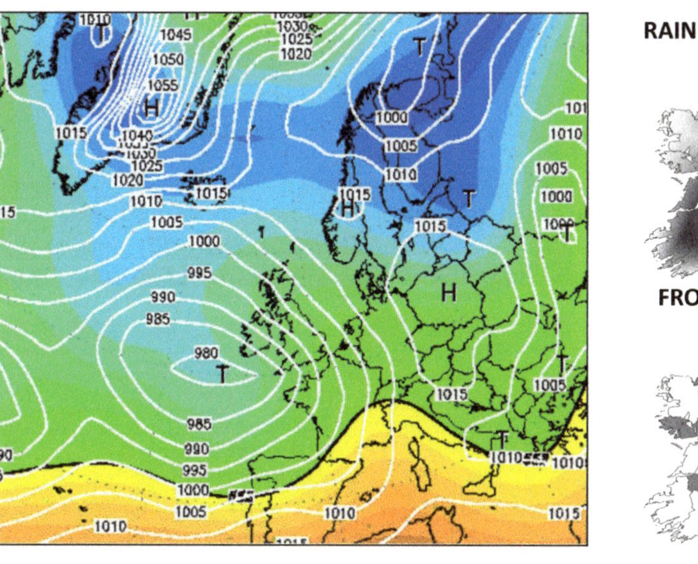

RAIN POTENTIAL

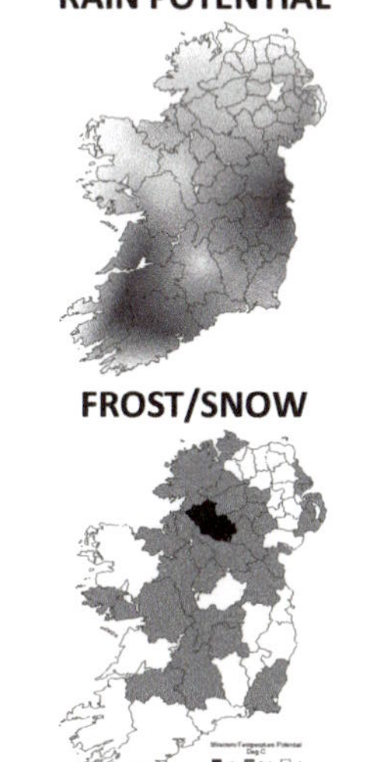

FROST/SNOW

APRIL **4** THURSDAY

A large depression fills over Ireland easing winds and variable in direction. Dull and gloomy with widespread scattered showers in most districts. Some brief sunny spells likely at first mostly in the north and east. Chance of overnight pockets of fog.

Belfast: Cloudy, milder, strong winds at times, squally showers
Dublin: Scattered showers, some sunny spells, milder, strong winds, occasional gusts
Cork: Overcast, rain, stormy air, strong winds with possible gale gusts
Galway: Overcast, rain, blustery

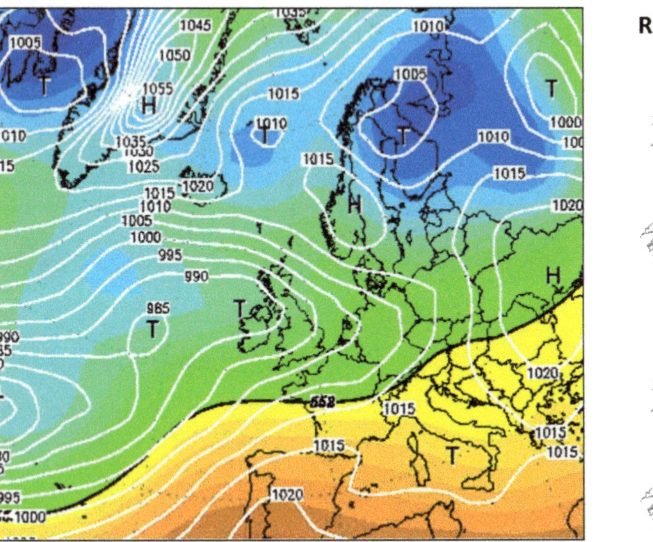

RAIN POTENTIAL

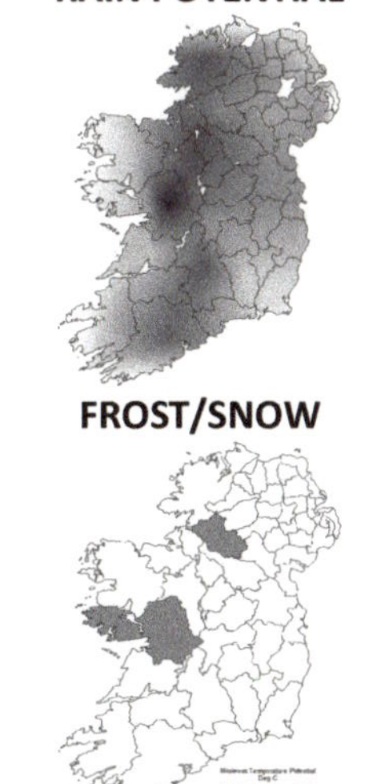

FROST/SNOW

APRIL 5 FRIDAY

A depression to the west of Ireland brings widespread cloud and rain throughout the country with winds strengthening and prevailing in a southeasterly backing southwest flow. Stormy air everywhere, with widespread cloud and heavy rain in many districts, and gusty winds that may reach gale to storm force for a time.

Belfast: Stormy air, strong winds, possible gales, dull and gloomy, rain, warmer
Dublin: Overcast, warmer, brief drizzly showers, stormy air with strong winds and gusts
Cork: Stormy air, overcast, heavy rain, gales to storm force at times in gusts
Galway: Strong gusty winds, dull and gloomy air, heavy rain

RAIN POTENTIAL

FROST/SNOW

APRIL 6 SATURDAY

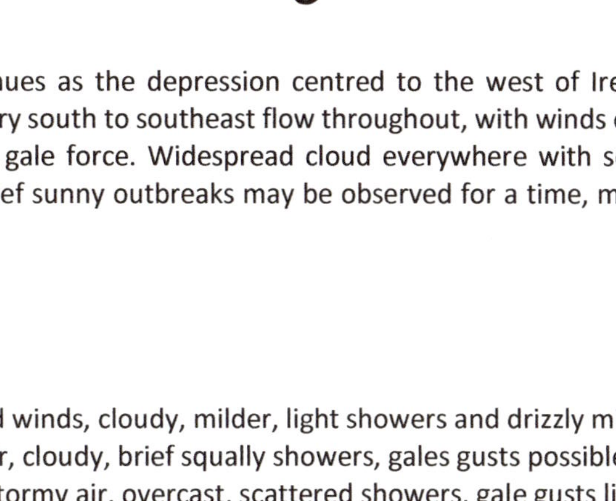

Stormy air continues as the depression centred to the west of Ireland maintains a strong and blustery south to southeast flow throughout, with winds often gusty and at times may reach gale force. Widespread cloud everywhere with squally showers in most districts. Brief sunny outbreaks may be observed for a time, mostly in the north and east.

Belfast: Unsettled winds, cloudy, milder, light showers and drizzly mists
Dublin: Stormy air, cloudy, brief squally showers, gales gusts possible
Cork: Unsettled stormy air, overcast, scattered showers, gale gusts likely
Galway: Overcast, scattered showers, strong winds with possible gale gusts

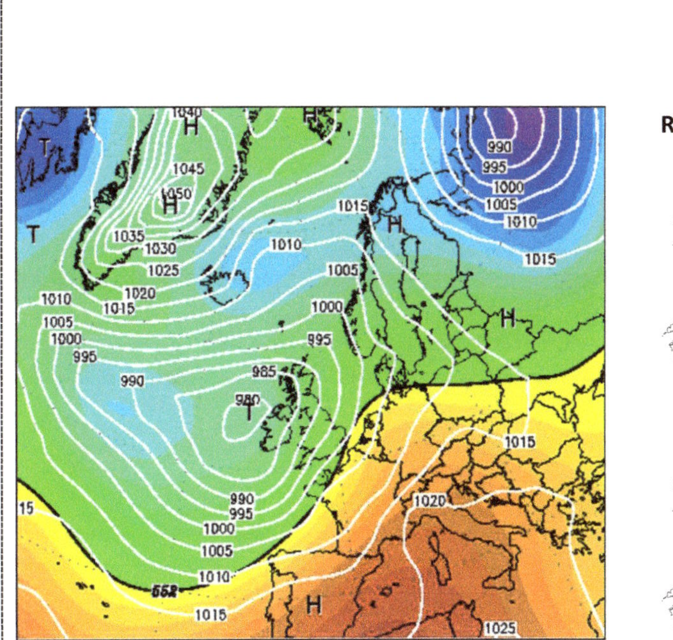

RAIN POTENTIAL

FROST/SNOW

APRIL **7** SUNDAY

A large depression slowly drifts over Ireland from the west to the north of the country with strong and blustery southwest winds prevailing, easing later with a northwest change. Dull and gloomy everywhere with widespread cloud and light squally showers in many districts. Strong winds may reach gale to storm force for a time, especially around the coasts. Chance of late light snowfall in some western parts.

Belfast: Overcast, brief showers, strong and unsettled winds with gale gusts likely
Dublin: Cooler, dull and gloomy, drizzly showers, stormy air with gales likely
Cork: Strong blustery winds, overcast, brief squalls, cooler
Galway: Dull and gloomy air, dry with passing threats of rain, stormy winds

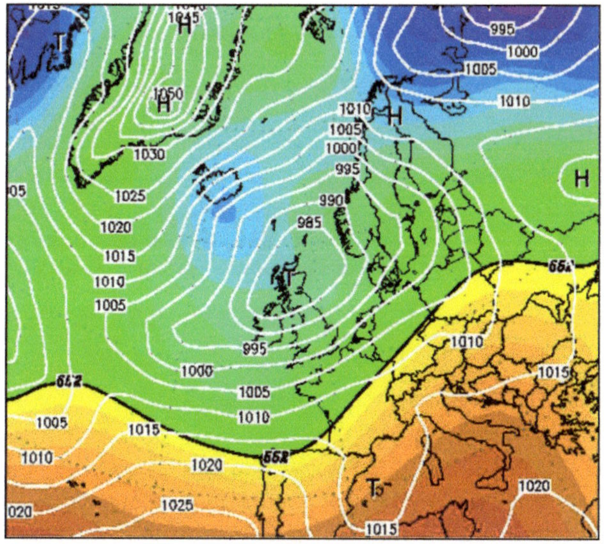

RAIN POTENTIAL

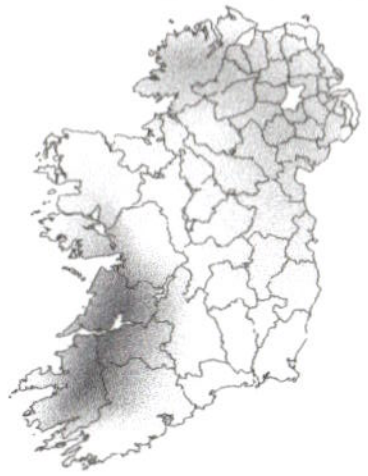

FROST/SNOW

APRIL **8** MONDAY

A large depression continues to bring widespread unsettled conditions throughout the country with variable to westerly winds prevailing, very strong at times and possibly with occasional gale gusts. Stormy everywhere. Wet and windy in all districts, and cold with the chance of overnight frosts or light snow in inland parts. Chance of some brief sunny spells later in the day.

Belfast: Unsettled, changeable, stormy, overnight heavy rain, some sunny spells, colder
Dublin: Cold overnight with squally showers, some sunny spells, winds easing for a time
Cork: Overcast, rain, winds easing to light for a time
Galway: Colder, stormy air, heavy rain, chance of frosts and overnight snow

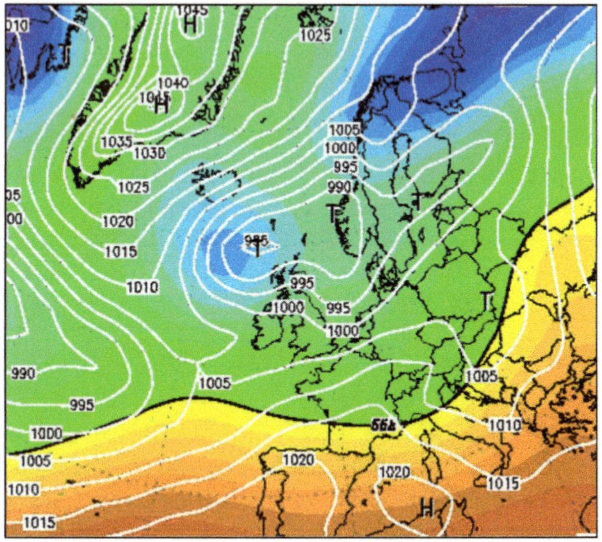

RAIN POTENTIAL

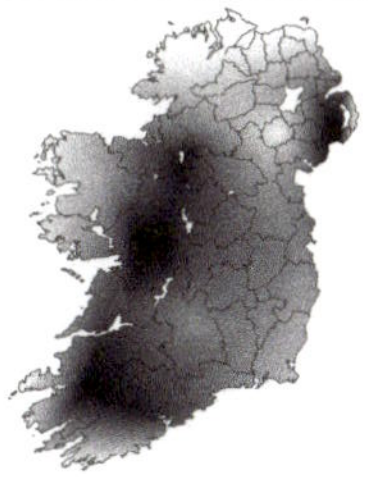

FROST/SNOW

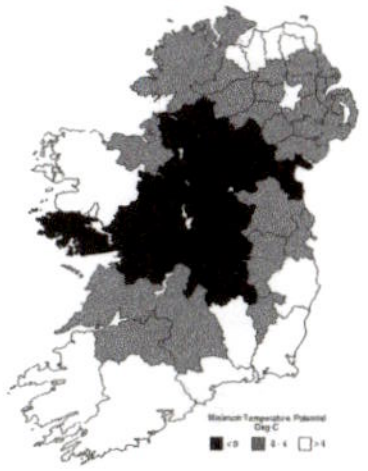

APRIL 9 TUESDAY

Stormy and low pressure conditions continue to prevail. Winds blustery in a south to southwest flow, occasionally up to gale force at times. Widespread rain and drizzle throughout, with the chance of some brief sunny breakthroughs. Milder.

Belfast: Cloudy, slightly milder, rain, fluctuating winds
Dublin: Mostly cloudy, isolated showers, milder, fluctuating winds
Cork: Unsettled and blustery, milder, overcast, showers of light rain
Galway: Cloudy, blustery, milder, drizzly showers

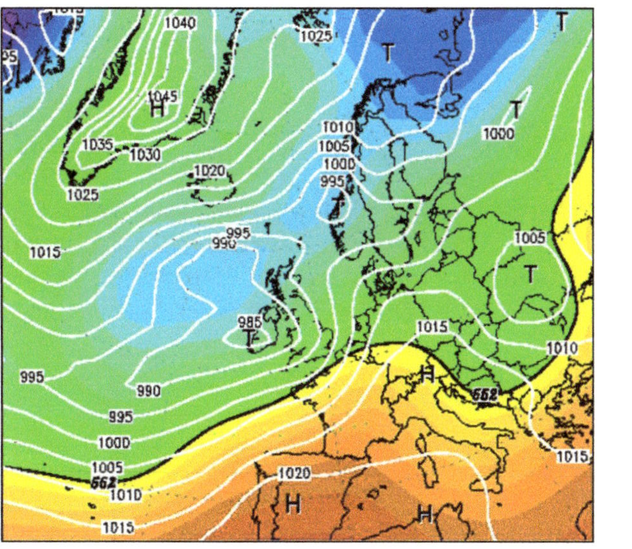

RAIN POTENTIAL

FROST/SNOW

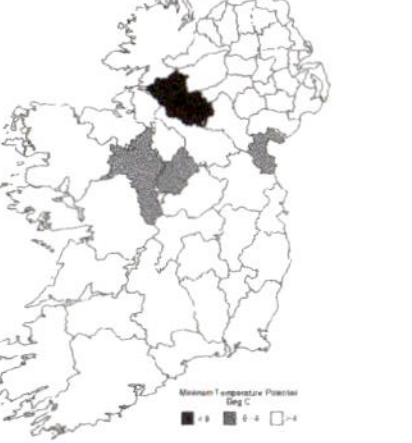

APRIL 10 WEDNESDAY

Low pressure and a stormy air persists over Ireland, with moderate to fresh southwest winds prevailing, backing southeasterly for a time then returning to southwesterly and up to gale force at times. A day mixed with some sunny spells and thundery rain bands crossing the country.

Belfast: Mix of brief sunny spells and showers, fluctuating winds, slightly milder
Dublin: Unsettled and changeable, light showers, some sunny spells, fresh breezes
Cork: Windy, mix of brief sunny spells and spotty showers, mild day cold night
Galway: Fluctuating winds, brief sunny spells, rain

RAIN POTENTIAL

FROST/SNOW

APRIL 11 THURSDAY

A depression continues to linger over Ireland directing a fresh south to southwest flow, stronger and more blustery at times. Another mixed day of some sunny spells interspersed with bands of showers, some heavy, in most districts, as well as the chance of scattered thunderstorms at times, particularly in northern parts.

Belfast: Changeable, occasional showers, chance of thunderstorms and gale gusts
Dublin: Partly cloudy, passing showers, cooler, breezy at times
Cork: Partly cloudy, light showers, windy with occasional gusts
Galway: Partly cloudy, light showers, fresh winds

RAIN POTENTIAL

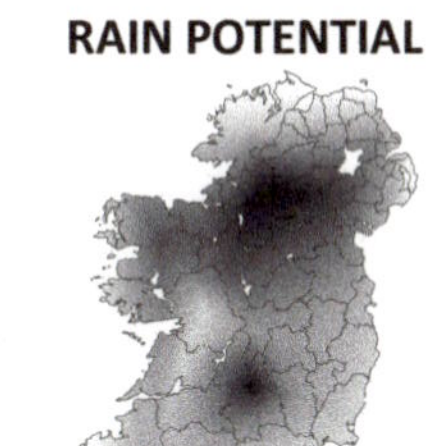

FROST/SNOW

APRIL 12 FRIDAY

A depression continues to linger over Ireland directing changeable winds both in direction and strength, prevailing primarily westerly, and with gusty winds occasionally reaching gale force at times. Overnight rain with isolated showers, heavy at times, and possible passing thunderstorms and hail showers. Some good sunny spells throughout the day.

Belfast: Partly cloudy, fluctuating winds with occasional gusts, spotty showers
Dublin: Overnight rain, some sunny spells, strong blustery winds with gusts
Cork: Sunny, mostly dry, cold overnight, windy at times
Galway: Dry, sunny, mild, unsettled winds with occasional gusts

RAIN POTENTIAL

FROST/SNOW

APRIL **13** SATURDAY

Low pressure drifts to the northeast of Ireland while high pressure ridges to the west. Winds generally in a west to northwest flow, still blustery at times in the north, while moderating to light elsewhere. Mostly dry and sunny with scattered frosts in many districts during the day with spotty showers overnight. Colder temperatures may bring the chance of some brief snow flurries mostly in the west and higher inland reaches.

Belfast: Sunny, brief overnight showers, mild day cold night, fresh breezes
Dublin: Sunny, colder, isolated showers, fresh breezes
Cork: Sunny, dry, mild day cold night, moderate breezes
Galway: Sunny, dry, cooler, frosts likely, moderate winds, chance of snow flurries

RAIN POTENTIAL

FROST/SNOW

APRIL **14** SUNDAY

High pressure ridges over Ireland, directing a very cold, light and variable flow over the country. Chance of scattered fogs here and there, and severe frosts overnight in many parts over the northern half of the country. Very sunny and dry throughout, although some isolated showers may briefly sprinkle around the west and southwest coasts.

Belfast: Very sunny, dry, cool, frosts, light breezes to mostly calm air
Dublin: Sunny, cold with frosts, dry, gentle northerly breezes
Cork: Sunny, dry, cooler, light shifting breezes
Galway: Very sunny, dry, cool with frosts likely, gentle northerly breezes

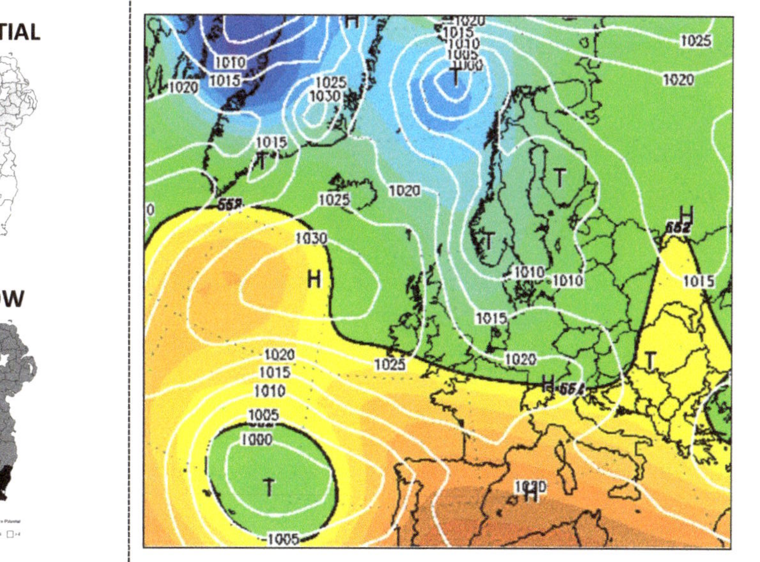

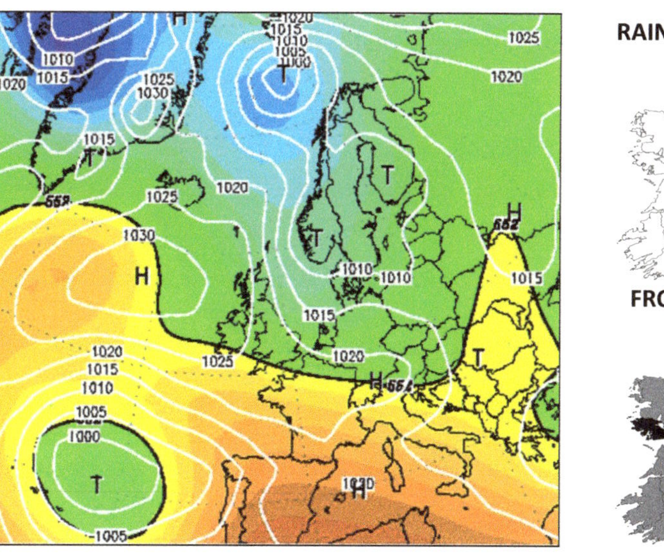

RAIN POTENTIAL

FROST/SNOW

APRIL **15** MONDAY

A large anticyclonic system lingers over Ireland while a depression passes to the south of the country. Winds generally light throughout in variable directions. Very cold with widespread frosts and sunny throughout. Mostly dry except for some pockets of showers in the north and west and chance of brief overnight showers in the central east.

Belfast: Sunny, cold and frosty, dry, light northwest breezes
Dublin: Sunny, dry, cool with severe frosts overnight, light breezes to mostly calm air
Cork: Very sunny, dry, cool, overnight frost, light breezes to mostly calm air
Galway: Sunny, cool, frosts, isolated showers, gentle breezes

APRIL **16** TUESDAY

The high pressure ridge drifts to the south of Ireland while low pressure moves into the north. Northwest winds prevail throughout, lighter in the south while blustery in the north. Overnight frosts in many districts with the chance of some light overnight snowfalls, mostly in central parts from the west to the east. Changeable throughout the day with a mix of sunny spells and scattered showers most districts.

Belfast: Mostly cloudy, cold, brief showers, windy with occasional gusts
Dublin: Windy spells, mostly cloudy, isolated showers, chance of snow flurries
Cork: Light breezes, sunny, brief morning showers, cold overnight
Galway: Partly cloudy, brief showers, chance of overnight frosts or light snow

APRIL **17** WEDNESDAY

An intense depression to the north of Ireland directs a west to southwest flow over the country, more blustery and accompanying storms over the northern parts. A changeable day with a mix of scattered showers and some sunny spells in most districts as well as the chance of some light overnight snow here and there.

Belfast: Stormy air, partly cloudy, squally showers, blustery northwesterlies at times
Dublin: Partly cloudy, light showers, moderate winds, chance of overnight light snow
Cork: Mostly sunny, light showers, moderate breezes
Galway: Partly cloudy, occasional showers, light breezes, chance of overnight snow

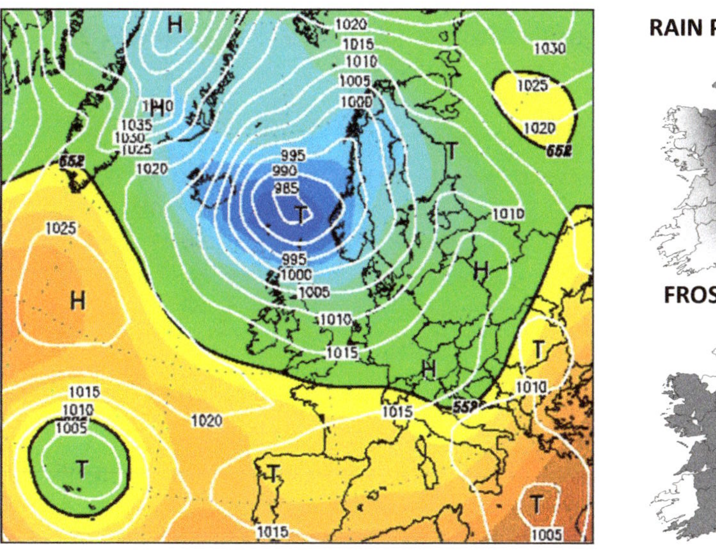

RAIN POTENTIAL

FROST/SNOW

APRIL **18** THURSDAY

Low pressure continues to direct a moderate northerly airflow over the country, with scattered frosts in sheltered parts and cold wintry showers overnight in most districts, including the chance of snow flurries in the west. Changeable day with good sunny spells interspersed with passing bands of showers. Winds blustery at times particularly in the north and along the western coasts.

Belfast: Changeable, sunny and cold, occasional showers, strong winds with gusts
Dublin: Sunny, windy, isolated showers, possible frost in sheltered pockets
Cork: Sunny, brief morning showers, cold overnight, fresh breezes at times
Galway: Unsettled winds, sunny, isolated showers, chance of frosts and snow flurries

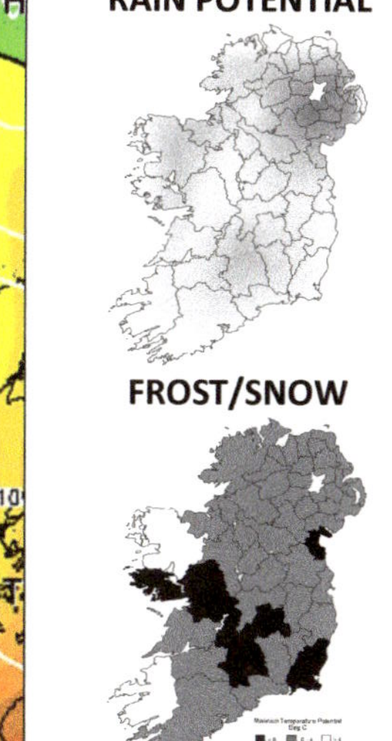

RAIN POTENTIAL

FROST/SNOW

APRIL 19 FRIDAY

High pressure ridges to the west of Ireland while low pressure flows to the east, with a cold northerly flow over the country. Changeable day with a mix of sunny spells and wintry showers with the passage of a wet front moving in from the northwest and later clearing to the south. Some showers may be heavy for a brief time in the northwest. Chance of light snow once more in some western parts.

Belfast: Sunny, isolated showers, cold overnight with chance of frost pockets
Dublin: Mostly sunny, cool, isolated showers, cold northerly air, chance of frost
Cork: Sunny, dry, light northerly breezes
Galway: Sunny, brief showers, cold northerly winds, chance of overnight snow

RAIN POTENTIAL

FROST/SNOW

APRIL 20 SATURDAY

A high pressure ridge over Ireland directs a very cold northerly air over the country with widespread frosts throughout, and light to moderate northwest breezes. Another mixed day of sunny spells and scattered showers in most districts, with showers turning to more steady rain in the west later in the day while remaining mostly dry in the far south and eastern coastal fringes. Cold temperatures may see wintry showers in the west turning to overnight snow.

Belfast: Partly cloudy, light showers, frosts, freshening breezes later in the day
Dublin: Sunny, dry, cool, severe frost pockets likely, gentle breezes
Cork: Very sunny, dry, cold, overnight frosts, mostly calm air
Galway: Partly cloudy, rain, frosts, cold northerly flow with chance of overnight snow

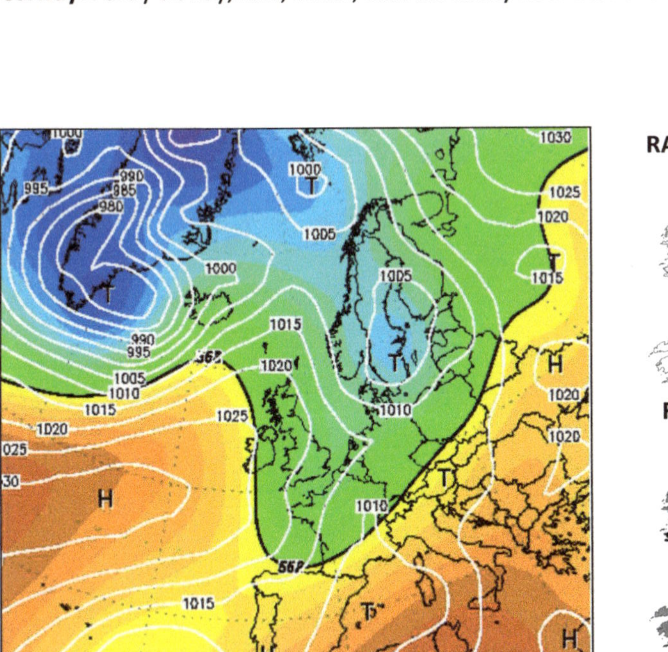

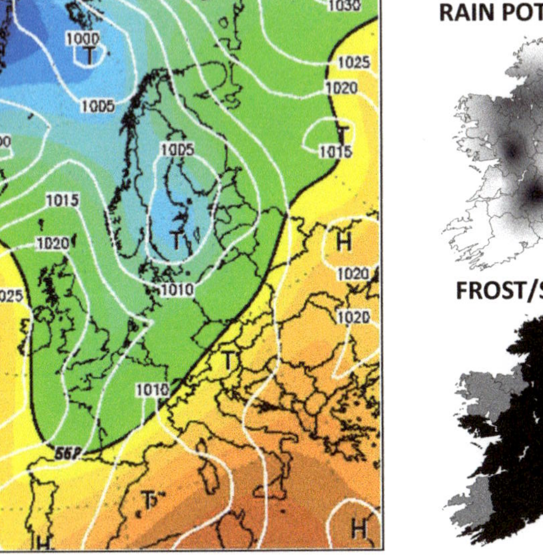

RAIN POTENTIAL

FROST/SNOW

APRIL **21** SUNDAY

High pressure to the south of Ireland continues to ridge over the country while an intense depression sits to the north. Fresh and blustery southwest winds at first, backing northwesterly later in the day. A wet front moves into the west at first, crossing to the eastern coasts during the day, followed by some sunny spells and further showery fronts. Mostly dry by evening.

Belfast: Mostly sunny and dry, milder, strong gusty winds with possible gale gusts
Dublin: Fluctuating breezes, milder, some cloudy spells, light showers
Cork: Partly cloudy, windy spells, light showers, warmer
Galway: Mostly sunny, milder, showers of light rain, strong winds with occasional gusts

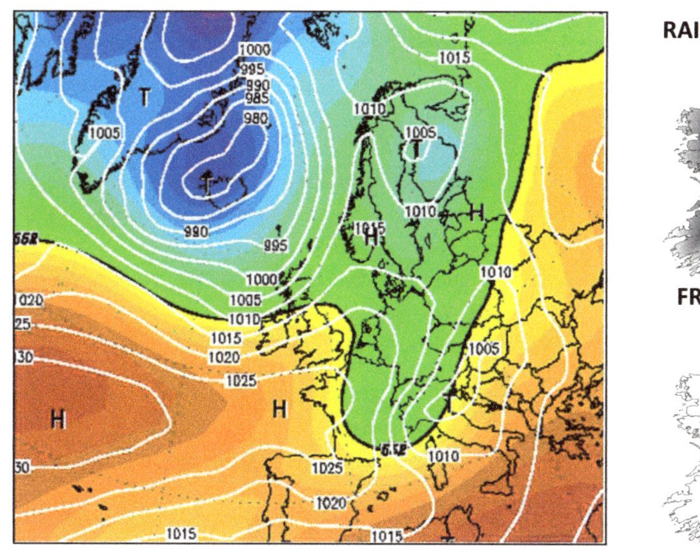

RAIN POTENTIAL

FROST/SNOW

APRIL **22** MONDAY

High pressure ridges to the southwest of Ireland while a low pressure trough flows over the country in a westerly flow, light at first but freshening to strong for a time before easing once more. Widespread cloud everywhere with isolated showers in most districts.

Belfast: Overcast, milder, dry with passing threats of rain, winds moderating for a time
Dublin: Cloudy, isolated showers, shifting breezy spells
Cork: Cloudy, dry, mild day cool night, shifting breezy spells
Galway: Stormy air, overcast, squally showers

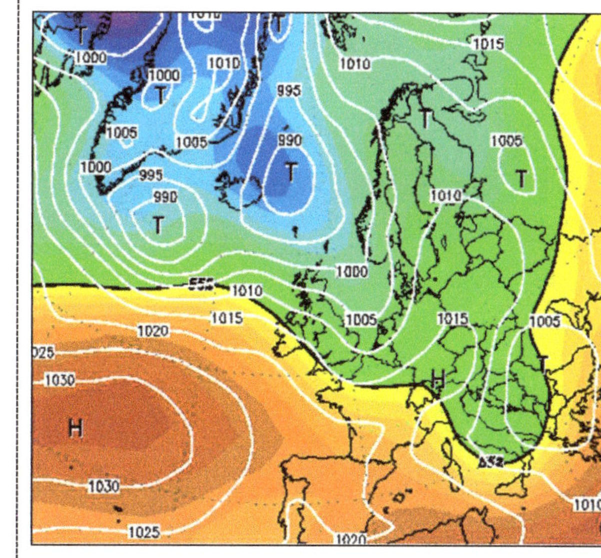

APRIL 23 TUESDAY

A depression to the north of Ireland moves down over the country directing very strong and blustery west to northwest winds throughout, with gale force gusts at times. Stormy air with a mix of light squally showers and sunny spells in most districts, although mostly dry along the eastern coasts.

Belfast: Mostly sunny, mild, isolated showers, strong winds to storm force possible
Dublin: Sunny, mostly dry, warmer, strong winds with chance of gales
Cork: Blustery winds, partly cloudy, isolated spotty showers
Galway: Partly cloudy, dry, windy with strong gusts, chance of gales

RAIN POTENTIAL

FROST/SNOW

APRIL 24 WEDNESDAY

Low pressure weakens over Ireland with westerly winds prevailing and easing, backing southeasterly later in the day. Mostly sunny everywhere, with isolated showers and patchy drizzles overnight and returning later in the day, mostly in the western half of the country.

Belfast: Sunny, windy, dry, colder overnight, winds easing, occasional gusts
Dublin: Sunny, mild and dry, windy
Cork: Partly cloudy, windy spells, light showers
Galway: Sunny, light showers, mild day cold night, windy at times

RAIN POTENTIAL

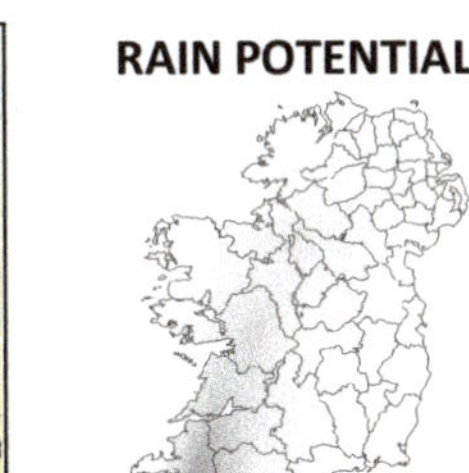

FROST/SNOW

APRIL 25 THURSDAY

Low pressure prevails with light and variable breezes throughout. Brief sunny spells and possible patches of fog in the south and east, with widespread cloud building everywhere, and very dull and gloomy in the south and west. Mostly dry in the east, with drizzly showers elsewhere.

Belfast: Cloudy, dry, mild, light breezes
Dublin: Cloudy, mostly dry, threats of rain, mild day cool night, light shifting breezes
Cork: Overcast, light drizzly showers, gentle breezes, chance of mist and fog patches
Galway: Overcast, light rain, mostly calm air

RAIN POTENTIAL

FROST/SNOW

APRIL 26 FRIDAY

A low pressure trough extends a gentle west to northwesterly flow over the country. Some light showers and drizzle patches over the western half of the country, while mostly dry in the east and far southern coastal regions. Occasional sunny spells most districts, with showers returning later in the evening.

Belfast: Partly cloudy, mild and mostly dry, odd threat of rain, moderate breezes
Dublin: Cloudy, dry, mild, light shifting breezes
Cork: Partly cloudy, dry, gentle breezes
Galway: Partly cloudy, gentle breezes, mild day cold night, some light drizzle patches

RAIN POTENTIAL

FROST/SNOW

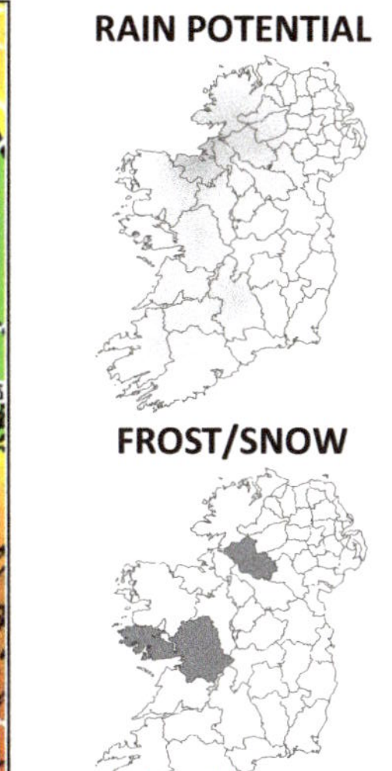

APRIL **27** SATURDAY

Low pressure trough maintains a stable west to northwest flow over the country while high pressure ridges south of Ireland. A changeable day with a mix of sunny spells and brief showers in most districts, with some heavy falls later in the day. Showers becoming more persistent in some western and northwestern districts for a time. Winds sometimes gusty around the northern coasts.

Belfast: Changeable, mix of sunny spells and light showers, breezy occasional gusts
Dublin: Changeable, mostly sunny, some showers of light rain, freshening breezes
Cork: Mostly sunny, dry, mild day cold night, light breezes
Galway: Dry, some sunny spells, milder, fresh breezes at times

RAIN POTENTIAL

FROST/SNOW

APRIL **28** SUNDAY

High pressure ridges to the south of Ireland while a low pressure trough lingers over the country with light to moderate westerly winds prevailing. Dull and gloomy with widespread cloud and drizzly showers throughout, with showers moving from the northwest and down through the country as the day progresses. Showers may be particularly heavy for a time in the far southwest.

Belfast: Overcast, scattered showers, cooler, chance of misty drizzles
Dublin: Cloudy, warmer, passing showers, windy spells
Cork: Overcast, showers of light rain, light breezes
Galway: Overcast, light showers, moderate breezes easing later in the day

RAIN POTENTIAL

FROST/SNOW

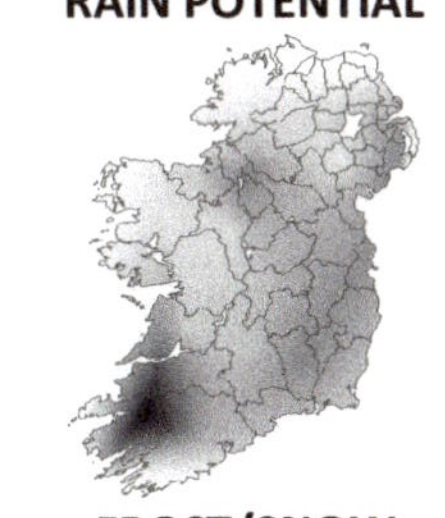

APRIL **29** MONDAY

Low pressure directs in a light variable flow over the country, with mist and fog patches in many districts overnight and chance of frost pockets in sheltered inland parts. Mostly dry and sunny day throughout, after light showers at first in the north, west, and south and again at the end of the day.

Belfast: Sunny, spotty showers, colder, mostly calm air, fog patches likely
Dublin: Mostly sunny, dry, mild day cold night, mostly calm air, chance of fog patches
Cork: Partly cloudy, warmer, scattered showers, gentle breezes to mostly calm air
Galway: Sunny, light showers, cold overnight, calm air, shallow fog patches possible

RAIN POTENTIAL

FROST/SNOW

APRIL **30** TUESDAY

Low pressure continues to prevail directing a light northeast to southeast flow, slightly fresher at times in the west. Mostly dry with some sunny spells and overnight fog patches in the east and north, with widespread cloud and showers in the west, and with more persistent rain and drizzle in the south of the country.

Belfast: Partly cloudy, milder, dry, light breezes to mostly calm, possible fog patches
Dublin: Cloudy, dry, milder, gentle breezes, possible overnight fog patches
Cork: Overcast, rain, colder, light breezes
Galway: Cloudy, dry, light breezes

RAIN POTENTIAL

FROST/SNOW

MAY

PHASES OF THE MOON

1st	Third Quarter
5th	Perigee #9
6th	Crossing Equator
8th	New Moon
11th	Northern Declination
15th	First Quarter
17th	Apogee
19th	Crossing Equator
23rd	Full Moon
25th	Southern Declination
30th	Third Quarter

1	2	3	4	5	6	7	8	9	10	11	12	13	14	15	16	17	18	19	20	21	22	23	24	25	26	27	28	29	30	31
Wed	Thur	Fri	Sat	Sun	Mon	Tue	Wed	Thur	Fri	Sat	Sun	Mon	Tue	Wed	Thur	Fri	Sat	Sun	Mon	Tue	Wed	Thur	Fri	Sat	Sun	Mon	Tue	Wed	Thur	Fri
3Q				P9	XhN		N			^				1Q		A		XhS				F		V					3Q	

MONTHLY SUMMARY

21st April -10th May Frontal systems cross the country in an easterly tract, bringing milder temperatures and outbreaks of showers in all districts, heavier falls at times in the west and north.

1st Chance of patches of fog. Warmer than normal especially in the west. Bright and sunny almost everywhere.

1st-9th Unusually warmer than normal daytime temperatures.

3rd Very sunny spell.

4th-5th High pressure prevails. Fog patches likely. Chance of a brush of rain in the west, with possible outbreaks of heavy rain in the Maam Valley region for a time.

7th-9th Very sunny spell, with many areas possibly seeing up to 12 hours of sunshine each day. Warmer than average.

8th-10th High pressure prevails.

10th Occasional outbreaks of heavy rain in the west and north of Ireland.

Over the next four weeks, low pressure will be the dominant weather features for much of the period, with high pressure only making an appearance in the last week of this outlook period. It will be wetter than average, with close to normal sunshine hours and mild temperatures that will be close to normal in the south and southwest, and above average elsewhere. Overall, rainfall amounts will be up to twice as much as is typical for this time of year, with particularly higher than normal rainfall in the west and southwest districts. Sunshine levels will be above normal overall in the northern half of the country by the end of this outlook period, particularly in the east and north, with average sunshine in the west and to below average sunshine hours in the south. On average overall, temperatures will be between half a degree to one degree higher than normal for this time of year in most parts except in the south and southwest, which will be closer to their norm. There will be a cooler than average spell however, in the second half of this outlook period, when maximum and minimum temperatures will be up to several degrees below the average for a few days, coming about when winds associated with a slow moving depression east of the country turn north to northwesterly in flow. During this outlook period, there is likely to be around 13 wet days in the northwest, (when 1mm or more rainfall is expected to be recorded), but up to 16-21 wet days elsewhere. Thunder may be heard on around 12 days, hail showers on 4 days, frosts and fogs on 5 days, gales on 2 days, but no snow.

11th-18th A depression near the northwest of Ireland extends a series of rain bands across the country, with some heavy falls at times. A generally southerly flow brings slightly warmer than normal temperatures and some blustery gusts at times.

12th A band of heavy rain crosses the country. Chance of hail showers.

13th Strong blustery winds at times.

13th-14th Chance of thunderstorms.

15th-17th Long sunny spells in most districts.

16th Some outbreaks of heavy rain likely.

17th Chance of thunderstorms.

18th Chance of heavy rain.

19th-22nd High pressure ridges over Ireland bringing a mostly dry and sunny spell with generally light to mostly calm winds. Warmer than normal, with many districts seeing daytime maximas above 20°C, except around the southern coastal fringes. Chance of fog patches.

20th-23rd Warmer than normal temperatures may generate scattered thunderstorms in many districts, with some heavy downpours at times. Very sunny almost everywhere.

23rd-6th June A very wet spell. A low pressure system moves onto Ireland during this period bringing rain or showers on most days, with some heavy falls in many areas, especially in the west and southwest, possibly resulting in some pockets of localised flooding.

24th-28th Chance of thunderstorms. Milder minimum temperatures with many districts seeing overnight lows around the 10-12°C range.

24th Heavy rain which may bring some localised flooding.

26th Unsettled winds may cause tornado activity, particularly in the northwest.

27th Heavy rain. Chance of localised flooding.

27th-28th Depression deepens as it moves close to the northwest coast. Winds strengthen throughout, possibly reaching gale force at times in the northwest, with strong gale gusts.

30th-31st A very deep depression sits to the southwest of the country bringing heavy rain in many districts, possibly causing some localised flooding. Colder north to northwest flow prevails.

IRELAND RAINFALL ESTIMATES
May 2024

d=mainly dry, n= nonrecordable, l=light shrs, s=significant shrs, r=rain, h=hvy falls

		ULSTER PROVINCE					CONNAUGHT PROVINCE						LEINSTER PROVINCE				MUNSTER PROVINCE						
		Alderg	Hillsbo	Armag	Loughl	Malin H	Sligo	Ardtarı	Belmul	Drums	Galway	Roscoı	Markre	Edende	Kilkenı	Dublin	Shannı	Newpo	Tralee	Killarne	Cork	All	Moon
1st	Wed	d	d	d	d	d	d	d	d	d	d	d	d	d	d	d	d	d	d	d	d	0	3Q
2nd	Thu	n	l	d	l	s	l	l	s	s	s	l	l	l	l	d	l	s	s	l	d	51	
3rd	Fri	l	d	l	d	d	d	d	n	d	d	d	d	d	d	d	d	d	d	d	d	3	
4th	Sat	n	l	d	s	l	s	l	l	s	s	s	l	l	l	l	l	s	s	l	l	53	
5th	Sun	l	d	l	s	l	l	s	l	n	d	n	l	d	d	d	n	l	n	l	l	17	P9 XhN
6th	Mon	d	d	d	d	l	l	l	l	l	l	s	n	d	d	d	d	n	d	d	d	10	
7th	Tue	d	d	d	d	d	d	d	d	d	d	d	l	d	d	d	d	d	d	d	d	1	
8th	Wed	d	d	d	d	d	d	d	d	d	d	d	d	d	d	d	d	d	d	d	d	0	N
9th	Thur	d	d	d	d	d	d	d	d	d	d	d	l	d	d	d	d	d	d	d	d	2	
10th	Fri	s	r	d	r	r	r	r	r	r	s	s	r	r	l	s	s	r	s	s	s	159	
11th	Sat	s	l	s	s	l	s	s	l	s	l	s	r	l	l	d	n	s	l	h	d	102	^
12th	Sun	l	s	l	l	r	r	s	h	r	r	r	s	r	s	l	r	s	h	h	s	182	
13th	Mon	s	l	s	n	n	l	d	l	l	n	l	l	l	s	d	l	n	s	s	n	38	
14th	Tue	s	r	l	l	l	l	l	s	l	l	l	n	s	r	l	l	l	l	s	l	62	
15th	Wed	l	d	l	d	d	l	l	s	n	s	s	l	n	l	d	l	l	r	h	l	69	1Q
16th	Thu	d	d	d	l	d	l	d	h	l	l	l	l	l	s	l	s	s	r	r	s	77	
17th	Fri	d	d	d	l	d	l	l	l	s	s	s	l	d	s	d	d	l	s	l	d	36	A
18th	Sat	r	r	s	s	d	r	s	s	s	s	r	r	l	l	r	l	l	n	d	s	124	XhS
19th	Sun	l	l	s	d	d	d	d	d	d	d	s	d	d	d	d	d	d	d	d	d	12	
20th	Mon	d	d	d	d	d	d	d	d	d	d	d	d	s	d	d	d	d	d	d	d	1	
21st	Tue	d	d	d	d	d	d	d	d	r	d	r	l	d	d	d	d	d	d	d	d	24	
22nd	Wed	s	l	s	d	d	d	d	d	l	d	s	d	s	n	h	l	d	l	s	l	48	
23rd	Thur	l	s	d	l	d	l	l	d	l	s	l	l	s	l	d	s	d	r	r	h	75	F
24th	Fri	s	r	l	r	s	r	r	r	r	r	h	r	h	l	s	s	r	r	s	s	203	
25th	Sat	s	l	r	s	s	s	s	l	l	l	l	s	l	s	r	s	s	h	l	s	112	V
26th	Sun	s	s	l	r	r	s	s	h	s	s	s	l	s	l	d	d	s	l	r	l	99	
27th	Mon	s	r	s	r	r	r	r	r	h	r	h	r	s	r	s	s	h	h	h	h	250	
28th	Tue	s	l	s	s	s	s	l	l	l	l	s	s	h	r	s	r	r	s	r	s	133	
29th	Wed	r	h	n	s	s	r	r	l	r	r	h	r	s	l	s	r	r	s	l	l	162	
30th	Thur	s	d	r	d	d	l	d	l	d	s	n	d	l	s	l	d	d	l	r	d	51	3Q
31st	Fri	l	s	d	r	l	r	r	d	h	h	h	r	h	h	h	h	r	h	h	h	305	
Estimate:		93	107	78	100	78	126	106	129	146	122	161	116	146	91	105	101	130	192	210	124	2460	
Average:		57	59	58	68	58	81	83	70	71	75	82	81	69	60	60	65	95	89	59	82	1423	
Trend:		wtr	wtr	wtr	wtr	wtr	wtr	wtr	wtr	wtr	wtr	wtr	wtr	wtr	wtr	wtr	wtr	wtr	wtr	wtr	wtr	wtr	

IRELAND SUNSHINE ESTIMATES

May 2024

F=fine (8-12 hours of sunshine), pc= partly cloudy (4-7 hours), c=cloudy (1-3 hours), o=overcast (0 hours)

		ULSTER PROVINCE					CONNAUGHT PROVINCE						LEINSTER PROVINCE				MUNSTER PROVINCE						
		Alderg	Hillsbo	Armag	Loughl	Malin H	Sligo	Ardtari	Belmul	Drums	Galway	Rosco	Markre	Edende	Kilkenn	Dublin	Shann	Newpo	Tralee	Killarne	Cork	All	Moon
1st	Wed	F	F	F	F	F	F	F	F	F	F	F	F	F	F	F	F	F	pc	pc	pc	218	3Q
2nd	Thu	c	c	c	c	c	o	c	o	c	o	o	c	c	c	c	o	c	c	c	c	24	
3rd	Fri	F	F	F	F	F	pc	F	pc	F	F	F	F	F	F	F	F	F	F	F	F	210	
4th	Sat	c	c	c	c	c	o	c	o	c	o	o	c	c	c	c	o	o	o	o	o	23	
5th	Sun	pc	pc	pc	pc	pc	pc	pc	pc	pc	pc	pc	pc	pc	F	pc	pc	pc	F	F	F	134	P9 XhN
6th	Mon	c	c	c	c	c	o	c	o	c	o	o	c	c	c	c	c	c	c	c	c	37	
7th	Tue	F	F	F	F	F	F	F	F	F	F	F	F	F	F	F	F	F	F	F	F	210	
8th	Wed	F	F	F	F	F	F	F	F	F	F	F	F	F	F	F	F	F	F	F	F	236	N
9th	Thur	pc	pc	pc	pc	pc	pc	pc	pc	pc	pc	pc	pc	pc	pc	pc	pc	pc	pc	pc	pc	117	
10th	Fri	o	o	o	o	o	o	o	o	o	o	o	o	o	o	o	o	o	o	o	o	0	
11th	Sat	F	F	F	F	F	pc	F	pc	pc	F	F	F	pc	pc	F	F	pc	pc	pc	pc	150	^
12th	Sun	c	c	c	c	c	c	c	c	c	c	c	c	c	c	c	c	c	c	c	c	46	
13th	Mon	F	F	F	F	F	F	F	F	pc	pc	F	F	pc	pc	F	pc	pc	o	o	o	133	
14th	Tue	o	o	o	o	o	c	o	c	o	pc	c	o	o	o	o	pc	c	o	o	o	29	
15th	Wed	F	F	F	F	F	F	F	F	F	F	F	F	F	pc	F	F	F	pc	pc	pc	180	1Q
16th	Thu	pc	pc	pc	pc	pc	pc	pc	pc	pc	pc	pc	pc	pc	pc	pc	pc	pc	pc	pc	pc	104	
17th	Fri	F	F	F	F	F	F	F	F	F	F	F	F	F	F	F	F	F	F	F	F	223	A
18th	Sat	o	o	o	o	o	o	o	o	o	o	o	o	o	o	o	c	c	c	c	c	8	XhS
19th	Sun	pc	pc	pc	pc	pc	pc	pc	pc	pc	pc	pc	pc	pc	pc	pc	pc	pc	F	F	F	130	
20th	Mon	F	F	F	F	F	F	F	F	F	F	F	F	F	F	F	F	F	F	F	F	250	
21st	Tue	F	F	F	F	F	F	F	F	F	F	F	F	F	F	F	F	F	F	F	F	215	
22nd	Wed	pc	pc	pc	pc	pc	pc	pc	pc	pc	pc	pc	pc	pc	pc	pc	c	c	c	c	c	106	
23rd	Thur	pc	pc	pc	pc	pc	F	pc	F	pc	pc	pc	pc	pc	pc	pc	c	c	c	c	c	112	F
24th	Fri	o	o	o	o	o	o	o	o	o	o	o	o	o	o	o	o	o	o	o	o	5	
25th	Sat	o	o	o	o	o	o	o	o	c	c	c	o	c	c	c	pc	c	pc	pc	pc	36	V
26th	Sun	c	c	c	c	c	o	c	o	c	c	o	c	c	c	pc	c	c	c	c	c	50	
27th	Mon	o	o	o	o	o	o	o	o	o	o	o	o	o	o	1.3	o	o	o	o	o	11	
28th	Tue	c	c	c	c	c	c	c	c	c	c	c	c	c	c	pc	pc	c	o	o	o	59	
29th	Wed	c	c	c	c	c	c	c	c	c	c	c	c	c	pc	pc	c	c	c	c	c	64	
30th	Thur	pc	pc	pc	pc	pc	pc	pc	pc	pc	pc	pc	pc	pc	pc	pc	c	c	c	c	c	108	3Q
31st	Fri	o	o	o	o	o	o	o	o	o	o	o	o	o	c	o	o	o	c	c	c	9	
Estimate hours:		169	169	169	169	169	151	169	151	167	156	153	169	167	165	187	161	163	144	144	144	3235	
Average hours:		189	189	165	191	192	189	188	190	153	180	183	172	176	170	189	179	168	180	180	187	3611	
Trend:		less	less	av	less	less	less	less	less	more	less	less	av	av	av	av	less	av	less	less	less	less	

MAY 1 WEDNESDAY

A weak depression to the south of Ireland extends a generally light to moderate easterly flow. Mostly dry and sunny throughout with scattered frosts and fogs in central to northern parts. Chance of some drizzly patches briefly around the southern coasts.

Belfast: Sunny, dry, mild day cold night, gentle breezes, chance of misty patches
Dublin: Dry, mild and sunny, fluctuating breezes
Cork: Warmer, partly cloudy, dry, breezy spells, chance of overnight fog patches
Galway: Sunny, dry, warmer, fluctuating breezes

RAIN POTENTIAL

FROST/SNOW

MAY 2 THURSDAY

A high pressure system to the southwest of Ireland extends a ridge over the country while low pressure sits to both the north and south. Southwest winds prevail, light for a while, but with occasional gusts at times. Widespread cloud throughout with a weak front crossing the country in a lazy southwesterly tract, starting in the northwest and eventually bringing drizzly showers into most districts, although remaining mostly dry around the eastern and southern coastal fringes.

Belfast: Cloudy, warmer, isolated showers, freshening winds later in the day
Dublin: Cloudy, mostly dry, threats of rain, warmer day cold night, moderate breezes
Cork: Cloudy, mostly dry, cooler, mostly calm air, chance of misty patches
Galway: Overcast, cooler, occasional showers, intermittent breezy spells

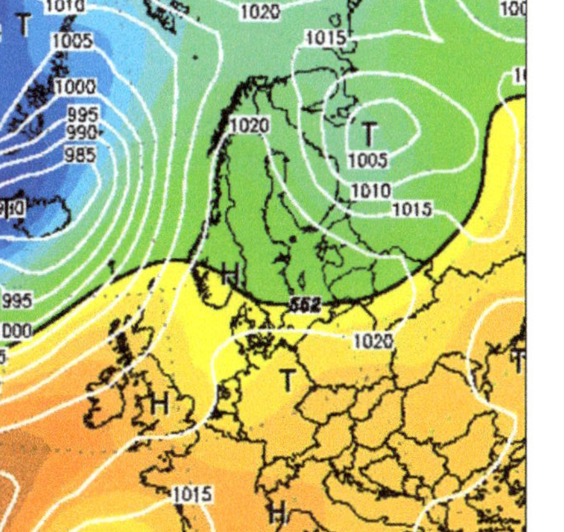

RAIN POTENTIAL

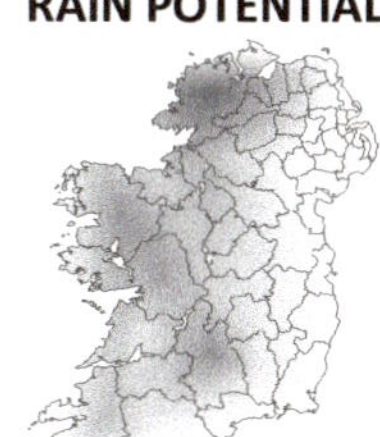

FROST/SNOW

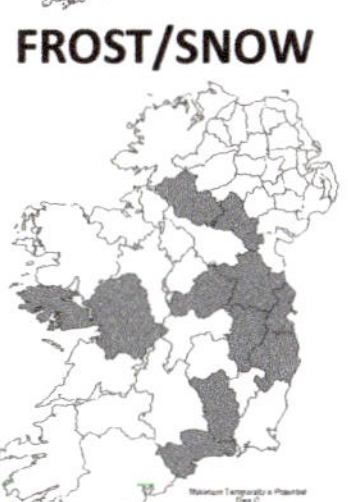

MAY 3 FRIDAY

Depressions to the north of Ireland extend a low pressure trough across the country while a weak high pressure ridge drifts to the west. Winds generally light throughout in west to variable flow. Showers clear the east of the country overnight bringing a mainly dry and sunny day throughout the country, and with the chance of some misty patches or light fog mostly in eastern parts and patches of frost in inland parts.

Belfast: Sunny, dry, mild day cold night, shifting breezy spells
Dublin: Sunny, dry, cooler, shifting windy spells, chance of isolated misty patches
Cork: Sunny, dry, mild day cooler night, light shifting breezes
Galway: Sunny, dry, colder overnight, light breezes

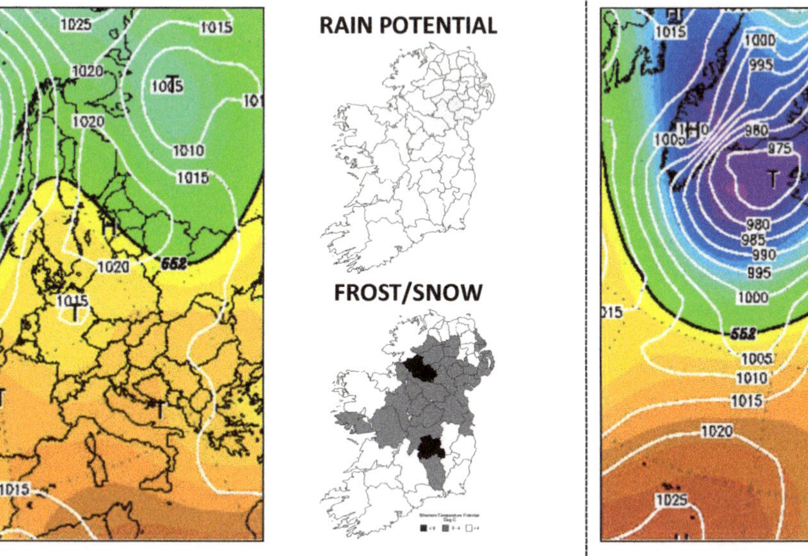

MAY 4 SATURDAY

A low pressure trough lingers over the country bringing light to moderate southerly breezes at first, but fresher at times in the west and north. Mostly dry at first with scattered frost and fog patches. Later, a wet front crosses the western coast and slowly moves across the country in a westerly tract to near the eastern coasts later in the day spreading light drizzly showers into many districts.

Belfast: Cloudy, mild, brief light showers, light breezes strengthening later in the day
Dublin: Mostly cloudy, isolated showers, gentle breezes, chance of late frost pockets
Cork: Overcast, colder, spells of light rain, calm air, possible shallow fog patches
Galway: Overcast, occasional showers, cooler, chance of overnight frost pockets

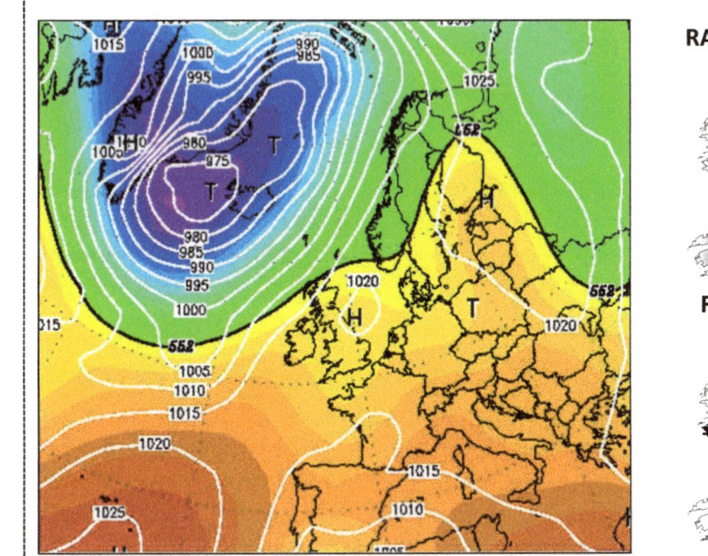

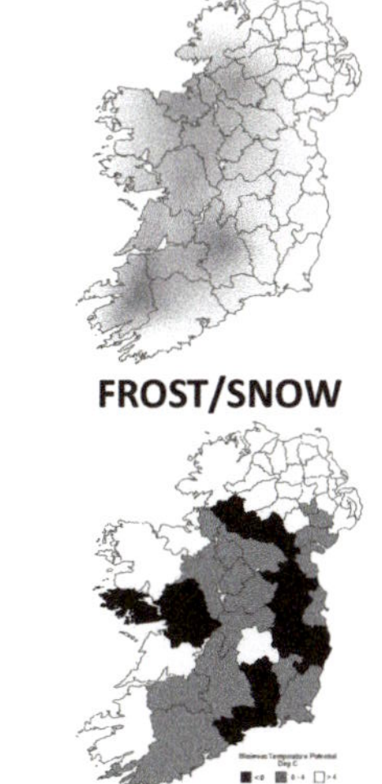

MAY **5** SUNDAY

An intense depression to the northwest of Ireland descends onto the country directing very strong southwest winds throughout, occasionally reaching gale force at times particularly in the north and west. A changeable day with a mix of sunny spells and the occasional showery spells in most districts although remaining mostly dry in the east and far southern parts.

Belfast: Partly cloudy, dry, strong winds, possible gales at times
Dublin: Partly cloudy, dry, warmer, blustery winds
Cork: Sunny, warmer, isolated showers, freshening winds at times
Galway: Partly cloudy, dry, milder, strong stormy air, gales likely

RAIN POTENTIAL

FROST/SNOW

MAY **6** MONDAY

Unsettled and blustery conditions over northern parts from the depression north of the country, while high pressure ridges to the south with winds generally light and in a westerly flow. Mostly dry with frequently cloudy spells throughout, except for some isolated drizzly spells mostly in the northwest. Chance of fog patches in the south of the country.

Belfast: Changeable, cloudy, mostly dry with odd threat of rain, drizzle patches possible
Dublin: Unsettled and blustery winds at times, dry, mostly cloudy
Cork: Cloudy, dry, winds easing to light, chance of overnight fog patches
Galway: Overcast, isolated showers, winds easing

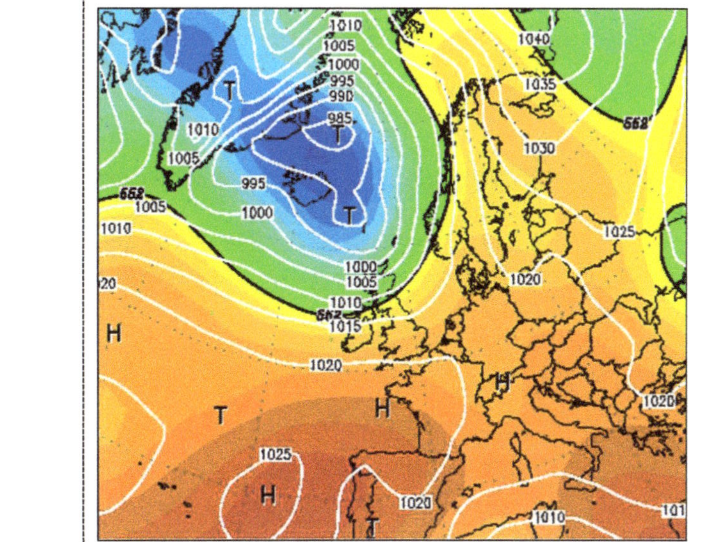

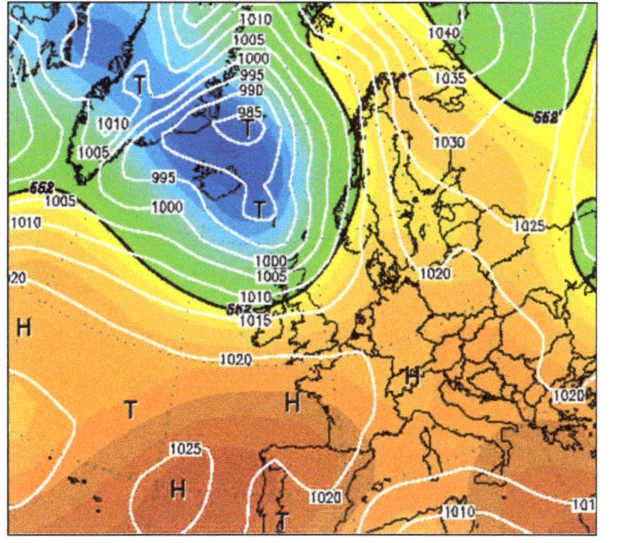

RAIN POTENTIAL

FROST/SNOW

MAY **7** TUESDAY

A large anticyclonic system moves up onto Ireland in a light to moderate northwest flow. Very sunny and dry throughout, with fog patches likely mostly in the east and south. Chance of some ground frosts overnight mostly in sheltered inland places.

Belfast: Sunny, dry, mild day cool night, gentle breezes
Dublin: Very bright and sunny, dry, warm day cool night, mostly calm and serene air
Cork: Sunny, dry, warmer day cooler night, mostly calm air, chance of fog
Galway: Sunny, dry, gentle shifting breezes, chance of shallow fog

RAIN POTENTIAL

FROST/SNOW

MAY **8** WEDNESDAY

A large anticyclonic system continues to strengthen as it lingers to the west of Ireland directing a light to moderate northwesterly flow throughout. Sunny and dry everywhere. Cool overnight with chance of some patches of frost and scattered fog mostly in northern and inland parts.

Belfast: Sunny, dry, light northerly breezes
Dublin: Pleasant air, very sunny, dry, gentle breezes, cold overnight
Cork: Very sunny and dry, mild day cool night, gentle breezes to mostly calm air
Galway: Sunny, dry, warmer, light breezes to mostly calm air

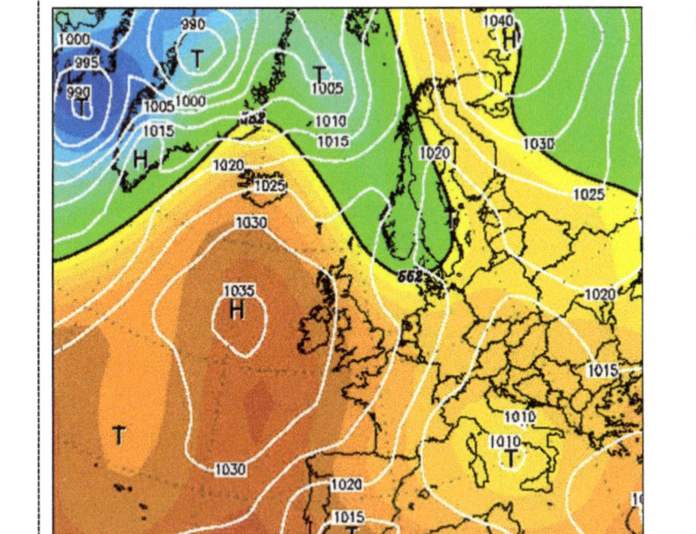

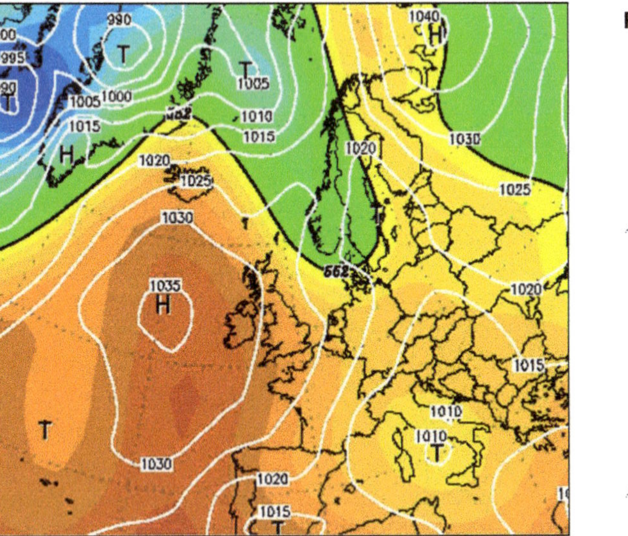

RAIN POTENTIAL

FROST/SNOW

MAY 9 THURSDAY

A high pressure ridge maintains a stable light northerly flow over the country and dry conditions throughout. Very sunny in southwestern parts with increased scattered cloud building elsewhere to mostly cloudy in the northwest and northern coastal fringes. Milder temperatures throughout.

Belfast: Increasing scattered cloudy spells, dry, milder, light breezes
Dublin: Partly cloudy, cooler, dry, light breezes
Cork: Partly cloudy, dry, cooler, light breezes to mostly calm air
Galway: Partly cloudy, light breezes, dry, mild temperatures

RAIN POTENTIAL

FROST/SNOW

MAY 10 FRIDAY

The high pressure ridge slips to the southwest of Ireland while a trough associated with a depression to the northwest moves down over the country in a light to moderate westerly flow, shifting to southwest for a time and then northwest. Cloud thickens to mostly overcast everywhere, with patches of light rain or drizzle in most districts, becoming more persistent and heavier especially in western parts.

Belfast: Mostly calm, overcast, rain
Dublin: Overcast, cooler, spells of passing rain, light breezes
Cork: Overcast, brief showers, cooler, light breezes to mostly calm air
Galway: Overcast, light breezes to mostly calm air, passing showers

RAIN POTENTIAL

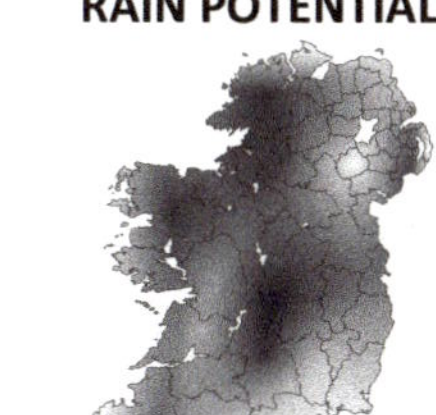

FROST/SNOW

MAY **11** SATURDAY

Low pressure prevails directing fresh to strong northwest winds, easing later with a southwesterly change. A changeable day with a mix of sunny spells and bands of showers in most districts with showers contracting to western parts before returning overnight.

Belfast: Sunny, cooler, brief showers, windy spells
Dublin: Unsettled windy spells, dry, sunny, coolish, chance of brief misty patches
Cork: Partly cloudy, dry, mild day cool night, freshening winds
Galway: Sunny, brief showers, cold overnight, windy at times

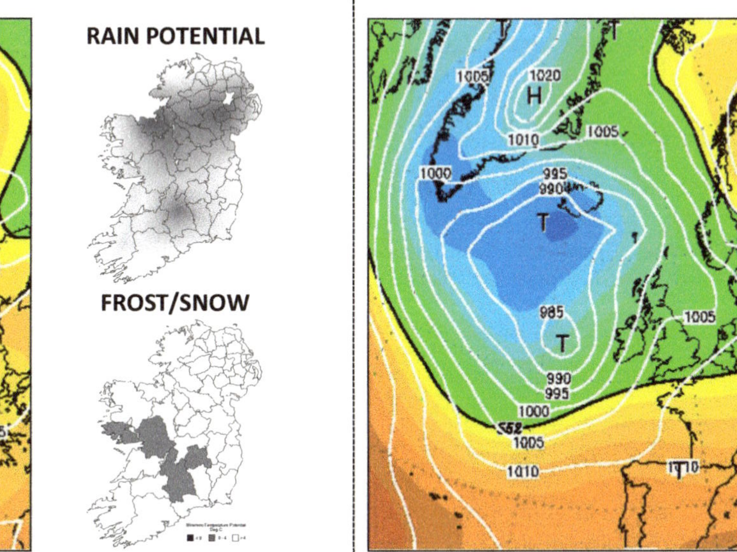

RAIN POTENTIAL

FROST/SNOW

MAY **12** SUNDAY

A large depression moves onto Ireland directing fresh to strong south to southwest winds throughout, with the chance of gale gusts for a time, particularly in the northwest. A trough of rain moves across the country in an easterly track, bringing widespread rain into all districts, including heavy falls for a time in western parts. Some brief sunny breakthroughs likely now and then, mostly in the far north and far southwest.

Belfast: Cloudy, scattered showers, windy with occasional gusts
Dublin: Cloudy, windy, brief showers
Cork: Cloudy, windy with occasional gusts, passing showers and drizzle spells
Galway: Cloudy, rain, cooler day milder night, blustery

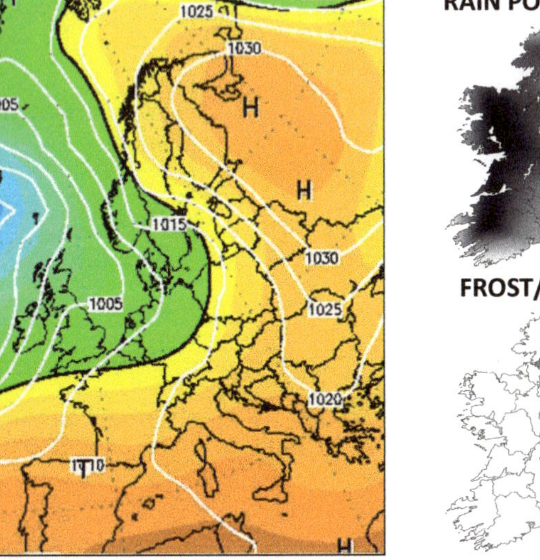

RAIN POTENTIAL

FROST/SNOW

MAY 13 MONDAY

Depression to the northwest of Ireland strengthens directing strong and blustery south to southwest winds over the country, with a stormy air and winds up to gale force at times. A frontal trough crossing the country brings the passage of squally light rain in the west at first, moving into the east later in the day, with sunny spells in between. Chance of thunderstorms, particularly in the west.

Belfast: Stormy air overnight, squally showers, sunny day, cooler
Dublin: Sunny, dry, warm day cool night, strong gusty winds, possible gales at times
Cork: Stormy air, overcast, isolated drizzly showers, gusty winds to gale force at times
Galway: Blustery winds, mostly sunny, milder, chance of thunderstorms and gales

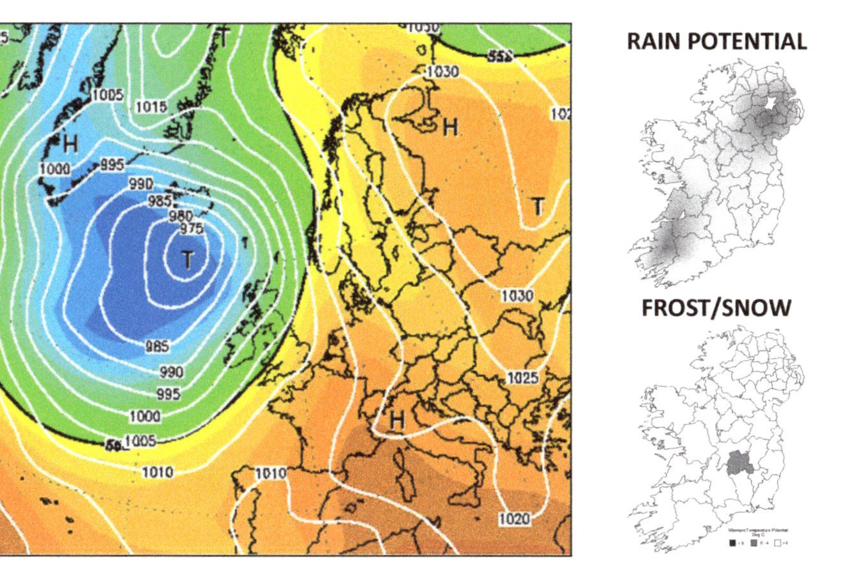

RAIN POTENTIAL

FROST/SNOW

MAY 14 TUESDAY

Low pressure continues to direct blustery front troughs over the country with south to southeast winds prevailing, strong at first but easing later in the day. Widespread cloud everywhere except for some brief sunny breakthroughs in the west. Bands of showers cross the country, sometimes accompanied by thunderstorm activity, with precipitation more persistent but lighter in the west, but heavier in the east.

Belfast: Overcast, rain, blustery with gusts, chance of thunderstorms and gale winds
Dublin: Overcast, cooler, light showers, windy spells
Cork: Overcast, light squally showers, windy, cold overnight
Galway: Blustery, mix of sunny spells and brief showers, warmer

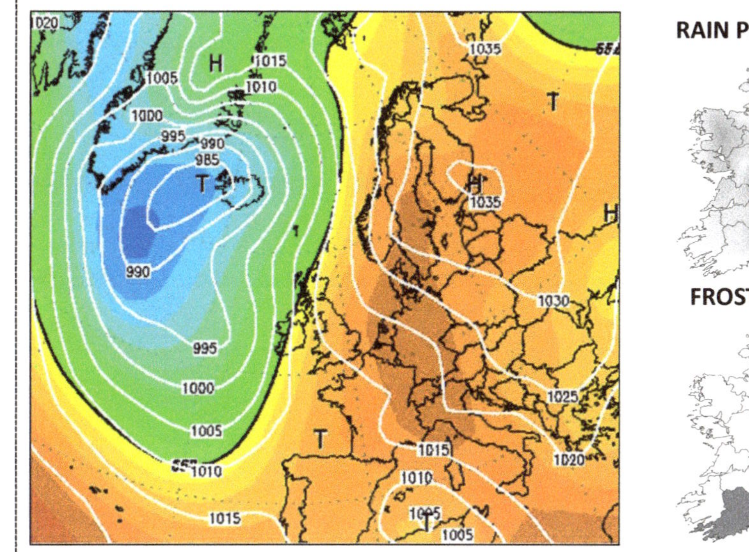

RAIN POTENTIAL

FROST/SNOW

MAY **15** WEDNESDAY

Depression to the west of Ireland weakens with southwest to southerly winds easing. A showery trough moves up over the country in a northeasterly track at first, followed by mostly dry and sunny conditions later in the day, particularly in the north and east. Chance of some misty patches overnight in eastern parts. Some heavy showers likely to become confined to the southwest later in the day.

Belfast: Sunny, dry, cooler, strong winds
Dublin: Sunny, dry, warm day cool night, blustery at times
Cork: Partly cloudy, brief showers, mild day cold night, windy with occasional gusts
Galway: Sunny, overnight showers, mild day cold night, blustery at times

RAIN POTENTIAL

FROST/SNOW

MAY **16** THURSDAY

A weak depression continues to linger over Ireland directing an unstable southerly flow throughout. A cold front crosses the country bringing heavy rain into western parts, and lighter scattered showers elsewhere throughout the day. Some sunny spells in most districts.

Belfast: Partly cloudy, mostly dry, threats of rain, warmer, fresh southerly breezes
Dublin: Partly cloudy, isolated showers, cooler, breezy spells
Cork: Mix of showers and sunny spells, cooler, fresh winds
Galway: Windy spells, cooler, mix of sunny spells and brief showers

RAIN POTENTIAL

FROST/SNOW

MAY 17 FRIDAY

A depression west of Ireland moves onto the country in a strong southerly flow, backing southeasterly later in the day. Mostly sunny day with long dry spells throughout. A band of thunderstorms is likely to cross the country, however, bringing outbreaks of heavy rain overnight, particularly in the southwest but also thundery showers in most districts south of Ulster, where only isolated drizzle patches can be expected overnight.

Belfast: Sunny, dry, windy, chance of brief overnight drizzle patch
Dublin: Sunny, dry, mild, moderate southerly flow
Cork: Sunny, mostly dry, coolish, moderate breezes
Galway: Sunny, overnight showers, blustery at times

RAIN POTENTIAL

FROST/SNOW

MAY 18 SATURDAY

A decaying low pressure system continues to prevail over Ireland with winds easing to a light east to northeast flow. Possible misty patches overnight before a slow-moving frontal system embedded with thunderstorms crosses the country bringing widespread cloud and rain throughout.

Belfast: Overcast, windy, rain
Dublin: Overcast, rain, cooler, moderate breezes, chance of overnight misty patches
Cork: Cloudy, occasional showers, light breezes to mostly calm air
Galway: Overcast, fluctuating winds, showers, cooler

RAIN POTENTIAL

FROST/SNOW

MAY **19** SUNDAY

A trough associated with a decaying depression to the west of Ireland brings a gentle southerly air with winds light and variable throughout, and fog and mist in many parts, especially in the southeast. A mostly sunny and dry day prevails, except for some drizzly showers likely for a time in the northeast of the country.

Belfast: Partly cloudy, warmer, isolated showers, light shifting breezes
Dublin: Partly cloudy, dry, warm day cool night, gentle breezes, chance of fog patches
Cork: Sunny, warmer, dry, light breezes to mostly calm air
Galway: Partly cloudy, dry, mostly calm air, chance of shallow fog in places

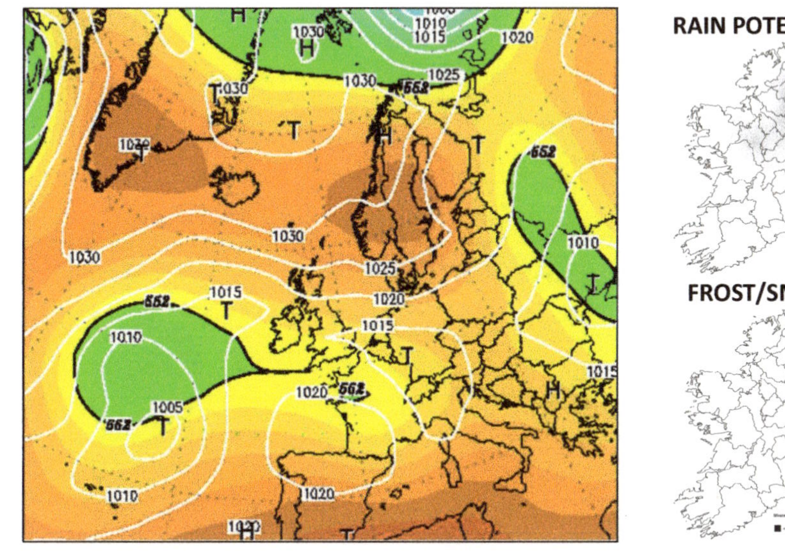

RAIN POTENTIAL

FROST/SNOW

MAY **20** MONDAY

An anticyclonic system north of Ireland ridges over the country bringing mostly calm, dry and sunny conditions throughout. Scattered fogs likely in many districts. Chance of isolated thunderstorms overnight in the far northwest.

Belfast: Warmer, very sunny, dry, pleasant serene atmosphere
Dublin: Sunny, dry, warm day cool night, pleasant serene atmosphere, chance of fog
Cork: Very sunny, dry, warmer, serene air, chance of misty patches overnight
Galway: Warmer, very sunny, dry, pleasant calm air, chance of overnight fog patches

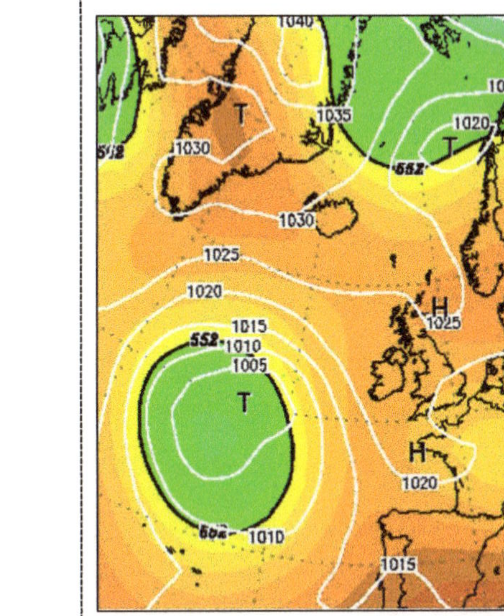

RAIN POTENTIAL

FROST/SNOW

MAY 21 TUESDAY

High pressure continues to ridge over Ireland maintaining a dry and mostly calm and serene air throughout with long sunny spells everywhere. Overnight fog likely in many districts. A low pressure trough to the west may extend a thundery front overnight, with some isolated thunderstorms that may bring an outburst of showers in northern midlands districts.

Belfast: Sunny, dry, warm day cool night, pleasant serene air
Dublin: Sunny, warmer, dry, calm and pleasant air, chance of fog overnight
Cork: Very sunny, dry, warmer, gentle breezes to calm air, fog possible
Galway: Sunny, warmer, dry, gentle breezes

MAY 22 WEDNESDAY

A wide trough lingers over Ireland bringing light breezes to a mostly calm air throughout with scattered fogs in many districts overnight, some taking a long time to clear. Cloudy in the east, while mostly sunny elsewhere. A band of thunderstorms is likely to cross the country, with heavy falls in the east for a time, while remaining mostly dry and pleasant in the far south and northwest.

Belfast: Calm, warm and humid, mostly sunny, brief showers likely
Dublin: Warm, mostly sunny, heavy overnight rain with chance of thunderstorms
Cork: Cloudy, calm, light showers, cooler, chance of fog patches
Galway: Partly cloudy, dry, cooler, calm atmosphere, chance of fog patches

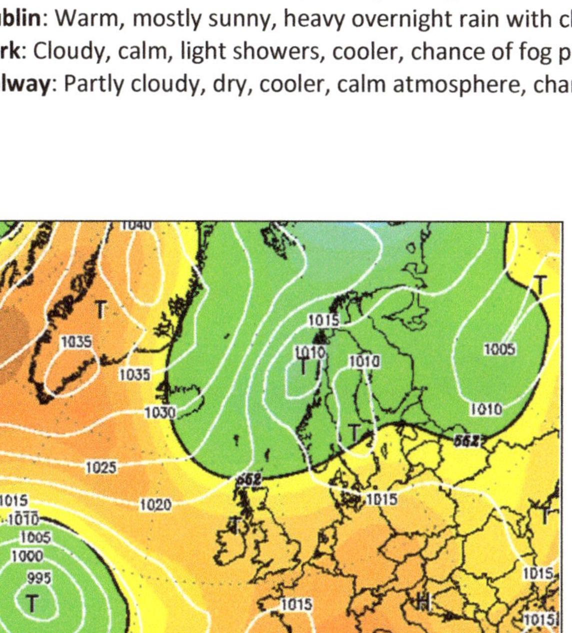

MAY **23** THURSDAY

High pressure ridges to the northwest and another to the far south, while low pressure prevails elsewhere. Winds generally light everywhere, in a gentle northerly flow, backing east to southeast later in the day. A changeable day with a mix of sunny spells and showers over the southern half of the country and in the far north, while mostly dry with longer sunny spells in the east. Chance of thunderstorms, particularly in the far south, accompanying some heavier falls of rain.

Belfast: Partly cloudy, cooler, brief spotty showers, light shifting breezes
Dublin: Mostly calm and dry, partly cloudy, cooler
Cork: Mostly cloudy, heavy rain, gentle breezes, chance of thunderstorms
Galway: Changeable, mix of sunny spells and showers, cooler, mostly calm air

MAY **24** FRIDAY

Low pressure trough maintains a light southeasterly flow over the country. Wet fronts within the trough bring widespread cloud and scattered rain activity everywhere, as well as the chance of overnight fog patches in the far northern and southern areas. Thunderstorms likely to accompany the rain band, particularly in the west.

Belfast: Overcast, rain, cooler, light breezes, chance of misty patches
Dublin: Light breezes to mostly calm air, overcast, occasional showers
Cork: Overcast, passing showers, gentle breezes, chance of overnight fog
Galway: Overcast, rain, cool, moderate breezes, chance of thunderstorms

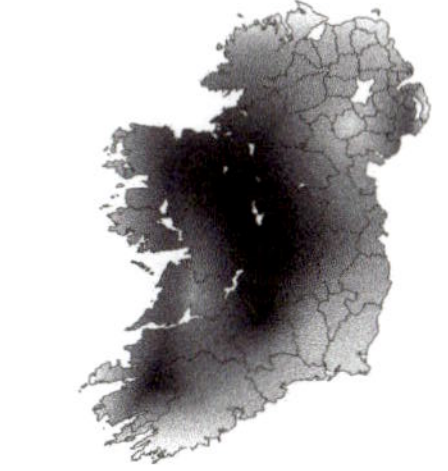

MAY **25** SATURDAY

A large depression to the west of Ireland moves onto the country in a moderate southerly flow. Widespread cloud in most districts, with outbreaks of rain throughout. Rain likely to be heavy at times in the south and southwest as well as around the coastal fringes in the east. Thunderstorms likely in the afternoon.

Belfast: Overcast, cooler, brief showers, breezy spells
Dublin: Cloudy, rain, occasional windy spells
Cork: Mix of sunny spells and showers, warmer, fresh breezes, chance of fog patches
Galway: Cloudy, brief drizzly showers, cool, moderate breezes

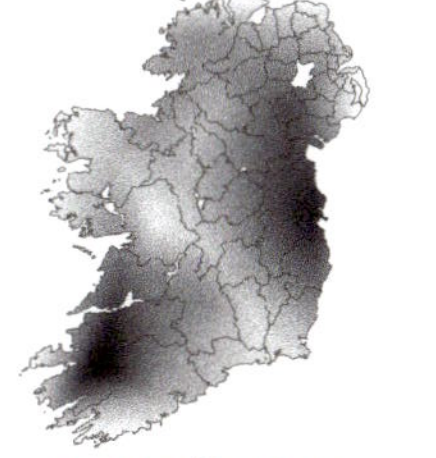

RAIN POTENTIAL

FROST/SNOW

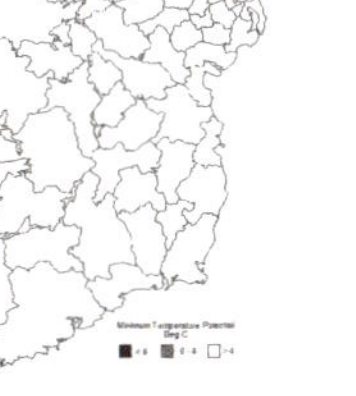

MAY **26** SUNDAY

Low pressure from a depression to the west of Ireland continues to prevail directing an unsettled south to southwest flow throughout. Widespread scattered cloud everywhere with showers starting in the west and spreading into most other parts as the day progresses. Showers likely to die out by evening allowing for scattered fog patches to develop particularly in some western parts.

Belfast: Mostly cloudy, rain, moderate breezes
Dublin: Mostly dry, cloudy spells, warmer day cooler night, shifting breezy spells
Cork: Cloudy, moderate breezes, brief showers
Galway: Cloudy, brief showers, fluctuating breezes, chance of fog patches

RAIN POTENTIAL

FROST/SNOW

MAY 27 MONDAY

A large depression to the west of Ireland continues to direct a southeasterly flow over the country, light at first before becoming fresh and blustery, with a stormy air later. Dull and gloomy with heavy cloud and frequent rain everywhere moving up from the southwest at first and progressing steadily up into the north of the country. Rain likely to be heavy and persistent at times, and occasionally accompanied by thunderstorms and gale gusts, especially in the western half of the country.

Belfast: Overcast, warmer, rain, stormy air at times
Dublin: Cooler, cloudy, occasional showers, fresh breezes at times
Cork: Overcast, heavy rain, windy spells
Galway: Overcast, rain, strengthening winds, chance of thunderstorms and gale gusts

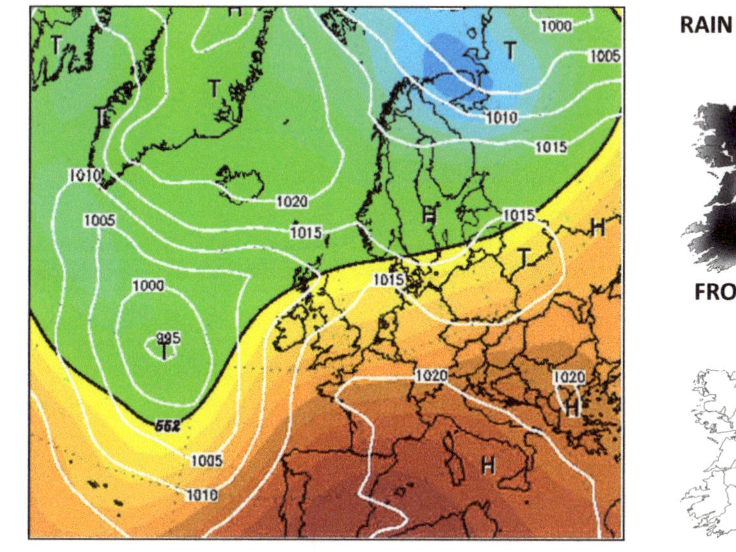

RAIN POTENTIAL

FROST/SNOW

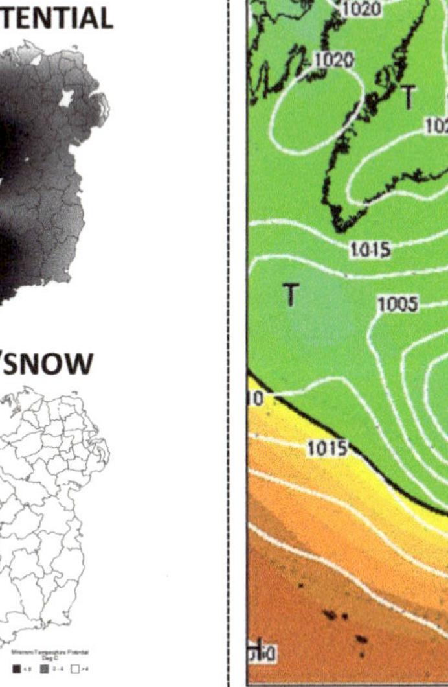

MAY 28 TUESDAY

The depression to the west of Ireland deepens directing a stormy air throughout the country with strong southwest winds prevailing, up to gale force at times. Widespread cloud and scattered showers everywhere, some accompanying thunderstorm activity, particularly in the west.

Belfast: Cloudy, cooler, light showers, chance of thunderstorms and gale winds at times
Dublin: Mix of showers and sunny spells, strong gusty winds, possible gale gusts
Cork: Overcast, cooler, blustery with winds to gale force at times, driving showers
Galway: Stormy air, showers, thunderstorms possible, blustery with gales possible

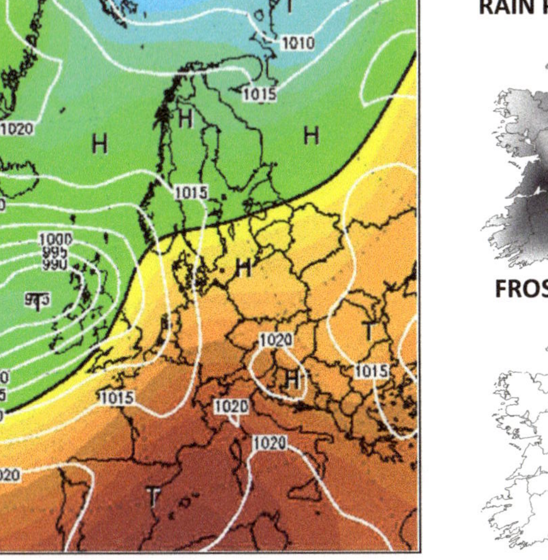

RAIN POTENTIAL

FROST/SNOW

MAY **29** WEDNESDAY

An intense depression continues to bring widespread storm activity and rain throughout the country, with heavy falls in places, scattered thunderstorms, and strong west to northwest winds prevailing, up to gale force at times, particularly around the southern coasts.

Belfast: Stormy air, heavy rain, brief sunny spells, chance of thunderstorms and gales
Dublin: Strong gusty winds, mix of sunny spells and showers, gales possible
Cork: Cloudy, light drizzly showers, strong gusty winds, possible gales
Galway: Unsettled and stormy air with gales, cloudy, rain, possible thunderstorms

RAIN POTENTIAL

FROST/SNOW

MAY **30** THURSDAY

A depression over Ireland brings very strong and fluctuating winds, possibly up to storm force at times. Widespread overnight rain throughout easing to showers later in the day, before another front brings a return of rain back into the southwest during the evening. Occasional sunny spells likely to break through in most districts, especially in western parts.

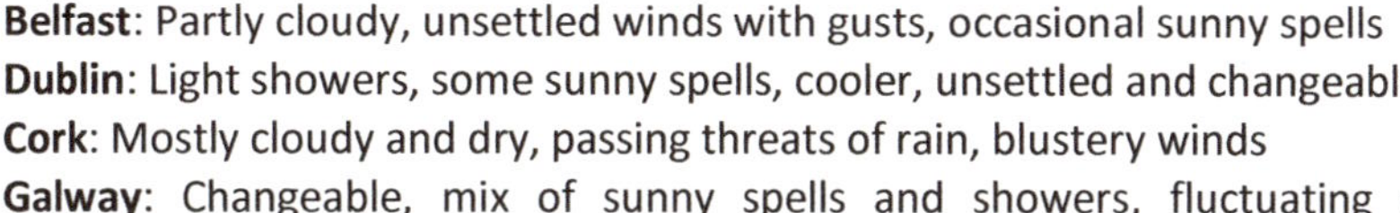

Belfast: Partly cloudy, unsettled winds with gusts, occasional sunny spells
Dublin: Light showers, some sunny spells, cooler, unsettled and changeable winds
Cork: Mostly cloudy and dry, passing threats of rain, blustery winds
Galway: Changeable, mix of sunny spells and showers, fluctuating winds with occasional gusts

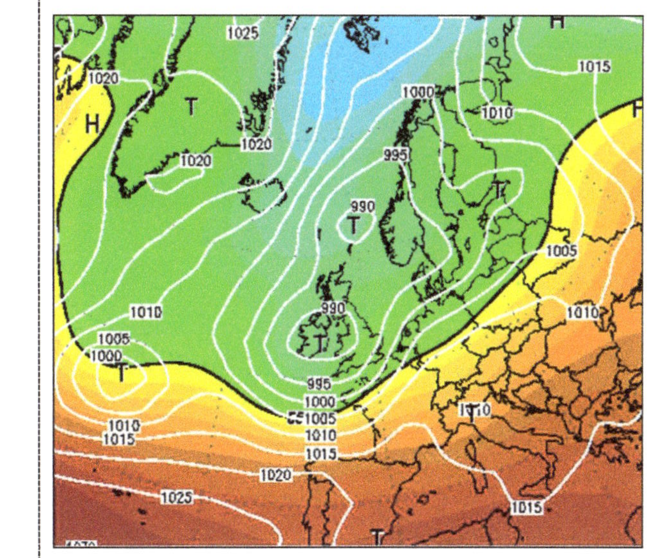

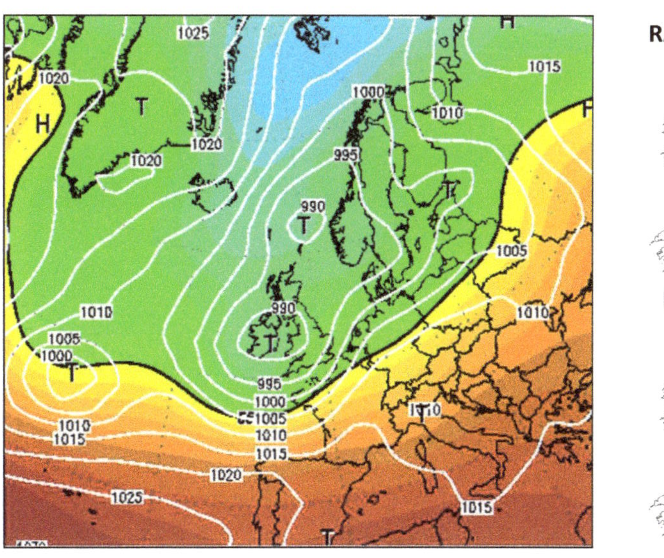

RAIN POTENTIAL

FROST/SNOW

MAY 31 FRIDAY

As the depression over Ireland continues to track away to the southwest of the country, a large trough between two low pressure centres brings easing winds despite a stormy air throughout. Widespread cloud carry bands of rain across the country with heavy falls in many areas, particularly over the southern half of the country. Rain likely to ease late in the day.

Belfast: Overcast, winds easing, colder, scattered showers
Dublin: Storm, overcast, heavy rain, cooler
Cork: Stormy, winds easing, cooler, heavy rain, brief sunny breakthroughs likely
Galway: Overcast, heavy rain, cooler, gusty winds easing later in the day

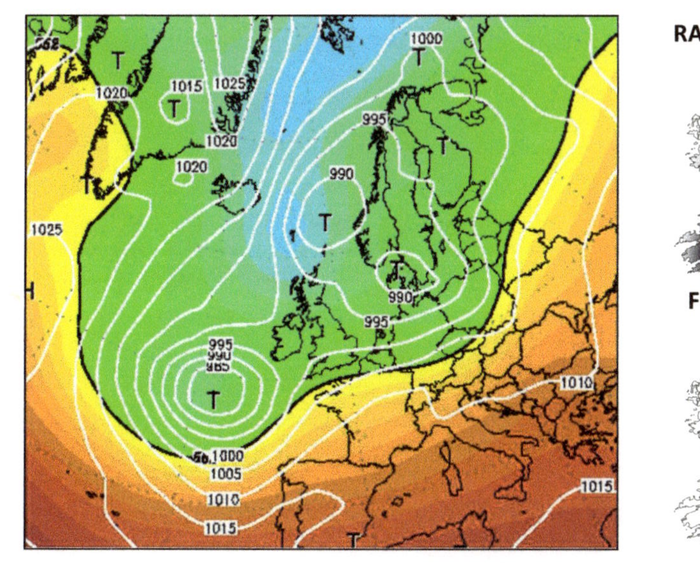

RAIN POTENTIAL

FROST/SNOW

JUNE

PHASES OF THE MOON

2nd	Perigee #12
2nd	Crossing Equator
6th	New Moon
7th	Northern Declination
14th	First Quarter
14th	Apogee
15th	Crossing Equator
22nd	Full Moon
22nd	Southern Declination
27th	Perigee #13
28th	Third Quarter
29th	Crossing Equator

1	2	3	4	5	6	7	8	9	10	11	12	13	14	15	16	17	18	19	20	21	22	23	24	25	26	27	28	29	30
Sat	Sun	Mon	Tue	Wed	Thur	Fri	Sat	Sun	Mon	Tue	Wed	Thur	Fri	Sat	Sun	Mon	Tue	Wed	Thur	Fri	Sat	Sun	Mon	Tue	Wed	Thur	Fri	Sat	Sun
	P12 XhN				N	^							1Q A	XhS							F V					P13	3Q	XhN	

MONTHLY SUMMARY

23rd May - 6th June A very wet spell. Low pressure system moves onto Ireland during this period bringing rain or showers on most days, with some heavy falls in many areas, especially in the west and southwest, possibly resulting in some pockets of localised flooding.
30th May-4th June Depression east of Ireland directs a colder north to northwest flow over the country. Colder everywhere with fairly widespread scattered ground frosts.
1st-3rd Cooler spell. Chance of hail showers.
2nd Chance of thunderstorms.
5th-6th Chance of patches of fog.
7th-10th A large anticyclonic system moves onto the country directing a lighter west to northwest flow, later backing northerly, and bringing mostly dry and sunny conditions throughout with warm daytime temperatures and colder overnight with pockets of frost in inland parts.
7th-27th Long dry spell likely in the Munster Province.
8th-19th Very bright and sunny spell throughout the country. Many districts likely to see up to 12-14 hours of sunshine a day.
9th-10th High pressure prevails. Cold spell with possible widespread frost patches. Dry everywhere and very sunny, especially in the west.

Over the next four weeks a large anticyclone situated to the west of Ireland brings fairly settled conditions for the first half of the outlook period, before pressure systems return to lower levels more typical of this time of year for the second half. As such, lighter winds, more sunshine than normal and mostly dry conditions will prevail over most of the first couple of weeks, to be followed for the remainder of the outlook period with increased cloud, fresher winds and increased precipitation events. Overall, the next four weeks will be drier, warmer, and sunnier than is usual for this time of year. Rainfall amounts will be around normal in the far north, while most other parts of the country can only anticipate around a quarter to half their usual amount of rain. Most of the rain that comes midway through this outlook period will be associated with passing thunderstorm activity. It will be extremely sunny over the first couple of weeks, with many areas enjoying up to 10 hours or more of sunshine on most days, and by the end of this outlook period, sunshine hours will average between 20-50% more than normal. The drier and sunnier than normal conditions will also reflect in warmer temperatures overall, with many districts maximum temperatures warmer by between 1-2°C above the norm, and up to 2°C warmer in the east and south. Minimum temperatures however, will be below normal at times during this outlook period, but no frosts are expected. Many districts will record maximum temperatures above 20°C during the first half of this outlook period, before temperatures return to what is typical for this time of year. Generally, winds will be lighter than what is normal for this time of year, particularly in southern parts. During this outlook period, there is likely to be only about 4-8 days with rain (days with 1mm or more rainfall), thunderstorms may be heard on 4 days, fogs seen on 6 days and possible light frosts on 6 days. No hail, snow or gales are expected.

11th-19th Anticyclonic. Light winds, warm, mostly dry and sunny spell as a large anticyclone lingers of the southeast of the country. Chance of some overnight fog patches here and there, particularly in east and southern parts. Very sunny.
11th-13th Unusually cool overnight minimum temperatures with chance of isolated pockets of light frost.
14th Overnight minimum temperatures rise back to normal levels.
15th-17th Chance of scattered fog patches.
17th-20th Very sunny and warm spell with many districts seeing daily maximum temperatures of around 22-25°C.
18th-19th Very warm spell with maximum temperatures over 25°C in many districts.
19th-20th Band of thunderstorms and rain brushes the south of the country before moving up over the western districts. Overnight minimum temperatures unusually milder than normal.
20th Winds will be fresher than they have been of late.
20th-26th South to southeast flow develops as the high drifts away in an easterly tract. Mostly dry spell with cloud increasing throughout and cooler temperatures.
24th-25th Unusually cool overnight minimum temperatures with chance of isolated pockets of light frost, particularly in the north and west.
26th Chance of scattered thunderstorms.
27th-10th July A series of Atlantic depression near the west and northern coasts bring a period of unsettled weather with fresh to strong southwest to westerly winds at first, later backing southerly and easing. Spells of rain or showers can be expected on most days, although amounts will generally be light with only a couple of exceptions. Generally cooler than normal at first before temperatures return to normal later in the period.
28th A band of rain crosses the country bringing a spell of heavy rain in northern parts, and occasional showers elsewhere. Chance of thunderstorms.
30th Very low pressure crosses the northwest of Ireland. Chance of heavy rain in some western districts, especially in the Maam Valley.

IRELAND RAINFALL ESTIMATES

June 2024

d=mainly dry, n= nonrecordable, l=light shrs, s=significant shrs, r=rain, h=hvy falls

		ULSTER PROVINCE					CONNAUGHT PROVINCE						LEINSTER PROVINCE				MUNSTER PROVINCE						
		Alderg	Hillsbo	Armag	Loughl	Malin H	Sligo	Ardtari	Belmul	Drums	Galway	Roscoi	Markre	Edende	Kilkenn	Dublin	Shannc	Newpo	Tralee	Killarne	Cork	All	Moon
1st	Sat	s	l	r	l	s	s	d	l	d	l	l	n	l	s	l	l	s	l	l	d	44	XhN
2nd	Sun	l	l	d	s	s	s	s	s	s	s	l	s	s	s	s	s	r	s	s	l	102	P12
3rd	Mon	l	d	s	d	l	d	d	l	d	d	d	l	d	l	s	l	l	s	s	s	38	
4th	Tue	s	r	d	r	l	r	r	s	r	h	r	s	r	s	l	r	r	s	s	s	151	
5th	Wed	l	d	s	n	s	l	l	l	d	l	n	d	l	l	s	s	l	s	s	d	60	
6th	Thu	d	d	d	d	l	d	d	n	d	n	d	d	l	n	l	s	d	r	l	s	33	N
7th	Fri	l	d	l	l	d	l	l	n	l	d	d	l	d	l	d	d	s	n	l	d	17	^
8th	Sat	n	d	l	d	l	d	d	n	d	d	n	l	d	d	d	d	d	d	d	d	4	
9th	Sun	d	d	d	d	d	d	d	d	d	d	d	d	d	d	d	d	d	d	d	d	0	
10th	Mon	d	d	d	d	d	d	d	d	d	d	d	d	d	d	d	d	d	d	d	d	0	
11th	Tue	d	d	d	d	d	d	d	d	d	d	d	d	d	d	d	d	d	d	d	d	0	
12th	Wed	d	d	d	d	d	d	d	d	d	d	d	d	d	d	d	d	d	d	d	d	0	
13th	Thur	d	d	d	d	d	d	d	d	d	d	d	d	d	d	d	d	d	d	d	d	0	
14th	Fri	d	d	d	d	d	d	d	d	d	d	d	d	d	d	d	d	d	d	d	d	0	1Q A XhS
15th	Sat	d	d	d	d	d	d	d	d	d	d	d	d	d	d	d	d	d	d	d	d	0	
16th	Sun	d	d	d	d	d	d	d	d	d	d	d	d	d	d	d	d	d	d	d	d	0	
17th	Mon	d	d	d	d	d	d	d	d	d	d	d	d	d	d	d	d	d	d	d	d	0	
18th	Tue	d	d	d	d	d	d	d	d	d	d	d	d	d	d	d	d	d	d	d	d	0	
19th	Wed	d	d	d	d	d	d	d	d	d	d	d	d	d	d	d	d	d	l	l	l	5	
20th	Thu	d	d	d	n	d	l	n	d	s	d	s	d	d	l	d	d	d	s	s	l	18	
21st	Fri	n	d	l	d	d	d	d	l	d	d	d	d	d	d	d	d	d	s	l	d	11	
22nd	Sat	d	d	d	n	l	d	n	l	n	n	l	l	d	d	d	d	l	d	d	d	9	F V
23rd	Sun	d	d	d	l	d	d	l	d	d	d	d	l	d	d	d	d	d	d	d	d	2	
24th	Mon	d	d	d	d	d	n	d	d	d	d	d	l	d	d	d	d	d	d	d	d	2	
25th	Tue	d	n	d	l	d	s	l	d	l	n	l	l	d	d	d	d	s	d	d	d	22	
26th	Wed	l	n	l	l	l	l	l	l	l	n	l	s	d	d	d	d	l	d	d	d	23	
27th	Thur	l	s	s	r	s	s	r	s	s	s	s	s	l	l	d	d	s	n	d	d	75	P13
28th	Fri	s	s	s	r	h	s	r	r	r	r	r	s	s	l	s	s	r	s	l	l	170	3Q XhN
29th	Sat	l	d	s	l	d	l	l	n	l	l	l	l	d	l	d	n	s	l	n	d	23	
30th	Sun	s	s	d	l	s	l	l	s	s	s	s	s	l	l	l	l	l	s	s	l	58	
	Estimate:	35	30	38	56	54	45	55	46	45	60	38	48	31	26	31	43	60	62	41	22	866	
	Average:	62	66	58	71	70	82	84	72	74	80	82	82	74	61	67	70	90	108	72	81	1505	
	Trend:	drr	drr	drr	drr	drr	drr	drr	drr	drr	drr	drr	drr	drr	drr	drr	drr	drr	drr	drr	drr	drr	

IRELAND SUNSHINE ESTIMATES
June 2024

F=fine (8-12 hours of sunshine), pc= partly cloudy (4-7 hours), c=cloudy (1-3 hours), o=overcast (0 hours)

		ULSTER PROVINCE					CONNAUGHT PROVINCE						LEINSTER PROVINCE				MUNSTER PROVINCE						
		Alderg	Hillsbo	Armag	Loughl	Malin H	Sligo	Ardtarı	Belmul	Drums	Galway	Roscoı	Markre	Edende	Kilkenı	Dublin	Shannc	Newpo	Tralee	Killarne	Cork	All	Moon
1st	Sat	pc	pc	pc	pc	pc	F	pc	F	pc	F	F	pc	pc	pc	c	F	pc	F	F	F	140	XhN
2nd	Sun	pc	pc	pc	pc	pc	pc	pc	pc	pc	pc	pc	pc	pc	pc	pc	c	pc	pc	pc	pc	124	P12
3rd	Mon	F	F	F	F	F	F	F	F	F	pc	F	F	F	pc	F	c	pc	c	c	c	178	
4th	Tue	F	F	F	F	F	pc	F	pc	pc	pc	pc	F	pc	pc	F	c	c	o	o	o	119	
5th	Wed	c	c	c	c	c	o	c	o	pc	o	o	c	pc	pc	pc	o	c	pc	pc	pc	59	
6th	Thu	pc	pc	pc	pc	pc	pc	pc	pc	pc	pc	pc	pc	pc	c	pc	c	c	c	c	c	93	N
7th	Fri	c	c	c	c	c	c	c	c	c	c	c	c	c	c	c	c	c	o	o	o	29	^
8th	Sat	F	F	F	F	F	F	F	F	F	F	F	F	F	F	F	F	F	F	F	F	238	
9th	Sun	F	F	F	F	F	F	F	F	F	F	F	F	F	F	F	F	F	F	F	F	255	
10th	Mon	F	F	F	F	F	pc	F	pc	F	F	F	F	F	F	F	F	F	F	F	F	236	
11th	Tue	F	F	F	F	F	F	F	F	F	F	F	F	F	F	F	F	F	F	F	F	297	
12th	Wed	F	F	F	F	F	F	F	F	F	F	F	F	F	F	F	F	F	F	F	F	280	
13th	Thur	F	F	F	F	F	F	F	F	F	F	F	F	F	F	F	F	F	F	F	F	267	
14th	Fri	F	F	F	F	F	F	F	F	F	F	F	F	F	F	F	F	F	F	F	F	221	1Q A XhS
15th	Sat	F	F	F	F	F	F	F	F	F	F	F	F	F	F	F	F	F	F	F	F	261	
16th	Sun	F	F	F	F	F	F	F	F	F	F	F	F	F	F	F	F	F	F	F	F	274	
17th	Mon	F	F	F	F	F	F	F	F	F	F	F	F	F	F	F	F	F	F	F	F	264	
18th	Tue	F	F	F	F	F	F	F	F	F	F	F	F	F	F	F	F	F	F	F	F	296	
19th	Wed	F	F	F	F	F	F	F	F	F	F	F	F	F	F	F	F	F	F	F	F	251	
20th	Thu	c	c	c	c	c	c	c	c	c	c	c	c	c	c	c	o	c	o	o	o	47	
21st	Fri	F	F	F	F	F	F	F	F	F	F	F	F	F	F	F	F	F	F	F	F	202	
22nd	Sat	pc	pc	pc	pc	pc	c	pc	c	pc	c	c	pc	pc	pc	F	c	pc	c	c	c	105	F V
23rd	Sun	c	c	c	c	c	o	c	o	c	c	c	c	c	c	c	c	c	c	c	c	30	
24th	Mon	pc	pc	pc	pc	pc	o	pc	o	pc	pc	pc	pc	pc	F	pc	F	F	F	F	F	145	
25th	Tue	c	c	c	c	c	c	c	c	pc	pc	c	c	pc	F	pc	pc	pc	F	F	F	106	
26th	Wed	c	c	c	c	c	o	o	o	c	c	c	o	c	pc	c	pc	pc	F	F	F	65	
27th	Thur	c	c	c	c	c	o	c	o	pc	pc	c	c	pc	F	pc	pc	F	F	F	F	110	P13
28th	Fri	o	o	o	o	o	o	o	o	c	o	o	o	c	c	c	o	o	c	c	c	15	3Q XhN
29th	Sat	c	c	c	c	c	c	c	c	c	pc	pc	c	c	c	c	F	pc	pc	pc	pc	71	
30th	Sun	c	c	c	c	c	c	c	c	c	c	c	c	c	c	pc	c	c	c	c	c	49	
Estimate hours:		236	236	236	236	236	214	236	214	249	224	219	236	249	263	258	234	248	267	267	267	4826	
Average hours:		167	167	142	169	171	162	162	161	137	152	159	155	158	155	180	154	146	152	154	172	3173	
Trend:		more	more	more	more	more	more	more	more	more	more	more	more	more	more	more	more	more	more	more	more	more	

JUNE **1** SATURDAY

An intense depression sits over the country directing cool and blustery north to northeasterly winds throughout with winds reaching gale force at times in numerous districts. A changeable day with a mix of scattered showers and sunny spells, but remaining mostly dry in the east and southeast coastal fringes.

Belfast: Partly cloudy, cold, isolated showers, strong blustery winds, possible gale gusts
Dublin: Stormy air, brief showers, chance of gale gusts and possible thunderstorms
Cork: Sunny, dry, mild day cold night, strong blustery winds, gales likely
Galway: Sunny, some brief evening showers likely, gusty winds

RAIN POTENTIAL

FROST/SNOW

JUNE **2** SUNDAY

A weakening depression drifts northwards off the country allowing for a blustery west to southwest flow to prevail, possibly up to gale force for a time. Another changeable day with a mix of good sunny spells and bands of scattered showers everywhere, some heavy, and including the chance of thunderstorms. Cold overnight with chance of scattered frosts in a number of districts, mostly in central and eastern parts.

Belfast: Partly cloudy, light showers, mild day cold overnight, blustery
Dublin: Changeable, scattered showers, sunny spells, thunderstorms likely
Cork: Mostly sunny, mild day cold night, light showers, unsettled winds
Galway: Partly cloudy, scattered showers, blustery, chance of light snow overnight

RAIN POTENTIAL

FROST/SNOW

JUNE **3** MONDAY

Low pressure continues to prevail over the northern half of Ireland while a high pressure system ridges to the south. Winds generally moderate to fresh in a westerly flow, with chance of gale gusts mostly in the north and west. Mostly cloudy in the south with scattered showers, and sunnier elsewhere with occasional showers in most districts. Chance of thunderstorms overnight and possible wintry showers of hail or unseasonal snow flurries briefly in the west.

Belfast: Sunny, dry, windy, chance of gales at times
Dublin: Overnight showers, sunny day, windy, chance of overnight thunderstorms
Cork: Cloudy, occasional showers, windy
Galway: Mostly sunny and dry, cool, windy with strong gusts, chance of overnight snow

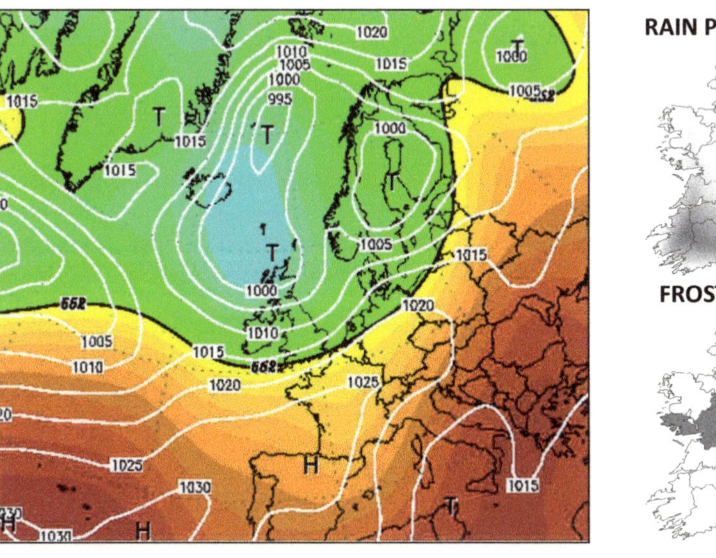

RAIN POTENTIAL

FROST/SNOW

JUNE **4** TUESDAY

Low pressure to the north of Ireland pushes the high pressure ridge off the south of the country in a light to variable northwest flow. Mostly sunny in the north and east, with cloud building in the west and south. Evening showers likely in most districts, heavier in the west and more drizzly in the south. Cooler overnight with the chance of some frost pockets in inland parts, and possible fog for a time mostly in the south.

Belfast: Sunny, evening rain, mild day cold night, light breezes
Dublin: Sunny, isolated evening showers, mild day cold night, mostly calm air
Cork: Overcast, drizzly showers, warmer, gentle breezes, chance of misty patches
Galway: Partly cloudy, heavy rain, cool, winds easing to light

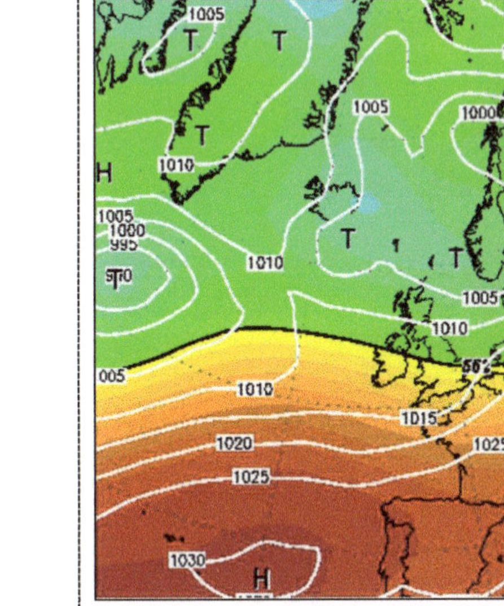

RAIN POTENTIAL

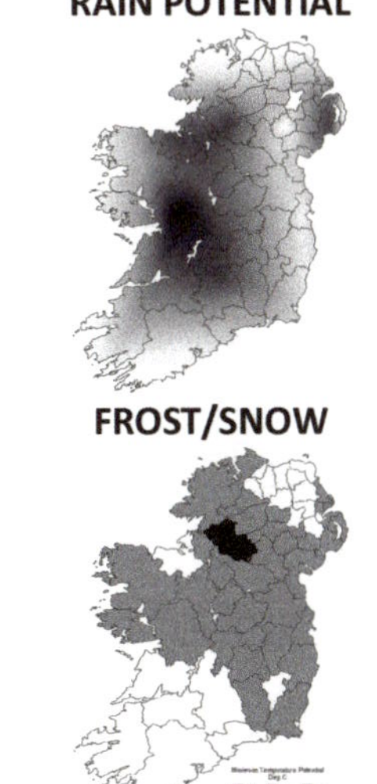

FROST/SNOW

JUNE **5** WEDNESDAY

Low pressure and moderate west to southwest flow prevails throughout the country. Mostly cloudy, particularly in the west, with a wet front bringing passing overnight showers, clearing later in the day. Chance of some overnight misty patches.

Belfast: Mostly cloudy and dry, warmer, moderate breezes
Dublin: Partly cloudy, outbreaks of rain, warmer, breezy spells
Cork: Warmer, partly cloudy, dry, fresh breezes at times
Galway: Overcast, fluctuating winds, light showers, milder

RAIN POTENTIAL

FROST/SNOW

JUNE **6** THURSDAY

Low pressure contracts to north of the country while a high pressure ridge moves back into the south. Westerly winds prevail, fresh and breezy at times. A cold front crosses southern parts bringing scattered showers for a time in the southern half of the country and isolated showers in the east and west, while remaining mostly dry in the north.

Belfast: Partly cloudy, warmer, dry, fresh breezes at times
Dublin: Partly cloudy, isolated showers, moderate breezes
Cork: Cloudy, cooler, scattered showers, windy at times
Galway: Partly cloudy, isolated showers, moderate breezes

RAIN POTENTIAL

FROST/SNOW

JUNE 7 FRIDAY

Low pressure to the north, high pressure ridging over the south. A showery trough with fresh southwest to westerly winds, later turning northwest prevails, and bringing some showers over the northern half of the country, while the high pressure ridge brings mostly dry and cloudy conditions with the odd patch of fog in southern parts.

Belfast: Slightly cooler, cloudy, dry, fresh breezes at times
Dublin: Cloudy, mostly dry with occasional threats of rain, cooler, moderate breezes
Cork: Dull, mostly dry, cooler, odd threat of rain, some overnight fog patches possible
Galway: Cloudy, dry, moderate northwest breezes

RAIN POTENTIAL

FROST/SNOW

JUNE 8 SATURDAY

A weak high pressure ridge directs a light to moderate northerly flow over the country. Mostly dry and sunny throughout with the chance of fresher winds and some isolated showers overnight in the west and north.

Belfast: Sunny, dry, mild day cool night, fresh breezes
Dublin: Sunny, dry, mild day cool night, light breezes
Cork: Sunny, dry, mild day cool night, gentle breezes
Galway: Sunny, dry, warmer day cooler night, moderate breezes with occasional gusts

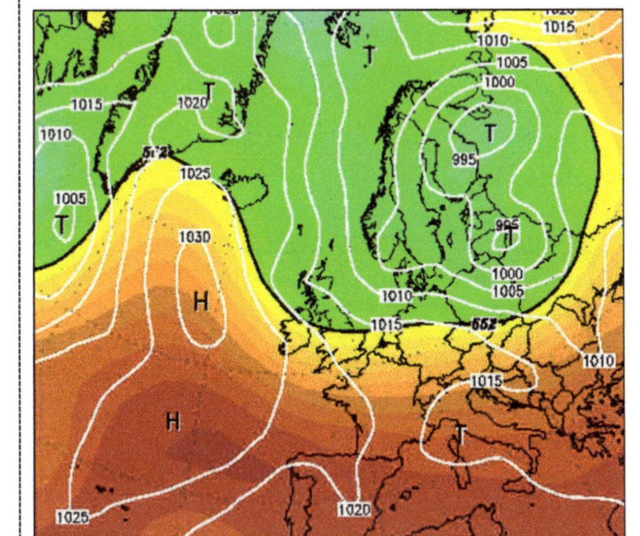

RAIN POTENTIAL

FROST/SNOW

JUNE **9** SUNDAY

High pressure ridging to the west of the country continues to strengthen with light north to northwest winds prevailing and cooler temperatures throughout. Very sunny and dry everywhere, with the chance of some isolated pockets of frost in the southeast.

Belfast: Very sunny, dry, slightly cooler, windy at times
Dublin: Very sunny, dry, mild day cool night, moderate breezes, chance of gusts
Cork: Very bright and sunny, dry, mild day cooler night, moderate breezes
Galway: Sunny, dry, mild day cool night, moderate northerly breezes

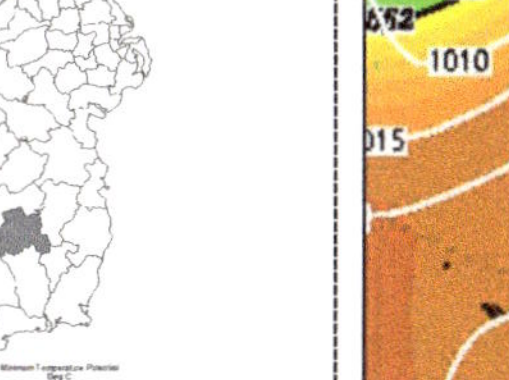

RAIN POTENTIAL

FROST/SNOW

JUNE **10** MONDAY

A large high pressure system settles over the country bringing light variable breezes and a pleasant northerly air throughout. Cool overnight with the chance of some light frosty patches mostly in the midland districts to the southeast. Very sunny and dry everywhere except for the chance of some brief showers around the northern coastal fringes.

Belfast: Mostly calm and serene air, sunny, dry, cold overnight
Dublin: Very bright and sunny, warmer day cold night, dry, mostly calm air
Cork: Very sunny, bright and dry, mild day cool night, calm and serene atmosphere
Galway: Sunny and dry, mild day cold night, light northerly air

RAIN POTENTIAL

FROST/SNOW

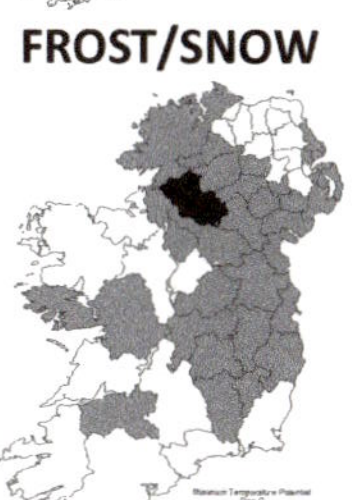

JUNE 11 TUESDAY

A large high sits over Ireland, with light to moderate westerly breezes prevailing, fresher at times mostly in the west and north. Very sunny, dry, and warm throughout with the chance of patches of fog mostly in the east and west.

Belfast: Warmer, bright and sunny, dry, gentle breezes
Dublin: Very sunny, warm and dry, chance of shallow fog patches, gentle breezes
Cork: Very bright and sunny, dry, warmer, light shifting breezes
Galway: Very bright and sunny, dry, milder, light breezes, chance of fog patches

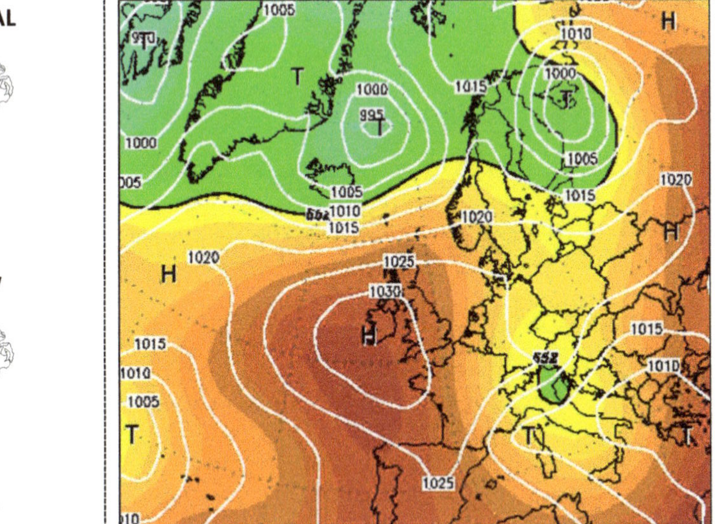

RAIN POTENTIAL

FROST/SNOW

JUNE 12 WEDNESDAY

A large high pressure ridge maintains a pleasant and warm air over the country with gentle winds prevailing. Very sunny and dry everywhere. Chance of fog patches mostly in central parts from the west of Ireland to the east.

Belfast: Very sunny, warmer, dry, light breezes
Dublin: Very bright and sunny, warm, humid, dry, calm air, fog patches possible
Cork: Calm, very sunny, mild and serene air, dry
Galway: Sunny, dry, warm day cool night, mostly calm air, chance of patches of mist

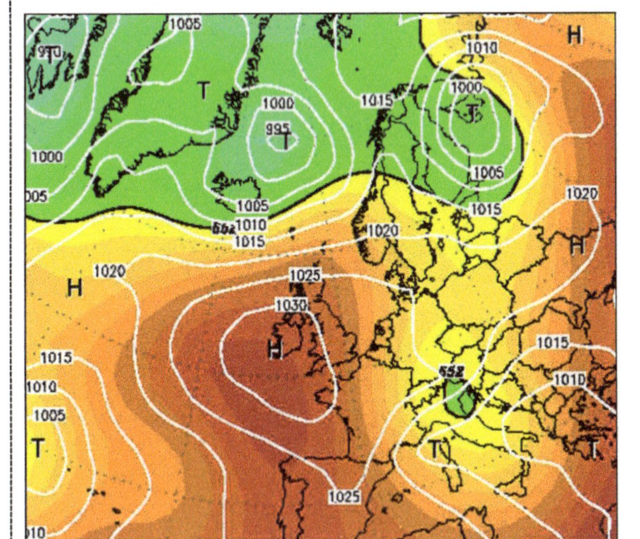

RAIN POTENTIAL

FROST/SNOW

JUNE **13** THURSDAY

Very sunny, dry, and warm conditions continue to prevail over Ireland as a large high lingers over the country. Generally a mostly calm and serene atmosphere everywhere, with the odd gentle breeze of various direction. Fog patches likely particularly in the north and west.

Belfast: Very sunny, warm, dry, cool overnight, mostly calm air, chance of fog patches
Dublin: Sunny, dry, cooler, calm air
Cork: Very sunny, warm and dry, calm atmosphere
Galway: Very sunny, dry and warm, cool overnight, serene air, fog possible

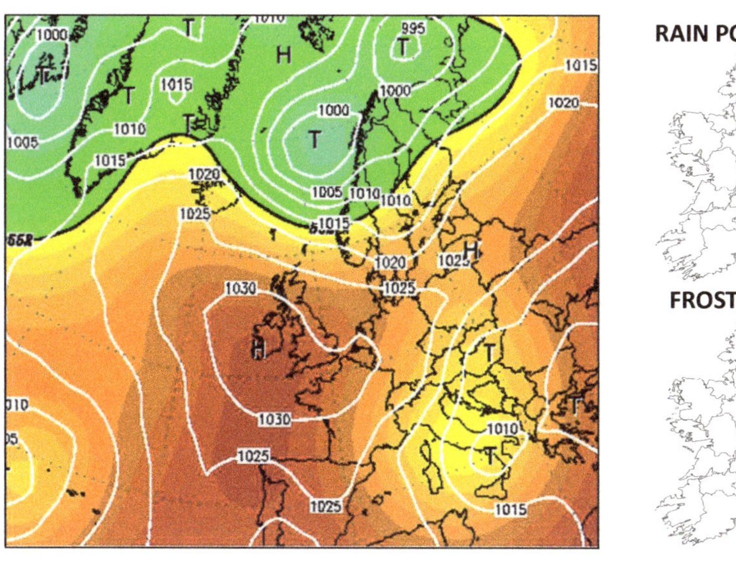

RAIN POTENTIAL

FROST/SNOW

JUNE **14** FRIDAY

A large anticyclone continues to linger over Ireland generating a pleasant and mostly calm and warm atmosphere throughout, with winds remaining light and in variable directions. Dry and sunny almost everywhere although some increased scattered cloud likely around the northern coasts. Chance of fog patches once more mostly in the west.

Belfast: Sunny, very warm and dry, calm and pleasant air
Dublin: Sunny, warm and dry, light shifting breezes to mostly calm air
Cork: Sunny, dry, warm, light shifting breezes
Galway: Warm, dry, sunny, light shifting breezes, chance of fog

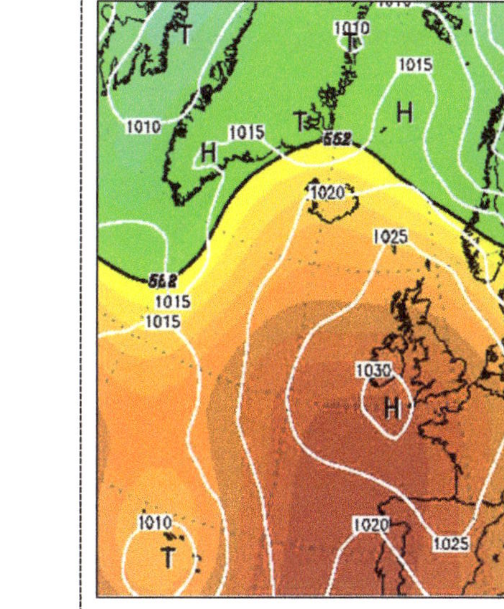

RAIN POTENTIAL

FROST/SNOW

JUNE **15** SATURDAY

The high pressure ridge continues to linger over the country. Warm, dry, and sunny everywhere, with patches of overnight fog likely in many districts, with fog patches likely to linger late into the morning in the north. A mostly calm air to light southerly flow prevails.

Belfast: Dry, warm, sunny, calm atmosphere, chance of overnight misty patches
Dublin: Sunny, dry, milder, calm and pleasant air, chance of fog patches
Cork: Dry, slightly cooler, very sunny, calm atmosphere, chance of fog patches
Galway: Very sunny, warm and dry, cool overnight, calm air, possible patches of fog

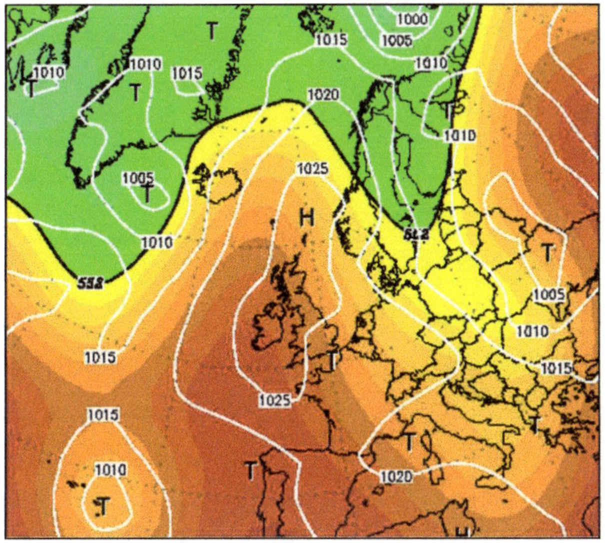

RAIN POTENTIAL

FROST/SNOW

JUNE **16** SUNDAY

Very sunny, dry, and warm throughout the country as high pressure continues to prevail. Mostly calm and serene air prevails with scattered patches of mist likely, especially in southern parts. Some gentle breezes in a westerly flow likely now and then.

Belfast: Very bright and sunny, dry, warm, calm and pleasant air
Dublin: Dry, warm and sunny, calm and serene air
Cork: Sunny, warm and dry, cooler overnight, calm and serene air, fog possible
Galway: Warm day cool night, dry, bright and sunny, calm and pleasant air

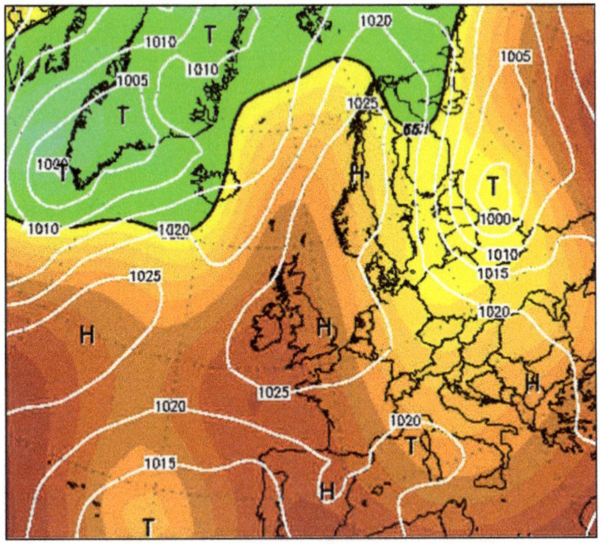

RAIN POTENTIAL

FROST/SNOW

JUNE 17 MONDAY

A large anticyclonic system continues to fill over Ireland bringing light breezes to a mostly calm and serene air throughout. Scattered fogs and mists in many districts clearing to a very sunny, dry, and hot day everywhere, before fog and misty patches return once more in the evening.

Belfast: Bright and sunny, hot day cool night, dry, calm atmosphere
Dublin: Very sunny, dry, slightly cooler, calm atmosphere
Cork: Sunny, dry, warmer, calm atmosphere
Galway: Sunny, dry, warm day cool night, calm and serene air, chance of misty patches

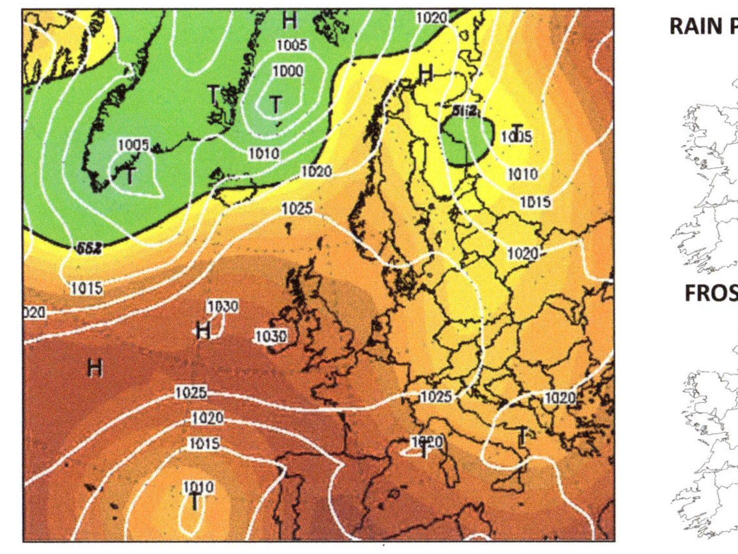

RAIN POTENTIAL

FROST/SNOW

JUNE 18 TUESDAY

The large high pressure system that has stalled over the country for a number of days now, finally begins to drift slowly westward towards Europe, with a low pressure trough west of Ireland. Very hot, dry, and sunny conditions continue everywhere, with fog overnight in many districts. Light easterly breezes to a mostly calm air prevails.

Belfast: Very sunny and dry, hot, mostly calm air, chance of overnight fog patches
Dublin: Very sunny, warmer, dry, serene atmosphere, chance of overnight fog patches
Cork: Dry, very sunny and warm, serene atmosphere
Galway: Hot, very sunny, dry, calm atmosphere, cool overnight, chance of shallow fog

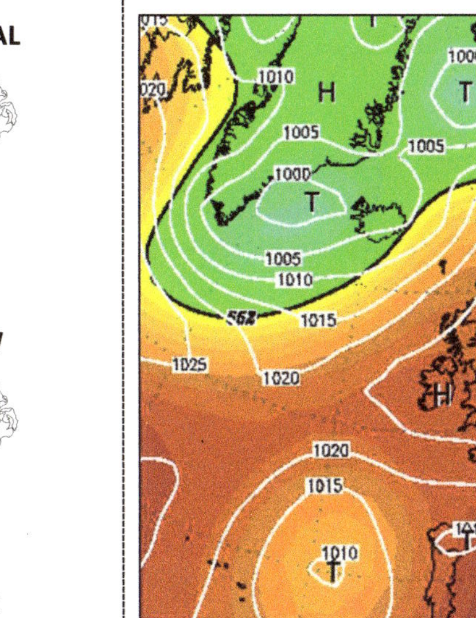

RAIN POTENTIAL

FROST/SNOW

JUNE **19** WEDNESDAY

High pressure moves to the east of Ireland while low pressure advances to the west, with a light to moderate east to southeast flow prevailing. Very hot and mostly dry and sunny conditions throughout most of the country, except for the chance of some thunderstorm activity bringing brief heavy showers in the southeast later in the day.

Belfast: Hot, dry, very sunny, cool overnight, light breezes
Dublin: Dry, warm, bright and sunny, light breezes to mostly calm air
Cork: Sunny, warm, light breezes, brief showers, chance of thunderstorms
Galway: Very warm, sunny, dry, moderate breezes

RAIN POTENTIAL

FROST/SNOW

JUNE **20** THURSDAY

A weak low pressure system sits to the west of Ireland directing freshening southeasterly breezes over the country, stronger and more blustery for a time in the west to southwest. Cloud increases everywhere, with bands of thunderstorms likely in the west and south, while remaining mostly dry over the east and north.

Belfast: Mostly cloudy, warm and dry, fresh breezes at times
Dublin: Cloudy and warm, dry, moderate breezes
Cork: Overcast and breezy, isolated showers, cooler, thunderstorms likely
Galway: Cloudy, warm and humid, mostly dry, breezy, chance of thunderstorms

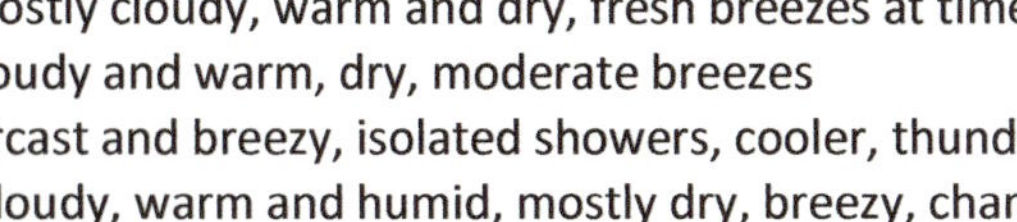

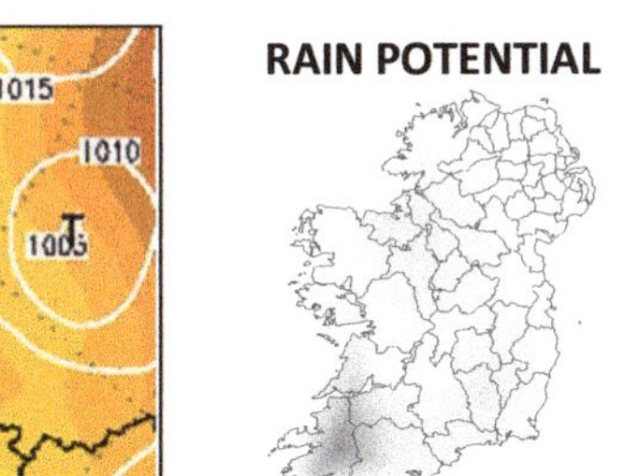

RAIN POTENTIAL

FROST/SNOW

JUNE 21 FRIDAY

An intense depression to the west of Ireland directs a weak frontal trough over the country in a fresh to strong southerly flow, with winds up to gale force with gusts at times and some light showers overnight mostly in the southwest. Mainly dry, warm, and sunny day throughout.

Belfast: Sunny, warm, dry, windy spells
Dublin: Sunny, warm and dry, occasional breezy spells
Cork: Sunny and dry, mild, freshening winds for a time
Galway: Sunny, cooler, dry, blustery

RAIN POTENTIAL

FROST/SNOW

JUNE 22 SATURDAY

The depression to the west of Ireland continues drifts in a northerly tract while directing a fresh to strong southerly flow over the western half of the country where increased cloud will bring some light scattered showers. A lighter, sunnier, and mostly dry southerly flow prevails in the east, associated with a high pressure system over Europe. Chance of some overnight misty patches for a time in some northern districts.

Belfast: Partly cloudy and dry, breezy spells, chance of brief misty patches
Dublin: Sunny and dry, warm, gentle breezes
Cork: Cloudy and cool, dry, windy at times
Galway: Unsettled and windy at times, cloudy, cooler, isolated showers

RAIN POTENTIAL

FROST/SNOW

JUNE 23 SUNDAY

A depression to the north of Ireland directs a fresh westerly flow over the northern half of the country, while a high pressure ridges over the southern half, with winds generally in a light southwesterly flow. Mostly cloudy and dry throughout with chance of fog patches in southern parts.

Belfast: Cloudy, warm, dry, windy at times
Dublin: Cloudy, cooler, dry, occasional breezy spells
Cork: Mostly cloudy, dry, gentle breezes to mostly calm air, possible fog patches
Galway: Cloudy, dry, mild, moderate breezes with occasional gusts likely

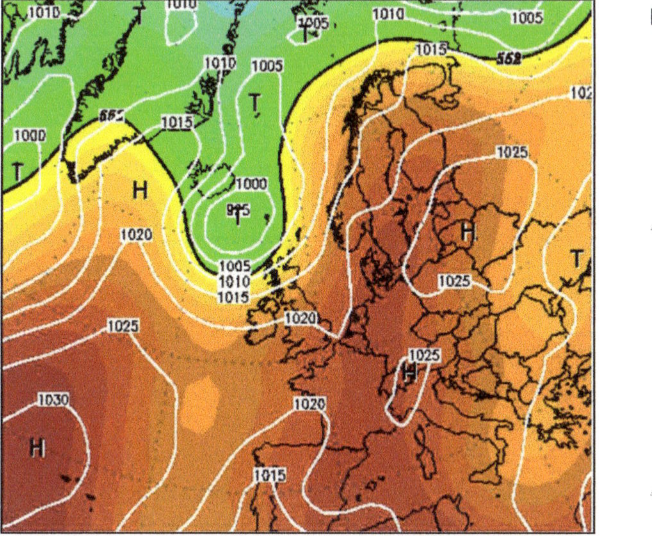

RAIN POTENTIAL

FROST/SNOW

JUNE 24 MONDAY

A high pressure ridge moves up over the country in a light north to northwest flow. Very sunny in the south of Ireland, with increased scattered cloud northwards but still mostly sunny through central parts, and a mostly cloudy patch in the far north. Dry, warm, and pleasant everywhere.

Belfast: Cooler, dry, cloudy spells, moderate breezes, chance of overnight fog patches
Dublin: Mostly sunny, mild and dry, moderate breezes
Cork: Very sunny, dry, mild day cool night, gentle breezes
Galway: Dry, partly cloudy, mild day cold night, shifting breezes

RAIN POTENTIAL

FROST/SNOW

JUNE 25 TUESDAY

High pressure maintains its ridge over Ireland with variable winds to a mostly calm air prevailing. Some isolated showers associated with the passage of a weak warm front moves through the north and west for a time, while it remains dry and sunnier in the south and east. Chance of fog patches overnight in sheltered districts.

Belfast: Mostly cloudy and cool, isolated showers, light breezes to mostly calm air
Dublin: Calm, partly cloudy, dry, mild
Cork: Sunny and mild, dry, calm and pleasant air
Galway: Calm and pleasant air, partly cloudy, mild, some isolated showers likely

RAIN POTENTIAL

FROST/SNOW

JUNE 26 WEDNESDAY

A complex of high and low pressure centres linger near the west of Ireland bringing a light, variable to southwesterly flow throughout, with some fresher breezes at times around the northwest coastal fringes. Mostly dry, sunny and pleasant in southern parts, while cloudier skies prevail elsewhere, and chance of some isolated overnight showers or drizzle patches likely over the northern half of the country.

Belfast: Light shifting breezes, widespread cloud, isolated showers, possible fog
Dublin: Cloudy, dry, warmer, light breezes to mostly calm air
Cork: Warm, sunny, dry, pleasant and serene air
Galway: Cloudy, isolated showers, mild, gentle breezes to mostly calm air

RAIN POTENTIAL

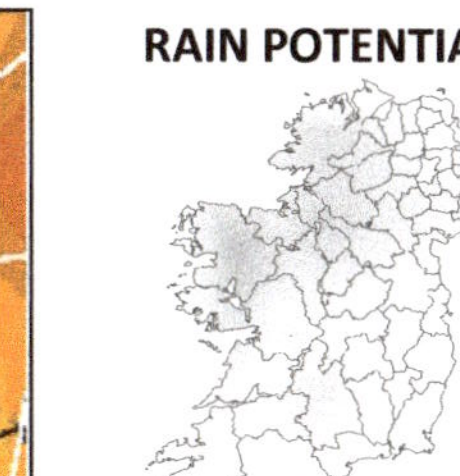

FROST/SNOW

JUNE **27** THURSDAY

High pressure drifts to the south of the country while a low pressure trough lingers over the northern half of the country. Mostly dry, warm, and sunny in southern parts and along the eastern coastlines, with increased cloud, light scattered showers, and a humid air elsewhere. Chance of overnight fog patches in central parts from the west to the east coasts.

Belfast: Cloudy, cooler, scattered showers, mostly calm air
Dublin: Pleasant and dry, warm, scattered cloud, light breezes, possible fog patches
Cork: Warm, sunny, dry, gentle shifting breezes
Galway: Partly cloudy, scattered showers, mild, gentle breezes, chance of misty drizzles

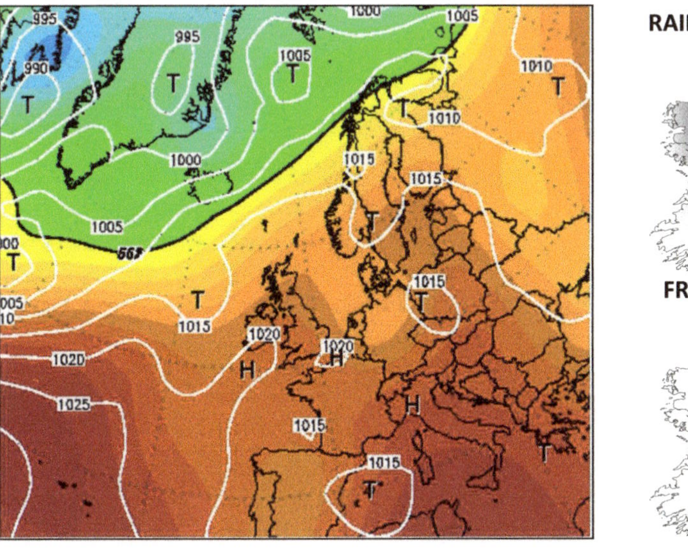

RAIN POTENTIAL

FROST/SNOW

JUNE **28** FRIDAY

A weak depression to the northwest of Ireland extends a wet front down over the country, with light southwest winds prevailing. Some overnight fog patches likely, mostly in the north. Cloud thickens and rain and drizzle outbreaks spread from the west to throughout the country as the day progresses, with some heavy falls at times in the north and west, while lighter in the far south.

Belfast: Light breezes, overcast, occasional showers, chance of misty drizzles
Dublin: Cloudy, rain, cooler, moderate breezes
Cork: Cloudy, breezy spells, brief showers, cooler
Galway: Overcast, cooler, rain, moderate breezes

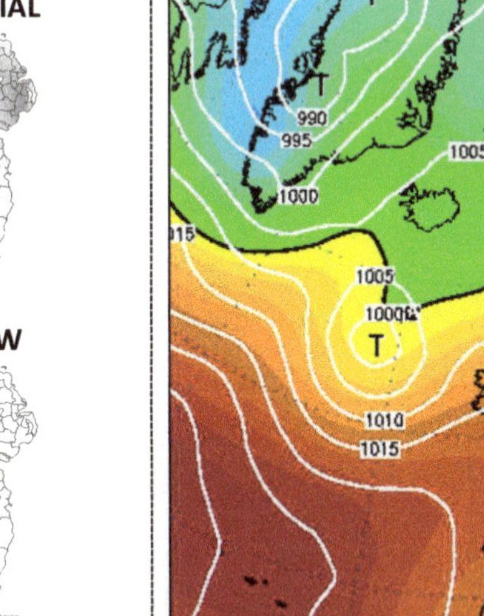

RAIN POTENTIAL

FROST/SNOW

JUNE **29** SATURDAY

A depression moves over the country directing unsettled west to southwest winds, backing northwesterly for a time, and blustery around the coastal fringes. Mostly cloudy with drizzle patches and isolated showers at first, followed by increasing sunny breakthroughs and drier conditions throughout.

Belfast: Cloudy, mostly dry with odd threat of rain, fluctuating windy spells
Dublin: Widespread cloud, dry, mild, shifting windy spells
Cork: Partly cloudy and mostly dry, mild, occasional breezy spells
Galway: Partly cloudy, passing brief showers, mild, unsettled and changeable winds

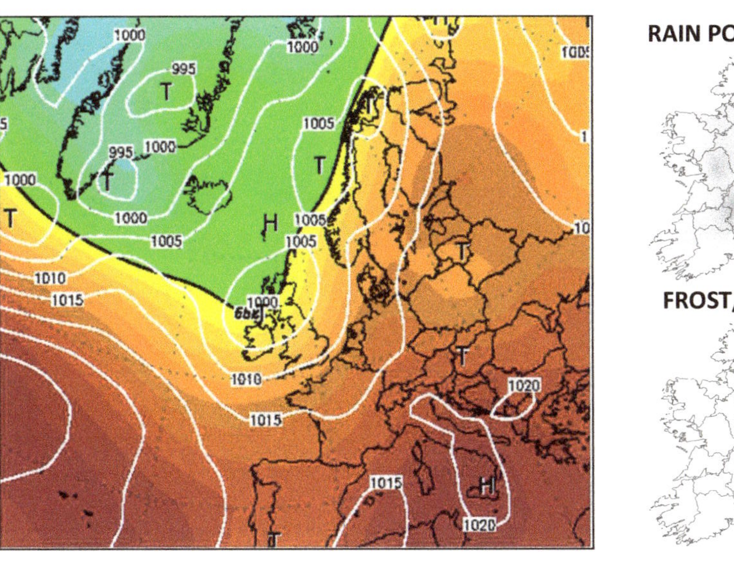

RAIN POTENTIAL

FROST/SNOW

JUNE **30** SUNDAY

The depression that has dominated the past few days drifts away from the country in a westerly path, directing a strong southwest flow at first, backing westerly later in the day. A couple of bands of rain move across the country during the day bringing scattered outbreaks of light rain spreading from the west to the east, with dry and sunnier spells between.

Belfast: Cloudy, light showers, mild, fluctuating breezes at times
Dublin: Partly cloudy and mild, brief showers, moderate breezes
Cork: Cloudy, light showers, cooler, fluctuating breezes
Galway: Cloudy, cooler, occasional showers, fresh winds at times

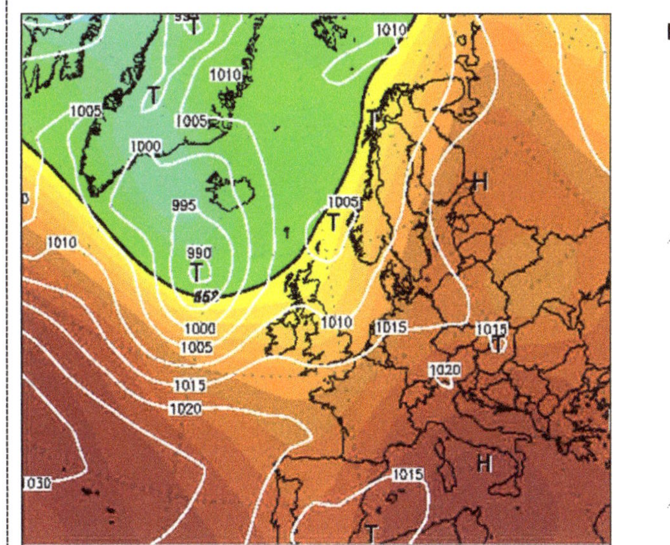

RAIN POTENTIAL

FROST/SNOW

JULY

PHASES OF THE MOON

5th	New Moon
5th	Northern Declination
12th	Apogee
13th	First Quarter
13th	Crossing Equator
19th	Southern Declination
21st	Full Moon
24th	Perigee #10
26th	Crossing Equator
28th	Third Quarter

1 Mon	2 Tue	3 Wed	4 Thur	5 Fri	6 Sat	7 Sun	8 Mon	9 Tue	10 Wed	11 Thur	12 Fri	13 Sat	14 Sun	15 Mon	16 Tue	17 Wed	18 Thur	19 Fri	20 Sat	21 Sun	22 Mon	23 Tue	24 Wed	25 Thur	26 Fri	27 Sat	28 Sun	29 Mon	30 Tue	31 Wed
				N ^							A	1Q XhS						V		F			P10		XhN		3Q			

MONTHLY SUMMARY

27th June-10th July A series of Atlantic depression near the west and northern coasts bring a period of unsettled weather with fresh to strong southwest to westerly winds at first, later backing southerly and easing. Spells of rain or showers can be expected on most days, although amounts will generally be light with only a couple of exceptions. Generally cooler than normal at first before temperatures return to normal later in the period.
1st Windy with occasional gusts, mostly in the northwest.
2nd-3rd Unusually cool overnight minimum temperatures with chance of isolated pockets of light frost.
5th-7th Very sunny and dry spell.
7th-9th Fog patches likely.
9th Spell of heavy rain in some places.

Over the next four weeks air pressure systems high pressure will be particularly dominant during the middle of this outlook period, and low pressure prevailing early and at the end of this look-ahead period. As such, the next four weeks will be mostly dry, sunny, and warmer than is typical for this time of year. In particular, the second and third weeks will be mostly dry. There will be some passing wet spells during this outlook period, but generally, it will be an unusually dry period for this time of year, resulting in some parts, especially in eastern counties, possibly only seeing around a third of their normal rainfall. In contrast, the northwest may have up to 50% more rainfall than normal. Sunshine hours will be above normal throughout the country during this outlook period with many districts, particularly in the east to the Midlands, possibly seeing up to 50% more sunshine hours than is typical for this time of year. Temperatures will be up to 2°C above average almost everywhere throughout the country, and up to 3°C warmer than normal in parts of the northern Midlands. Some counties in the east of the country can anticipate maximum temperatures in excess of 20°C for many days in a row, and possibly over 25°C on up to 10 days. During this outlook period, there is likely to be around 5-9 wet days (with 1mm or more recorded rainfall) in the east and south, and up to 12-16 days with some rain in most other areas. Thunderstorms may be heard on 10 days, fogs observed on 7 days, frosts on 2 days, but no hail, gales, or snow.

11th Very bright and sunny in the Leinster Province.
11th-17th High pressure prevails bringing a mostly dry and clear spell with light breezes to a mostly calm air. Cooler than normal overnight, while unusually hot during the day. Some weak fronts likely to cross the country bringing only the occasional patchy showery outbreaks.
18th A depression north of Ireland directs a strong southwesterly flow and a band of heavy rain across the country, with particularly heavy falls in the west and southwest. Generally hot and humid with widespread cloud.
19th-21st Mostly dry spell except for some lingering showers mostly in the north and west.
22nd-31st Mostly dry spell especially in the east and south. Very bright and sunny throughout, with many districts seeing up to 15 hours of sunshine on some days.
23rd-24th Depression to the west of the country extends a cooler air across the country. Chance of pockets of ground frost in some inland areas and eastern districts.
26th-30th Very hot spell, with daily maximas temperatures possibly over 25°C in many districts, and even above 30°C in some places, particularly in western and northern parts.
28th-29th Very hot spell, particularly in the west.

IRELAND RAINFALL ESTIMATES

July 2024

d=mainly dry, n= nonrecordable, l=light shrs, s=significant shrs, r=rain, h=hvy falls

		ULSTER PROVINCE					CONNAUGHT PROVINCE						LEINSTER PROVINCE				MUNSTER PROVINCE						
		Alderg	Hillsbo	Armag	Loughl	Malin H	Sligo	Ardtarı	Belmul	Drums	Galway	Roscoı	Markre	Edende	Kilkenı	Dublin	Shannc	Newpo	Tralee	Killarne	Cork	All	Moon
1st	Mon	l	l	l	l	r	l	l	l	s	l	l	d	l	d	l	l	l	d	d	d	35	
2nd	Tue	n	d	l	d	l	d	d	d	d	d	d	d	d	d	d	d	d	d	d	d	5	
3rd	Wed	d	d	d	d	n	d	d	d	d	d	d	d	l	l	d	d	d	l	l	d	4	
4th	Thu	l	l	d	d	d	d	d	d	d	l	d	d	l	l	s	l	d	l	l	n	15	^
5th	Fri	d	d	d	d	d	d	d	d	d	d	d	d	d	d	d	d	d	d	d	d	0	N
6th	Sat	d	d	d	d	d	d	d	d	d	d	d	d	d	l	l	d	d	d	d	l	5	
7th	Sun	d	d	d	d	d	d	d	d	d	d	d	l	d	d	l	n	d	d	d	d	2	
8th	Mon	d	d	d	l	l	l	l	l	l	l	l	l	d	d	d	d	l	s	l	n	20	
9th	Tue	l	l	l	s	r	s	s	l	s	s	s	r	s	r	s	s	s	h	s	s	127	
10th	Wed	n	d	l	d	d	d	d	d	d	d	d	s	d	d	s	d	d	d	l	d	13	
11th	Thur	d	d	d	d	l	n	d	l	d	d	n	l	d	n	d	d	l	l	d	d	8	
12th	Fri	l	l	d	d	d	d	d	n	d	d	d	d	l	l	n	d	d	d	d	d	7	A XhS
13th	Sat	n	d	l	d	d	d	d	d	l	d	d	d	s	r	d	d	d	d	d	s	22	1Q
14th	Sun	d	d	d	d	l	l	d	d	d	d	s	l	r	s	l	n	d	d	d	l	32	
15th	Mon	d	d	d	s	d	l	s	d	d	d	d	l	d	d	d	d	l	d	d	s	15	
16th	Tue	d	d	d	s	l	l	s	s	l	s	d	l	d	n	d	d	s	l	n	d	26	
17th	Wed	d	d	d	d	l	l	d	l	l	s	l	l	d	d	d	l	l	l	d	d	12	
18th	Thu	s	r	n	s	r	r	s	h	r	s	h	s	r	s	s	s	r	h	h	r	237	
19th	Fri	l	d	s	d	n	d	d	n	s	d	l	d	d	d	l	l	l	n	d	d	13	V
20th	Sat	l	s	l	s	s	l	s	l	l	n	l	l	d	d	d	d	l	l	d	d	29	
21st	Sun	s	d	s	d	d	n	d	d	d	d	l	n	d	d	d	d	n	d	d	d	5	F
22nd	Mon	d	d	d	l	d	n	l	n	l	l	l	d	d	d	d	l	d	l	d	d	8	
23rd	Tue	l	d	l	d	d	d	d	d	d	l	d	d	d	d	d	d	d	d	d	d	3	
24th	Wed	d	d	d	d	d	d	d	d	d	d	d	d	d	d	d	d	d	d	d	d	0	P10
25th	Thur	d	d	d	d	d	d	d	d	d	d	d	d	d	d	d	d	d	d	d	d	0	XhN
26th	Fri	d	d	d	d	d	d	d	d	d	d	d	d	d	d	d	d	d	d	d	d	0	
27th	Sat	d	d	d	d	d	d	d	d	d	d	d	d	d	d	d	d	d	d	d	d	0	
28th	Sun	d	d	d	d	d	d	d	d	d	d	d	d	d	d	d	d	d	d	d	d	0	3Q
29th	Mon	d	d	d	d	d	d	d	d	d	d	d	d	d	d	d	n	d	d	d	n	1	
30th	Tue	s	d	s	d	l	d	d	d	d	d	d	d	d	d	d	d	d	n	d	d	15	
31st	Wed	d	d	d	d	d	l	d	d	d	n	d	d	d	d	d	d	d	d	l	n	4	
	Estimate:	25	26	23	24	52	30	24	43	32	29	36	33	38	44	28	24	31	57	29	36	663	
	Average:	71	77	63	86	81	93	95	79	73	87	86	93	71	59	56	66	101	103	75	79	1593	
	Trend:	drr	drr	drr	drr	drr	drr	drr	drr	drr	drr	drr	drr	drr	drr	drr	drr	drr	drr	drr	drr	drr	

IRELAND SUNSHINE ESTIMATES

July 2024

F=fine (8-12 hours of sunshine), pc= partly cloudy (4-7 hours), c=cloudy (1-3 hours), o=overcast (0 hours)

		ULSTER PROVINCE					CONNAUGHT PROVINCE						LEINSTER PROVINCE				MUNSTER PROVINCE						
		Alderg	Hillsbo	Armag	Loughl	Malin H	Sligo	Ardtar	Belmul	Drums	Galway	Rosco	Markre	Edende	Kilkenn	Dublin	Shann	Newpo	Tralee	Killarn	Cork	All	Moon
1st	Mon	pc	pc	pc	pc	pc	c	pc	c	pc	pc	pc	pc	pc	pc	pc	pc	pc	pc	pc	pc	108	
2nd	Tue	c	c	c	c	c	o	c	o	pc	pc	c	c	pc	pc	pc	pc	pc	F	F	F	97	
3rd	Wed	F	F	F	F	F	pc	F	pc	pc	c	pc	F	pc	pc	F	c	c	o	o	o	115	
4th	Thu	c	c	c	c	c	c	c	c	c	pc	c	c	c	c	o	pc	c	c	c	c	49	^
5th	Fri	F	F	F	F	F	F	F	F	F	pc	pc	F	F	F	F	pc	pc	c	c	c	167	N
6th	Sat	c	c	c	c	c	pc	c	pc	pc	pc	pc	c	c	c	o	pc	pc	F	F	F	105	
7th	Sun	c	c	c	c	c	pc	c	pc	c	c	pc	c	c	c	c	o	c	pc	pc	pc	68	
8th	Mon	c	c	c	c	c	c	c	c	pc	pc	c	c	pc	pc	pc	pc	pc	F	F	F	104	
9th	Tue	o	o	o	o	o	c	o	c	o	o	o	o	o	o	o	o	o	o	o	o	10	
10th	Wed	pc	pc	pc	pc	pc	F	pc	F	pc	F	F	pc	pc	pc	pc	F	F	F	F	F	154	
11th	Thur	F	F	F	F	F	c	F	c	F	pc	c	F	F	F	F	F	F	F	F	F	169	
12th	Fri	pc	pc	pc	pc	pc	F	pc	F	pc	pc	pc	pc	pc	c	c	pc	pc	pc	pc	pc	104	A XhS
13th	Sat	F	F	F	F	F	F	F	F	pc	F	F	F	pc	o	c	c	c	o	o	o	119	1Q
14th	Sun	c	c	c	c	c	o	c	o	c	o	o	c	c	c	c	c	c	c	c	c	33	
15th	Mon	F	F	F	F	F	F	F	F	F	F	F	F	F	pc	F	F	pc	c	c	c	168	
16th	Tue	c	c	c	c	c	c	c	c	c	o	c	c	c	c	c	o	c	c	c	c	37	
17th	Wed	pc	pc	pc	pc	pc	pc	pc	pc	pc	pc	pc	pc	pc	pc	pc	pc	pc	F	F	F	120	
18th	Thu	o	o	o	o	o	o	o	o	o	o	o	o	o	o	o	o	o	o	o	o	5	
19th	Fri	pc	pc	pc	pc	pc	pc	pc	pc	pc	F	pc	pc	pc	pc	c	F	F	F	F	F	128	V
20th	Sat	pc	pc	pc	pc	pc	pc	pc	pc	pc	pc	pc	pc	pc	pc	pc	pc	pc	pc	pc	pc	105	
21st	Sun	F	F	F	F	F	F	F	F	F	F	F	F	F	F	F	F	F	F	F	F	181	F
22nd	Mon	pc	pc	pc	pc	pc	pc	pc	pc	pc	pc	pc	pc	pc	c	c	pc	c	c	c	c	97	
23rd	Tue	F	F	F	F	F	F	F	F	F	F	F	F	F	F	F	F	F	F	F	F	260	
24th	Wed	F	F	F	F	F	F	F	F	F	F	F	F	F	F	F	F	F	F	F	F	300	P10
25th	Thur	F	F	F	F	F	F	F	F	F	F	F	F	F	F	F	F	F	F	F	F	298	XhN
26th	Fri	F	F	F	F	F	F	F	F	F	F	F	F	F	F	F	F	F	F	F	F	230	
27th	Sat	F	F	F	F	F	F	F	F	F	F	F	F	F	F	F	F	F	F	F	F	272	
28th	Sun	F	F	F	F	F	F	F	F	F	F	F	F	F	F	F	F	F	F	F	F	286	3Q
29th	Mon	pc	pc	pc	pc	pc	pc	pc	pc	pc	pc	pc	pc	pc	pc	F	c	pc	c	c	c	118	
30th	Tue	pc	pc	pc	pc	pc	pc	pc	pc	pc	pc	pc	pc	pc	F	pc	pc	pc	F	F	F	131	
31st	Wed	pc	pc	pc	pc	pc	pc	pc	pc	pc	pc	pc	pc	pc	pc	pc	pc	pc	pc	pc	pc	106	
Estimate hours:		211	211	211	211	211	201	211	201	215	207	204	211	215	219	220	213	216	218	218	218	4244	
Average hours:		152	152	135	143	133	121	122	133	129	144	157	142	155	144	167	142	121	142	140	167	2841	
Trend:		more	more	more	more	more	more	more	more	more	more	more	more	more	more	more	more	more	more	more	more	more	

JULY **1** MONDAY

An intense depression near the northern coastline of Ireland directs a stormy air over the country with strong westerly winds prevailing, sometimes reaching gale force at times, particularly in the north and northwest. A changeable day with a mix of sunny spells and showers over the northern half of the country, while remaining mostly dry down through the southern half.

Belfast: Stormy winds, occasional sunny spells, brief showers
Dublin: Stormy air, brief squalls, occasional sunny spells, cooler
Cork: Partly cloudy, dry, mild day cool night, freshening winds
Galway: Partly cloudy, dry, cool, windy with gusts

RAIN POTENTIAL

FROST/SNOW

JULY **2** TUESDAY

High pressure ridges over the south of Ireland while a blustery depression lingers to the north with northwest winds prevailing throughout, stronger and more blustery over northern and western districts. Mostly dry throughout after isolated showers in the northwest clear. Cloudy in the north, with increasing sunny spells to the east and west, and sunny in the south.

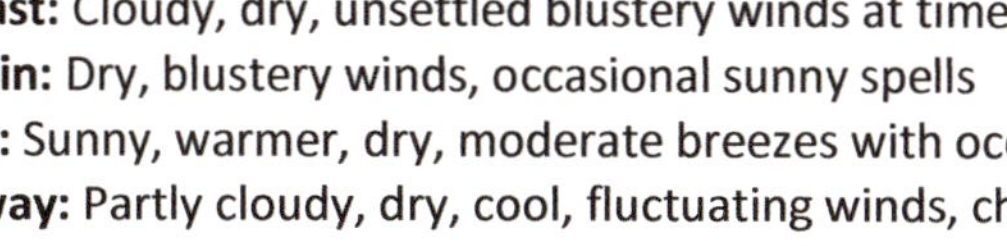

Belfast: Cloudy, dry, unsettled blustery winds at times
Dublin: Dry, blustery winds, occasional sunny spells
Cork: Sunny, warmer, dry, moderate breezes with occasional gusts
Galway: Partly cloudy, dry, cool, fluctuating winds, chance of gale gusts

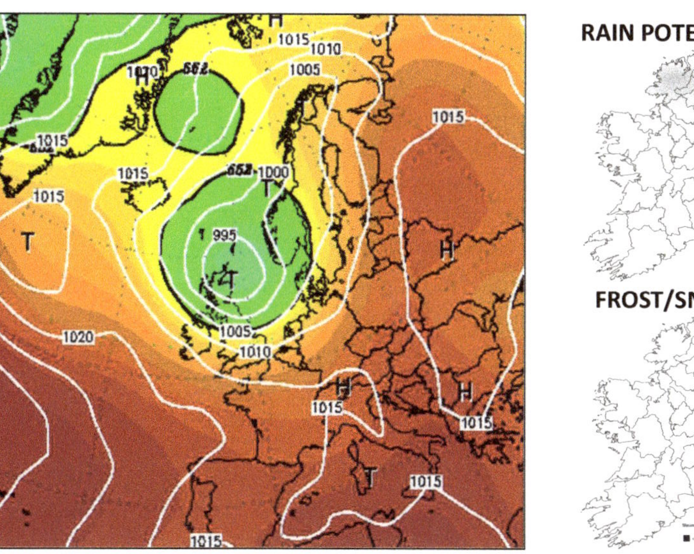

RAIN POTENTIAL

FROST/SNOW

JULY **3** WEDNESDAY

A weak high pressure system over Ireland brings light breezes to a mostly calm air throughout the country. Mostly cloudy with the occasional threat of rain in the west, while mostly sunny and dry elsewhere. Chance of isolated showers developing in the southwest in the evening.

Belfast: Sunny, cooler, dry, gentle breezes to mostly calm air
Dublin: Sunny, dry, milder, calm serene atmosphere
Cork: Overcast, dry with passing threats of rain, calm and serene atmosphere
Galway: Partly cloudy, dry, mild day cold night, calm air

JULY **4** THURSDAY

Low pressure trough lingers over Ireland, with light variable breezes prevailing. Scattered cloud throughout, thicker in the north and south, and brighter in the east and west. Outbreaks of drizzly rain in many districts during the day.

Belfast: Warmer, cloudy, brief showers, gentle breezes to mostly calm air
Dublin: Overcast, cooler, calm and misty air at times, chance of brief drizzle patches
Cork: Warm, frequent cloudy spells, isolated showers, calm atmosphere, chance of fog
Galway: Partly cloudy, milder, brief showers, calm atmosphere

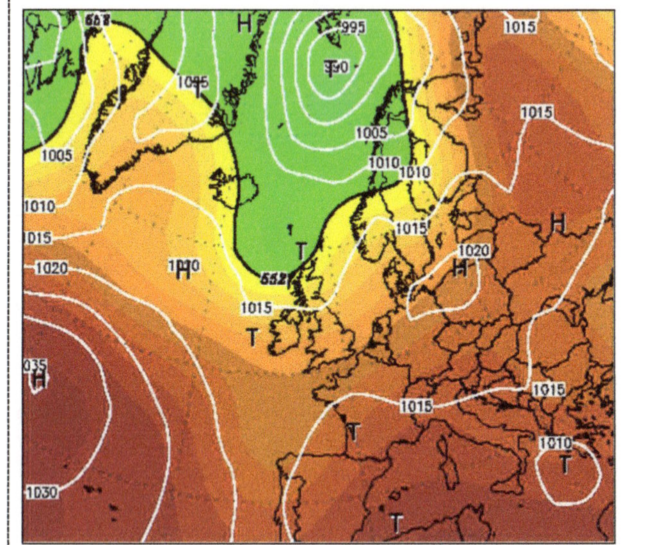

JULY **5** FRIDAY

High pressure ridges to the west of Ireland directing a light to moderate, mostly easterly flow. Overnight light showers in the west for a brief time. Dry and sunny throughout the day everywhere, and very sunny in northern parts. Some pockets of mists likely early morning in many districts.

Belfast: Sunny, dry, mild day cool night, mostly calm air
Dublin: Very bright and sunny, warmer, dry, gentle breezes
Cork: Mostly cloudy and dry, warm, gentle breezes
Galway: Partly cloudy, dry, slightly warmer, gentle breezes

RAIN POTENTIAL

FROST/SNOW

JULY **6** SATURDAY

A high pressure ridge moves over the west of the country with low pressure to the east, and a light to moderate northerly flow prevailing throughout. Mostly cloudy with passing outbreaks of showers over the eastern half of the country, while mostly sunny and dry in the western half.

Belfast: Cloudy, warmer, dry, gentle shifting breezes
Dublin: Overcast, cooler, light shifting breezes, passing drizzly showers
Cork: Sunny, mild, gentle breezes, isolated overnight showers mostly dry day
Galway: Dry, mild, gentle breezes, pleasant passing cloud

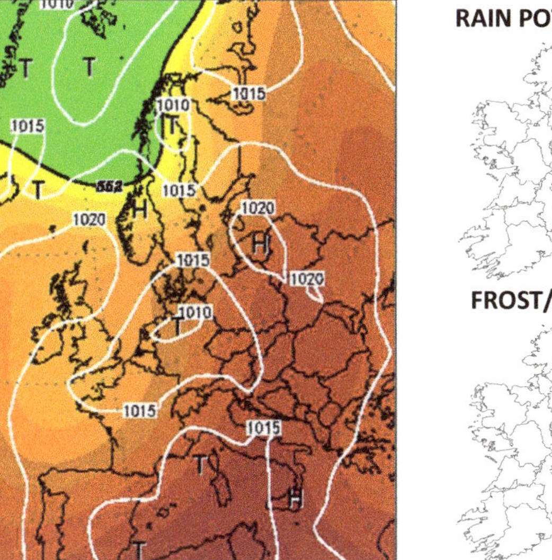

RAIN POTENTIAL

FROST/SNOW

JULY **7** SUNDAY

A weak high pressure ridge lingers over the country directing a light southeasterly flow throughout. Changeable in the eastern half of the country, with frequent cloud and a mix of mists, fog, and light showers, while sunnier and mostly dry with the odd patch of fog over the western half.

Belfast: Mostly cloudy, dry, calm air
Dublin: Cloudy, cool, spotty drizzle spells, light shifting breezes
Cork: Partly cloudy, dry, mild, light breezes, chance of overnight fog patches
Galway: Cloudy, dry, shifting breezes

RAIN POTENTIAL

FROST/SNOW

JULY **8** MONDAY

Low pressure trough sits to the northwest of Ireland, another to the southeast, and a high pressure ridge flows between with light to moderate southwesterly winds prevailing. Widespread and persistent fog and misty patches likely in many districts, particularly in the east and south. Mostly sunny and dry day unfolds in the east, while increased cloud and isolated showers develop elsewhere.

Belfast: Mostly cloudy, light breezes, dry, increasing threats of rain later in the day
Dublin: Partly cloudy, warmer, mostly dry, gentle breezes, chance of misty patches
Cork: Sunny, isolated showers, possible fog patches overnight, gentle shifting breezes
Galway: Mix of sunny spells and light showers, cooler, light southerly breezes

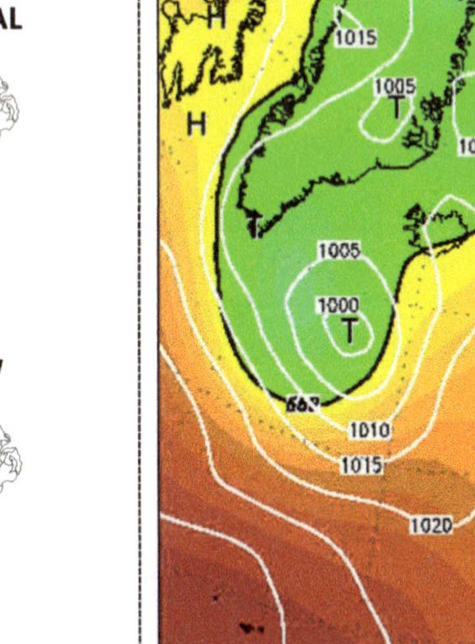

RAIN POTENTIAL

FROST/SNOW

JULY **9** TUESDAY

Depression to the northwest of Ireland moves onto the country with light south to southwest winds prevailing at first, freshening later in the day. Widespread cloud with occasional showers everywhere, with some outbreaks of heavy rain at times, particularly in the southwest and east. Chance of overnight fog in southern parts.

Belfast: Warm, humid, overcast, brief showers, moderate winds
Dublin: Overcast, occasional showers, moderate breezes
Cork: Overcast, mild, showers, chance of overnight fog and drizzle patches
Galway: Overcast, showers, moderate breezes

RAIN POTENTIAL

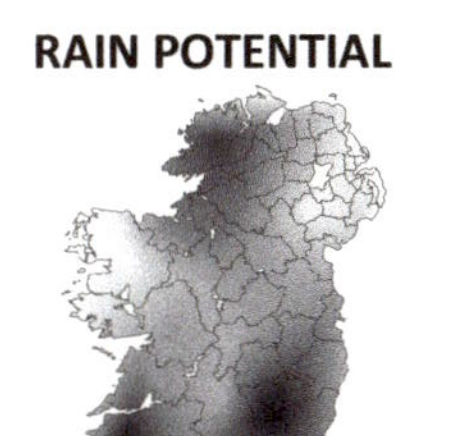

FROST/SNOW

JULY **10** WEDNESDAY

Depression to the northwest of Ireland maintains a moderate to fresh southwesterly flow over the country. Warm and humid everywhere, with long sunny spells developing, particularly in the south and west. Some showers in the east of the country, while dry elsewhere. Chance of pockets of fog in some northern and southern districts.

Belfast: Warm, mostly sunny and dry, breezy spells
Dublin: Partly cloudy, occasional showers, warm and humid, moderate breezes
Cork: Sunny, dry, warmer day cooler night, moderate breezes
Galway: Sunny, warmer, dry, moderate breezes

RAIN POTENTIAL

FROST/SNOW

JULY **11** THURSDAY

Low pressure trough directs a light to moderate southerly flow over the country. Dry, warm, and sunny almost everywhere, except for the chance of some fog in the south and brief overnight showers in the far north and western coastal areas.

Belfast: Sunny, warmer, dry, gentle breezes
Dublin: Very sunny, dry, warm day cool night, light shifting breezes
Cork: Sunny, mild, mostly dry, mostly calm air, chance of fog patches
Galway: Partly cloudy, dry, mild day cool night, light breezes

RAIN POTENTIAL

FROST/SNOW

JULY **12** FRIDAY

A high pressure system to the east of Ireland extends it ridge over the country with light east to northeast breezes prevailing. Mostly sunny and dry in the west and south, with overnight patches of fog likely in the south. Cloudier elsewhere with the passage of thunderstorms likely to bring some scattered light showers, mostly in the east and north.

Belfast: Calm, warmer, humid, cloudy spells, light showers, chance of misty patches
Dublin: Cloudy, isolated showers, light breezes, chance of thunderstorms overnight
Cork: Partly cloudy, dry, warmer, light shifting breezes, chance of overnight fog patches
Galway: Warm and mostly sunny, dry, gentle breezes

RAIN POTENTIAL

FROST/SNOW

JULY 13 SATURDAY

A weak high pressure ridges over the west of Ireland while low pressure contacts to the southeast, with a light to moderate northerly flow prevailing. Some thunderstorms and showers overnight in the east to the southeast of the country at first, clearing to odd showers between long dry spells. Mostly dry and sunny in the north and west with patches of morning mist likely.

Belfast: Sunny, warm and dry, light breezes
Dublin: Cloudy, warm, dry, occasional breezy spells
Cork: Overcast, cooler, light showers, occasional breezy spells
Galway: Warm, sunny, dry, fresh breezes at times

RAIN POTENTIAL

FROST/SNOW

JULY 14 SUNDAY

A weak high pressure ridge between two low pressure fronts moves up through the country in a light northerly flow. Mostly cloudy and humid throughout. Some thunderstorms and scattered showers likely in the east of the country, while mostly dry along the western coastal districts. Showers may be heavy for a time in some central districts to the east coasts.

Belfast: Cloudy, warm, dry, gentle shifting breezes
Dublin: Cloudy, cooler, light showers, light breezes
Cork: Cloudy, warm and humid air, brief showers, moderate breezes
Galway: Cloudy, dry, warm day cool night, mostly calm with odd breezy interlude

RAIN POTENTIAL

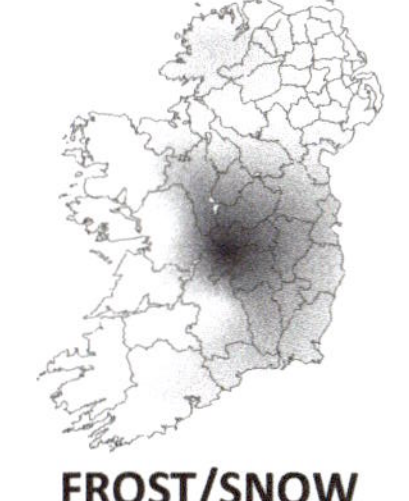

FROST/SNOW

JULY **15** MONDAY

A weak low pressure trough directs light breezes of variable directions over the country. Mostly sunny and dry over the northern half of the country, with cloudier conditions and passing showers in southern parts. An evening change to threats of rain along the western coasts with drizzle patches developing. Chance of fog in central parts and inland western districts.

Belfast: Sunny, hot, dry, shifting breezy spells
Dublin: Sunny, warm and dry, gentle shifting breezes, chance of shallow fog patches
Cork: Changeable, warm and humid, mostly cloudy, outbreaks of rain
Galway: Sunny and dry, mild temperatures, shifting breezes

RAIN POTENTIAL

FROST/SNOW

JULY **16** TUESDAY

A weak low pressure trough continues to drift over the country directing light southwesterly breezes, freshening to gusty at times later in the day. Warm, humid, and mostly cloudy everywhere. Mostly dry in the east and southeast coastal fringes, while outbreaks of drizzle or light showers and some fog patches likely elsewhere.

Belfast: Very hot and dry, frequent cloud, breezy spells
Dublin: Cloudy, warm, dry, gentle shifting breezes
Cork: Cloudy, dry, warm, humid, moderate shifting breezes
Galway: Overcast, cooler, showers, breezy spells, chance of overnight patches of fog

RAIN POTENTIAL

FROST/SNOW

JULY 17 WEDNESDAY

Low pressure with light to moderate winds prevail, winds gustier at times in the north. Mostly dry and sunny in the north and east of the country, while cloudier with scattered showers likely in the west.

Belfast: Hot, dry, partly cloudy, moderate breezes
Dublin: Mostly sunny and dry, warm, moderate breezes
Cork: Sunny, cooler, dry, breezy spells
Galway: Partly cloudy, mild, light showers, moderate breezes

RAIN POTENTIAL

FROST/SNOW

JULY 18 THURSDAY

Depressions to the north and west of Ireland direct light to moderate southwest winds, backing southeasterly and strengthening later in the day. Stormy conditions prevail throughout, with wet fronts crossing the country from the west and moving into the east as the day progresses, before easing later. Rain likely to be persistent and heavy for a time in many districts, particularly in the west.

Belfast: Overcast, rain, warm and humid air, fresh breezes at times
Dublin: Unsettled air, cooler, widespread cloud, scattered showers
Cork: Overcast, cooler, rain, fresh breezes at times
Galway: Overcast, cooler, rain, moderate breezes

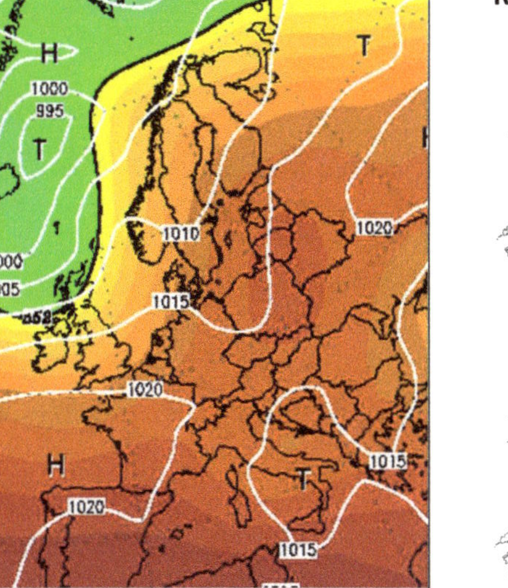

RAIN POTENTIAL

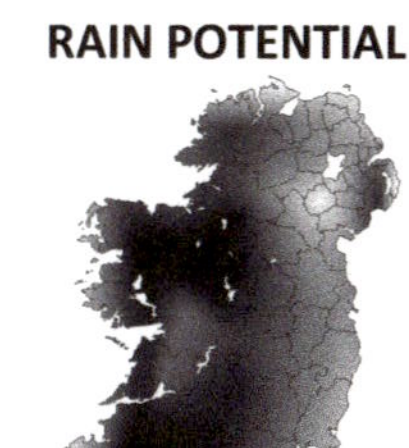

FROST/SNOW

JULY 19 FRIDAY

A depression moves over the north of the country overnight, with light rain and patchy drizzle prevailing in most districts for a time, before clearing to mostly dry with increasing sunny spells during the day. Fresh to moderate west to southwest winds prevail, easing later in the day when backing to south to southwest.

Belfast: Unsettled and blustery winds at times, partly cloudy, cooler and dry
Dublin: Unsettled and blustery at times, cloudy, isolated showers
Cork: Unsettled winds, sunny, warm and humid air, mostly dry
Galway: Sunny, slightly warmer, dry, unsettled and blustery winds

RAIN POTENTIAL

FROST/SNOW

JULY 20 SATURDAY

High pressure ridges to the south of the country while low pressure continues over the north with south to southwest breezes prevailing, fresh for a time particularly in the north and west. Mostly dry in the east and south, while a mix of sunny spells and showers likely in the west and north, with more frequent showers in the northwest for a time.

Belfast: Changeable, breezy, mix of showers and sunny spells, warmer, humid
Dublin: Partly cloudy, warm, mostly dry with odd threat of rain, moderate breezes
Cork: Partly cloudy, mostly dry, cooler, moderate breezes
Galway: Partly cloudy, isolated showers, fresh breezes at times

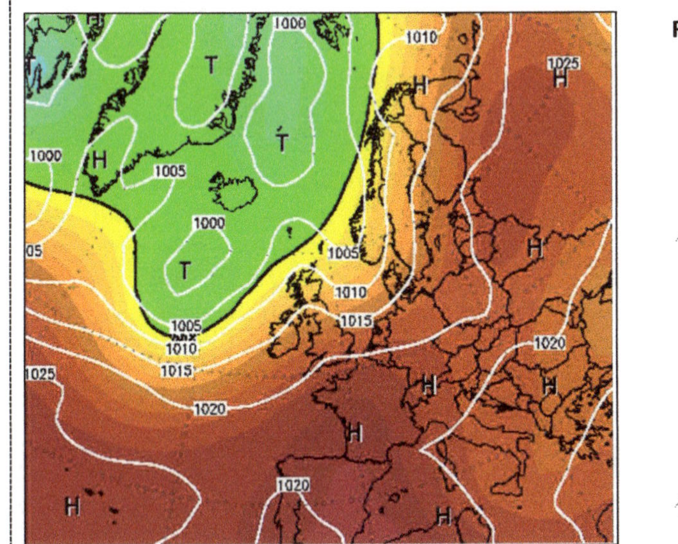

RAIN POTENTIAL

FROST/SNOW

JULY 21 SUNDAY

High pressure to the south of Ireland and low pressure to the north, with a light to moderate south to southwest flow prevailing. Mostly sunny and dry everywhere, except for the chance of some isolated showers in the northeast and brief drizzly spells in the northwest.

Belfast: Sunny, dry, fresh winds, slightly cooler
Dublin: Sunny, dry, warm day cool night, fresh breezes at times
Cork: Sunny and dry, mild day cool night, shifting breezy spells
Galway: Sunny and dry, mild day cool night, fresh breezes

RAIN POTENTIAL

FROST/SNOW

JULY 22 MONDAY

High pressure ridges over the southern half of Ireland while a depression remains stalled north of the country. Light westerly winds prevail, with the occasional breezier spell. Mostly cloudy and dry everywhere, except for the chance of some isolated overnight showers or drizzle patches in the west and north.

Belfast: Partly cloudy, mostly dry, mild, some windy spells, chance of spotty drizzles
Dublin: Cloudy and dry, milder, windy spells
Cork: Cloudy, dry, breezy
Galway: Partly cloudy, light showers, fresh winds

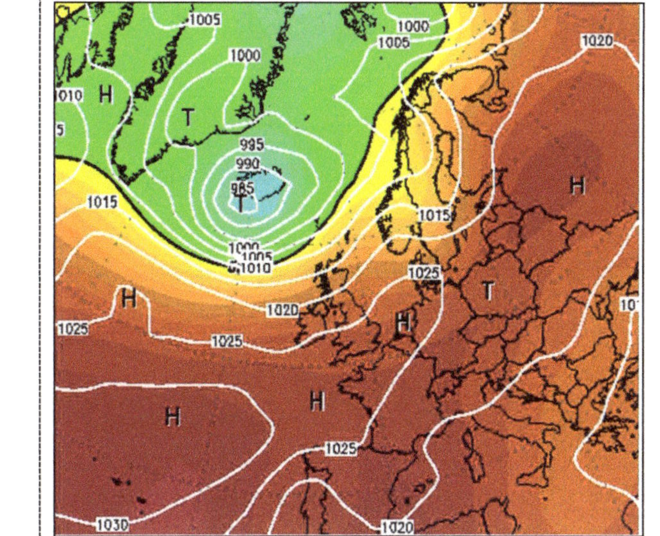

RAIN POTENTIAL

FROST/SNOW

JULY 23 TUESDAY

A large high to the southwest of Ireland continues to ridge over the country, directing a light northerly flow throughout. Very sunny and dry everywhere. Chance of some overnight patches of fog or brief drizzle breakouts in the west.

Belfast: Very sunny, dry, mild day cool night, moderate breezes
Dublin: Very sunny, dry, mild day cool night, gentle breezes to mostly calm air
Cork: Very sunny, dry, mild temperatures, mostly calm air
Galway: Sunny, isolated showers, light breezes, chance of overnight fog

RAIN POTENTIAL

FROST/SNOW

JULY 24 WEDNESDAY

A large anticyclonic system moves over the country bringing a mostly calm and serene air throughout. Sunny, dry, and warm everywhere. Chance of overnight misty patches in some sheltered western districts.

Belfast: Very bright and sunny, mild day cool night, calm atmosphere
Dublin: Very bright and sunny, mild day cool night, serene air
Cork: Very bright and sunny, dry, mild, gentle breezes to mostly calm air
Galway: Warm day cold night, very sunny, dry, calm air, chance of fog patches

RAIN POTENTIAL

FROST/SNOW

JULY **25** THURSDAY

Anticyclonic conditions continue to prevail throughout the country. Sunny, dry, and warm everywhere in a light to mostly calm air, with some pleasant sea breezes along the coastal fringes. Chance of fog patches overnight in sheltered districts.

Belfast: Very bright and sunny, warm day cool night, gentle breezes to mostly calm air
Dublin: Very sunny, dry, mild day cool night, calm air, chance of shallow fog patches
Cork: Very sunny, warm and dry, cool overnight, moderate breezes at times
Galway: Very sunny, warm and dry, cool overnight, light shifting breezes, humid

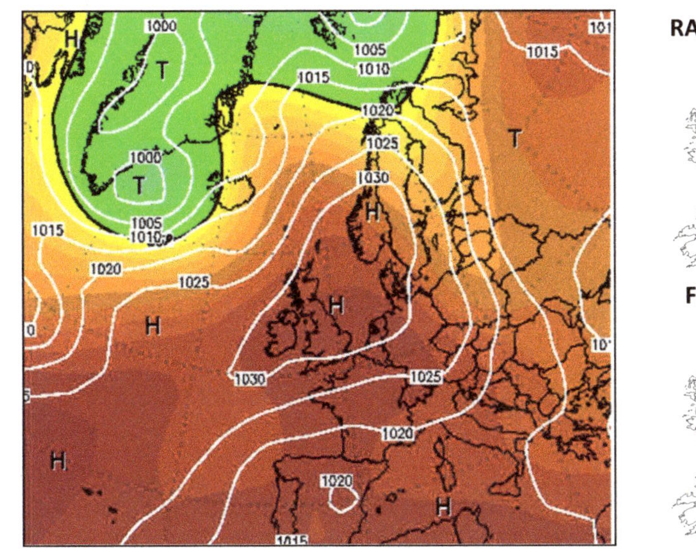

RAIN POTENTIAL

FROST/SNOW

JULY **26** FRIDAY

Anticyclonic conditions continue. Hot, dry, and sunny everywhere. Winds generally light to mostly calm and in a generally variable to easterly flow, except for the odd breezier spells around the northwest coasts. Chance of overnight fog patches in sheltered areas.

Belfast: Hot, sunny, dry, moderate southwesterly breezes
Dublin: Gentle southerly air, sunny, hot day cool night, dry, possible fog overnight
Cork: Sunny, hot and dry, gentle breezes to mostly calm air
Galway: Calm and serene air, dry, sunny, warm day cool night, possible shallow fog

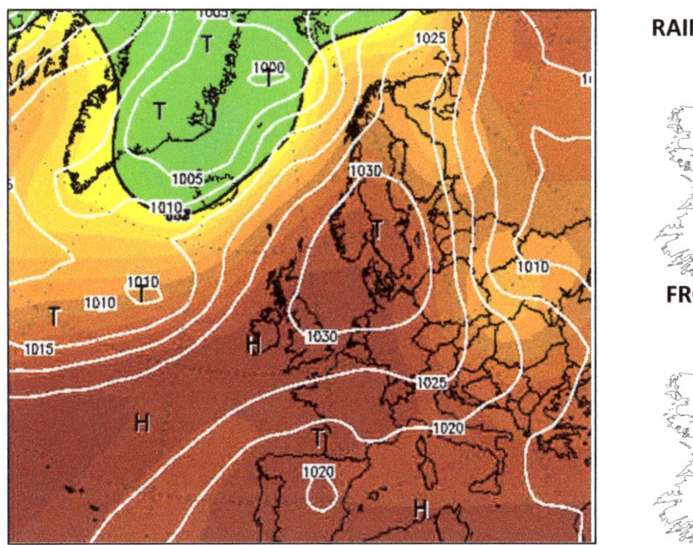

RAIN POTENTIAL

FROST/SNOW

JULY **27** SATURDAY

A large high continues to prevail over Ireland with light southeasterly breezes to a mostly calm air prevailing. Very hot, dry, and sunny everywhere, except for the chance of isolated overnight drizzle patches in the far northwest and brief pockets of fog in the west.

Belfast: Very hot and sunny, dry, moderate to light breezes
Dublin: Very hot, sunny and dry, calm air
Cork: Very sunny, hot, dry, calm atmosphere
Galway: Hot, humid, very sunny, dry, light breezes to mostly calm air, possible fog

RAIN POTENTIAL

FROST/SNOW

JULY **28** SUNDAY

High pressure remains stalled over Ireland maintaining a mostly calm and serene air over the country. Possible shallow fog patches overnight in many areas, clearing to a sunny and dry day everywhere. Very hot in many districts but milder along the coasts due to gentle sea breezes.

Belfast: Very hot, sunny, dry, calm atmosphere
Dublin: Calm and dry air, hot, very sunny
Cork: Very sunny, hot and dry, calm atmosphere
Galway: Very hot and sunny, dry, calm air, possible shallow fog

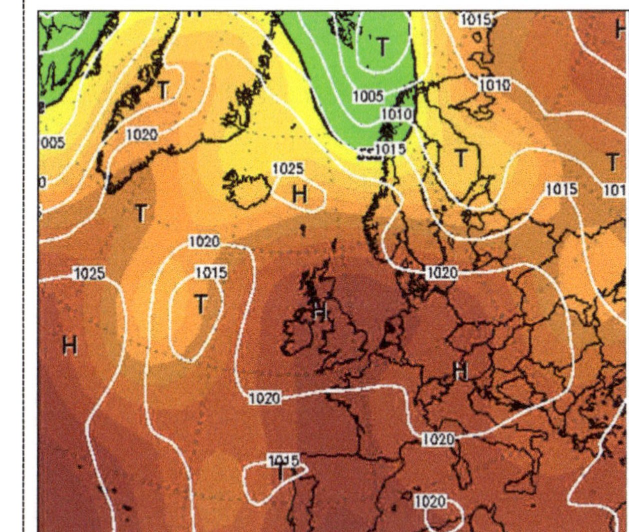

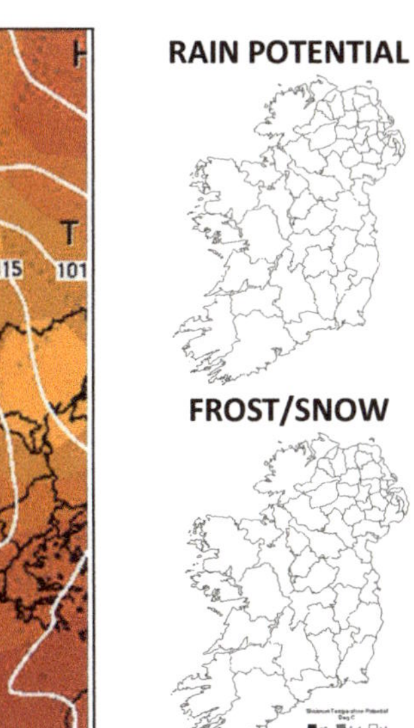

JULY **29** MONDAY

A weak low pressure system moves near the west of the country while the high pressure system drifts northwards but still ridging over the north. Winds generally light to moderate in a southeasterly flow. Overnight patches of mist or fog in many districts. Mainly dry, sunny, and hot throughout the day everywhere. Cloud likely to build up in the south of the country by the evening, with the chance of some light drizzle spells developing.

Belfast: Sunny, very hot, dry, calm and static atmosphere
Dublin: Gentle southerly air, sunny, very hot, dry, humid at times
Cork: Cloudy, warm and humid, isolated showers, offshore breezes
Galway: Very hot, dry, some cloudy spells, gentle breezes

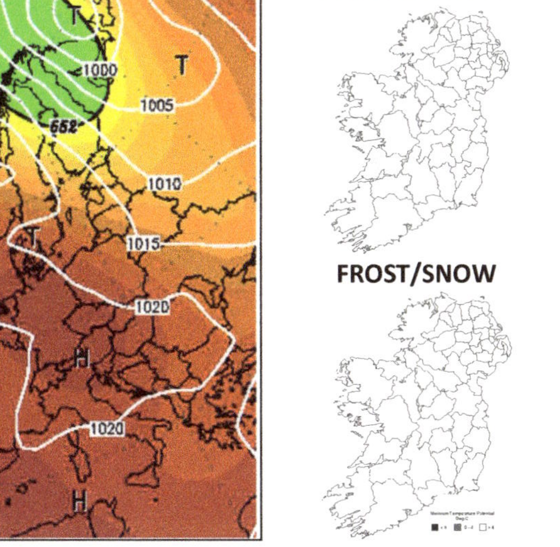

RAIN POTENTIAL

FROST/SNOW

JULY **30** TUESDAY

A weak low pressure system spreads over the country directing a light southeasterly flow, backing southwesterly later in the day, with breezier spells around the coasts. Hot, mostly sunny, and humid everywhere. Chance of some misty patches in the south and west overnight. Heat generated thunderstorms may be observed, particularly in northern parts and accompanied by showers, while remaining dry elsewhere.

Belfast: Very hot, mostly dry, occasional cloudy spells, calm air, possible thunderstorm
Dublin: Mostly sunny, hot and dry, gentle shifting breezes
Cork: Sunny, dry, warm day cool night, calm air, chance of fog patches overnight
Galway: Partly cloudy, dry, cooler, calm air, chance of mists and thunderstorms

RAIN POTENTIAL

FROST/SNOW

JULY **31** WEDNESDAY

A weak depression lingers to the southwest of Ireland, while high pressure ridges over Europe, with a low pressure trough crossing the country in a light southerly flow. Patches of mist or fog here and there overnight, clearing to a mostly dry, warm, humid, calm and sunny day, except for the chance of some isolated showers in the south and west.

Belfast: Partly cloudy, light southerly flow, hot, dry, chance of overnight fog patches
Dublin: Partly cloudy, slightly cooler, dry, calm air
Cork: Partly cloudy, warm, light breezes, chance of fog, isolated showers possible
Galway: Warm, humid, isolated showers, some sunny spells, chance of misty patches

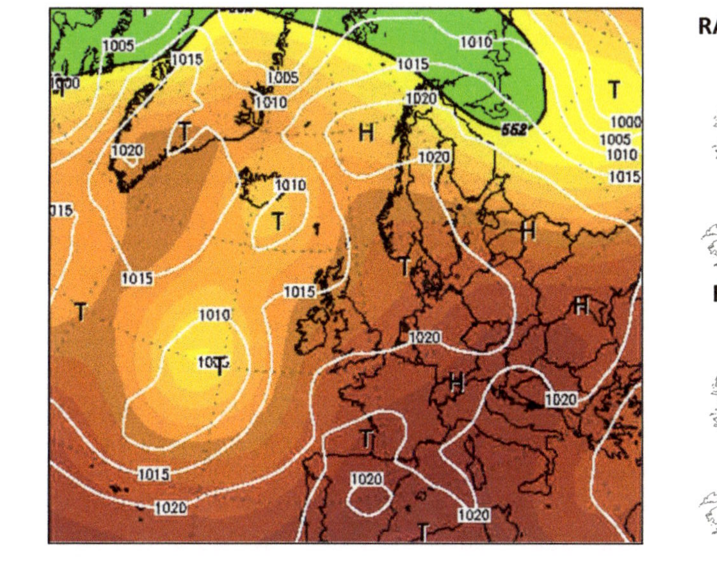

RAIN POTENTIAL

FROST/SNOW

AUGUST

PHASES OF THE MOON

1st	Northern Declination
4th	New Moon
9th	Apogee
9th	Crossing Equator
12th	First Quarter
15th	Southern Declination
19th	Full Moon
21st	Perigee #7
23rd	Crossing Equator
26th	Third Quarter
28th	Northern Declination

1	2	3	4	5	6	7	8	9	10	11	12	13	14	15	16	17	18	19	20	21	22	23	24	25	26	27	28	29	30	31
Thur	Fri	Sat	Sun	Mon	Tue	Wed	Thur	Fri	Sat	Sun	Mon	Tue	Wed	Thur	Fri	Sat	Sun	Mon	Tue	Wed	Thur	Fri	Sat	Sun	Mon	Tue	Wed	Thur	Fri	Sat
^			N					A XhS			1Q			V				F		P7		XhN			3Q		^			

MONTHLY SUMMARY

1st-6th August Mostly dry spell continues from last month especially in the east and south.
1st-10th Depressions near the west of Ireland extend low pressure over the country, with unsettled conditions prevailing everywhere. While mostly dry, warm, and humid at first, particularly in the east and south, unsettled southerly winds and showers develop elsewhere, spreading later throughout most districts and sometimes associated with thunderstorm activity.
4th-7th Very warm spell with daily maximas in the upper 20's in many districts.
8th Low pressure moves near the west of the country. Strong winds prevail, with occasional gusts, mostly in the northwest.
10th Chance of funnel formation winds in southern inland parts, possibly in the Tipperary to Clonmel region.

The next four weeks ahead will see low pressure briefly prevailing at first, then again around the middle of this outlook period, and with higher pressure mostly towards the end of the first week and into the second and again in the last week. Generally, much of the coming outlook period will be fairly dull and unsettled compared to normal, but also warm, humid and with spotty outbreaks of rain on many days, but only small amounts resulting in the overall expectation of below average rainfall in most districts. Rainfall will be generally less than normal for this time of year, despite some rainfall occurring on most days. Many places will record below average rainfall, especially in the northwest, where only around a third of their normal rainfall is expected. In contrast, thunderstorms in the second half of this outlook period may bring above average rain to some parts of Leinster and Ulster. Overall, however, most of the rain that does cross the country will be in the form of passing showers, and only thunderstorm bands bringing the occasional spell of heavy rain. Overall, sunshine hours will be around the norm for this time of year in the south of Ireland, while unusually dull in northern parts. The sunniest period during the next four weeks will be over the first week, after which widespread cloud settles for the remainder of this outlook period. The drier than normal conditions will also reflect in warmer than normal temperatures everywhere, with most districts likely to be around half to one degree warmer than average, and more than 1° C warmer in some parts of the south and southeast. Despite this, a cold spell midway through this outlook period will see some pockets of frosts. The first week of this outlook period and the last few days will be much warmer than normal, with the remaining days being milder and with little variation each day. During this outlook period there is likely to be around 9-11 wet days in the south of the country (with 1mm or more recorded rainfall likely), but up to 20 wet days in the northwest. Thunderstorms may be heard on 8 days, frosts on 1 day, and fogs on 7 days, but there are no days when hail, snow, or gales are expected.

11th A weak front crosses the country.
11th-15th High pressure to the west of Ireland ridges over country bringing mostly dry, warm and sunny conditions, especially in the west and southwest. Light westerly winds prevail, later becoming light and more variable.
14th-18th Chance of scattered fogs.
15th Unusually warmer than normal day with many districts recording maximas over 25°C, and overnight minimas around an average of 16-17°C especially in the southeast. Increased cloud in many districts, but remaining very sunny in the south.
16th Chance of a band of showers crossing the country.
16th-31st Low pressure lingers to the northeast of the country while high pressure is situated to the southwest, with a resulting northerly airflow prevailing. Very dull and cloudy spell with occasional showers on most days.
17th-20th Anticyclonic conditions move over the country. Very sunny in the southeast.
18th-31st Milder spell.
18th Chance of some light ground frosts in sheltered parts in inland districts.
19th Very blustery at times in the northwest.
19th-20th Very windy spell with the chance of gale gust in some districts.
20th Very sunny day.
21st-25th Cooler spell with daily maximas around the 15°C range in many districts.
25th-29th A slow moving depression extends wet fronts over the country, bringing more persistent rain and the occasional outbreak of heavy falls and possible thunderstorms, especially in the south and east of the country.
24th Chance of isolated pockets of overnight light frosts in inland districts.
26th Depression moves close to the southern coasts of Ireland.
26th-29th Chance of widespread thunderstorms and heavy rain in places, especially in the southeast. Unsettled winds may cause funnel formations in the southeast.
28th Very heavy rain likely in the southeast of the country.

IRELAND RAINFALL ESTIMATES

August 2024

d=mainly dry, n= nonrecordable, l=light shrs, s=significant shrs, r=rain, h=hvy falls

		ULSTER PROVINCE					CONNAUGHT PROVINCE						LEINSTER PROVINCE				MUNSTER PROVINCE						
		Alderg	Hillsbo	Armag	Loughl	Malin H	Sligo	Ardtarı	Belmul	Drums	Galway	Roscoı	Markre	Edende	Kilkenı	Dublin	Shannc	Newpo	Tralee	Killarne	Cork	All	Moon
1st	Thu	d	d	d	l	s	s	l	s	s	s	l	s	d	d	d	s	s	h	l	l	60	^
2nd	Fri	d	d	d	l	d	l	l	n	l	d	l	l	d	d	d	n	n	d	d	d	7	
3rd	Sat	d	d	n	l	s	l	l	s	d	r	d	s	d	d	d	d	r	s	d	d	48	
4th	Sun	d	d	d	r	d	s	r	l	l	l	d	s	d	d	d	d	l	l	d	d	40	N
5th	Mon	d	d	d	l	s	l	l	d	d	d	d	n	d	d	d	l	l	l	d	l	13	
6th	Tue	n	n	d	s	d	s	s	l	l	s	l	s	d	d	d	l	s	s	s	d	57	
7th	Wed	s	s	l	n	r	l	n	n	s	d	s	l	s	s	l	l	l	l	s	s	73	
8th	Thur	s	l	r	s	s	l	s	s	l	l	s	l	l	n	l	l	l	h	s	s	97	XhS
9th	Fri	s	s	n	l	s	s	l	s	l	r	s	l	s	s	l	l	l	r	r	s	113	A
10th	Sat	r	h	s	s	s	s	s	s	l	s	r	l	l	n	l	s	l	s	l	s	117	
11th	Sun	s	l	h	n	l	l	n	l	l	d	l	l	d	d	d	d	d	n	d	l	36	
12th	Mon	n	d	l	d	l	d	d	d	d	d	d	l	d	d	d	d	d	d	d	d	2	1Q
13th	Tue	d	d	d	d	d	d	d	d	d	d	d	l	d	d	d	d	d	d	d	d	1	
14th	Wed	l	l	d	l	s	d	l	d	n	l	l	l	d	d	n	d	l	l	d	d	22	
15th	Thu	l	l	l	l	l	d	l	l	l	s	s	d	l	d	d	d	s	l	s	d	40	V
16th	Fri	l	l	l	d	l	d	d	l	d	d	d	n	l	l	l	s	d	r	l	l	39	
17th	Sat	d	d	n	d	d	d	d	d	d	d	d	l	d	d	d	d	n	d	d	d	4	
18th	Sun	l	l	d	s	s	l	s	l	s	l	s	l	l	d	s	n	s	l	n	d	55	
19th	Mon	l	l	d	l	s	l	l	d	s	d	l	l	d	d	d	d	d	d	d	d	14	F
20th	Tue	s	l	l	l	l	l	l	d	d	d	l	r	s	d	l	d	d	d	d	d	31	
21st	Wed	l	d	l	d	l	d	d	d	d	d	d	d	d	d	s	d	d	d	d	d	4	P7
22nd	Thur	d	d	d	d	d	d	d	d	d	d	d	d	l	l	d	d	d	d	d	d	2	XhN
23rd	Fri	d	d	d	d	d	d	d	d	d	d	d	d	d	d	d	d	d	d	d	d	0	
24th	Sat	n	l	d	l	d	l	l	d	l	d	l	l	d	d	d	d	d	d	d	d	12	
25th	Sun	l	s	n	l	l	l	l	l	r	r	s	l	s	s	r	r	l	s	r	l	105	
26th	Mon	l	d	s	d	l	l	d	s	n	s	n	d	d	l	l	d	l	n	l	l	18	3Q
27th	Tue	l	s	d	d	d	l	d	l	l	d	l	d	l	s	n	d	d	d	l	r	38	
28th	Wed	s	s	l	s	n	s	s	d	s	l	r	s	h	h	s	l	d	l	l	h	141	^
29th	Thu	r	l	h	d	l	l	d	d	l	d	s	l	l	d	h	d	d	n	s	d	59	
30th	Fri	l	l	l	s	s	s	s	n	l	d	l	s	s	d	d	l	l	l	l	d	43	
31st	Sat	s	d	s	d	n	d	d	d	d	d	d	n	d	d	d	d	d	s	d	d	6	
	Estimate:	73	70	74	63	82	53	63	51	59	72	73	68	73	50	50	35	57	95	63	78	1299	
	Average:	84	87	76	101	95	102	104	102	88	108	109	102	86	72	73	82	133	106	77	97	1882	
	Trend:	drr	drr	av	drr	drr	drr	drr	drr	drr	drr	drr	drr	drr	drr	drr	drr	drr	drr	drr	drr	drr	

IRELAND SUNSHINE ESTIMATES

August 2024

F=fine (8-12 hours of sunshine), pc= partly cloudy (4-7 hours), c=cloudy (1-3 hours), o=overcast (0 hours)

		ULSTER PROVINCE					CONNAUGHT PROVINCE						LEINSTER PROVINCE				MUNSTER PROVINCE						
		Alderg	Hillsbo	Armag	Loughl	Malin I	Sligo	Ardtarı	Belmul	Drums	Galway	Roscoı	Markre	Edende	Kilkenı	Dublin	Shannc	Newpo	Tralee	Killarne	Cork	All	Moon
1st	Thu	pc	pc	pc	pc	pc	pc	pc	pc	pc	pc	pc	pc	pc	pc	pc	pc	pc	pc	pc	pc	105	^
2nd	Fri	pc	pc	pc	pc	pc	pc	pc	pc	pc	pc	pc	pc	pc	pc	F	pc	pc	pc	pc	pc	117	
3rd	Sat	pc	pc	pc	pc	pc	c	pc	c	F	c	c	pc	F	F	F	c	pc	F	F	F	132	
4th	Sun	pc	pc	pc	pc	pc	o	pc	o	pc	c	c	pc	pc	pc	F	pc	pc	pc	pc	pc	103	N
5th	Mon	pc	pc	pc	pc	pc	pc	pc	pc	pc	c	pc	pc	pc	pc	pc	c	pc	c	c	c	94	
6th	Tue	pc	pc	pc	pc	pc	c	pc	c	pc	c	c	pc	pc	pc	pc	c	c	c	c	c	72	
7th	Wed	c	c	c	c	c	o	c	o	c	o	o	c	c	c	c	o	c	c	c	c	24	
8th	Thur	pc	pc	pc	pc	pc	pc	pc	pc	pc	pc	pc	pc	pc	pc	pc	pc	pc	pc	pc	pc	104	XhS
9th	Fri	pc	pc	pc	pc	pc	pc	pc	pc	F	pc	pc	pc	F	F	F	pc	F	F	F	F	150	A
10th	Sat	c	c	c	c	c	pc	c	pc	c	c	c	c	c	c	o	o	c	c	c	c	50	
11th	Sun	pc	pc	pc	pc	pc	pc	pc	pc	pc	pc	pc	pc	pc	pc	F	pc	pc	c	c	c	122	
12th	Mon	F	F	F	F	F	pc	F	pc	F	c	pc	F	F	F	F	c	pc	pc	pc	pc	140	1Q
13th	Tue	pc	pc	pc	pc	pc	c	pc	c	F	c	c	pc	F	F	F	c	pc	F	F	F	134	
14th	Wed	o	o	o	o	o	o	o	o	c	o	o	o	c	c	o	o	c	pc	pc	pc	24	
15th	Thu	c	c	c	c	c	c	c	c	c	c	c	c	c	pc	c	c	pc	pc	pc	pc	77	V
16th	Fri	c	c	c	c	c	c	c	c	o	c	c	c	o	o	o	c	c	o	o	o	27	
17th	Sat	pc	pc	pc	pc	pc	c	pc	c	pc	pc	c	pc	pc	F	pc	F	F	F	F	F	141	
18th	Sun	c	c	c	c	c	o	c	o	pc	c	o	c	pc	pc	pc	c	pc	F	F	F	75	
19th	Mon	pc	pc	pc	pc	pc	pc	pc	pc	pc	pc	pc	pc	pc	F	pc	pc	F	F	F	F	146	F
20th	Tue	c	c	c	c	c	c	c	c	c	pc	pc	c	c	pc	o	F	F	F	F	F	94	
21st	Wed	c	c	c	c	c	c	c	c	c	o	c	c	c	o	c	o	o	o	o	o	23	P7
22nd	Thur	c	c	c	c	c	c	c	c	pc	pc	pc	c	pc	pc	c	pc	pc	pc	pc	pc	89	XhN
23rd	Fri	pc	pc	pc	pc	pc	pc	pc	pc	pc	pc	pc	pc	pc	F	pc	pc	pc	F	F	F	145	
24th	Sat	c	c	c	c	c	o	c	o	pc	c	o	c	pc	F	pc	c	pc	F	F	F	93	
25th	Sun	o	o	o	o	o	o	o	o	o	c	c	o	o	o	o	c	c	o	o	o	13	
26th	Mon	c	c	c	c	c	c	c	c	c	c	c	c	c	c	o	c	c	c	c	c	47	3Q
27th	Tue	pc	pc	pc	pc	pc	c	pc	c	pc	c	c	pc	pc	pc	pc	c	pc	pc	pc	pc	103	
28th	Wed	c	c	c	c	c	c	c	c	c	c	c	c	c	c	c	c	c	c	c	c	47	^
29th	Thu	o	o	o	o	o	o	o	o	c	o	o	o	c	c	o	o	o	c	c	c	19	
30th	Fri	pc	pc	pc	pc	pc	c	pc	c	pc	c	c	pc	pc	c	pc	c	c	c	c	c	70	
31st	Sat	pc	pc	pc	pc	pc	c	pc	c	pc	pc	c	pc	pc	pc	F	pc	pc	pc	pc	pc	117	
Estimate hours:		126	126	126	126	126	88	126	88	148	105	97	126	148	171	163	122	147	180	180	180	2696	
Average hours:		146	146	133	140	133	127	127	144	122	140	152	136	148	140	158	138	124	140	140	159	2793	
Trend:		less	less	av	less	av	less	av	less	more	less	less	av	av	more	av	less	more	more	more	more	average	

AUGUST 1 THURSDAY

A low pressure rain embedded trough associated with a depression to the west of Ireland maintains a light to moderate mostly southerly flow over the country. A changeable day with a mix of sunny spells and scattered showers in most parts, although remaining mostly dry in the east, where some patches of early morning fog are likely. Warm and humid.

Belfast: Partly cloudy, hot southwesterly breezes, dry
Dublin: Partly cloudy, dry and warm, mostly calm air, chance of overnight fog patches
Cork: Warm, scattered cloudy spells, brief showers of light rain, moderate breezes
Galway: Unsettled and breezy, cooler, some sunny spells, occasional showers

RAIN POTENTIAL

FROST/SNOW

AUGUST 2 FRIDAY

A weak depression moves onto the country with overnight showers clearing to a mostly dry and sunny day throughout. Generally, light south to southwesterly winds prevail, with the odd breezier spell now and then, particularly later in the day.

Belfast: Partly cloudy, dry, warm, light onshore breezes
Dublin: Mostly sunny, warmer, dry, gentle shifting breezes
Cork: Partly cloudy, dry with odd threat of rain, moderate breezes
Galway: Partly cloudy, mild and dry, moderate breezes

RAIN POTENTIAL

FROST/SNOW

AUGUST **3** SATURDAY

A low pressure trough laying off the west of Ireland directs a light southerly air through the country along the ridge line of a high pressure system to the east. Mostly dry and very sunny in the east and coastal southern districts, while a band of rain embedded with thunderstorms moves up through the country starting in the southwest and spreading up through western parts and into the northern districts.

Belfast: Partly cloudy, dry, warm, shifting breezes
Dublin: Very sunny, warm, dry, changeable breezes at times
Cork: Sunny, warmer, dry, fluctuating breezes
Galway: Cloudy, rain, fluctuating winds, warmer

RAIN POTENTIAL

FROST/SNOW

AUGUST **4** SUNDAY

The high pressure ridge moves further east of Ireland while a weak low pressure centre lingers to the west, with a warm and gentle southerly flow prevailing over the country. Mostly dry, hot, humid, and sunny at first, with the chance of scattered fogs mostly in the east. Later in the day another band of showers likely cross the western coasts and bring isolated showers from the southwest and up into the northwest.

Belfast: Hot and humid, dry, scattered cloudy spells, mostly calm air
Dublin: Sunny, warm, dry, mostly calm air, chance of light shallow fog patches
Cork: Warm, dry, scattered cloudy spells, gentle southerly air
Galway: Mostly cloudy, warmer, brief showers, light southerly flow

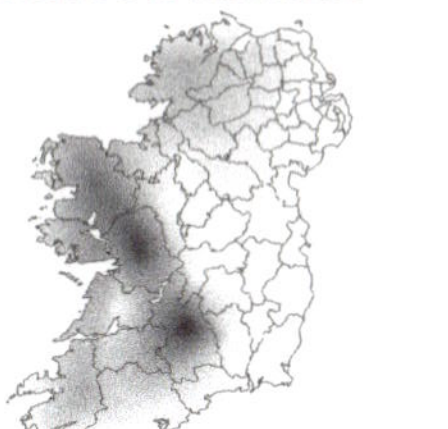

RAIN POTENTIAL

FROST/SNOW

AUGUST **5** MONDAY

A weak low pressure trough continues to direct a mostly calm to gentle southerly flow over the country. Some patches of mist or fog likely overnight mostly around the southern coasts. A hot and humid day with a mix of sunny spells and cloudy spells throughout. Some isolated showers likely for a brief time in the north, more drizzly in the northwest, while mainly dry elsewhere.

Belfast: Very hot, dry, humid, occasional cloudy spells, some breezy spells
Dublin: Partly cloudy, mostly calm air, dry, hot and humid
Cork: Cloudy, calm, humid, light drizzle, chance of fog patches
Galway: Mostly cloudy, dry, warm, shifting breezes

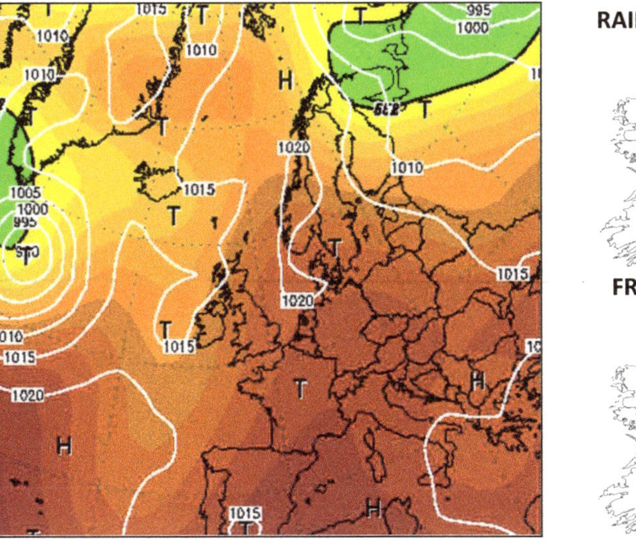

RAIN POTENTIAL

FROST/SNOW

AUGUST **6** TUESDAY

A significant depression to the northwest of Ireland moves closer towards the western coasts of the country, with low pressure prevailing throughout. Winds generally light in a southwest flow with the likelihood of patches of fog, mist, and drizzle here and there. A wet front crosses the southwest at first and moves northwards through the country bringing wet spells through most of the country except along the eastern and northern coastal fringes. Some outbreaks of rain may be heavy for a brief time, particularly in the midlands.

Belfast: Partly cloudy, hot, humid, isolated showers, light breezes
Dublin: Partly cloudy, dry, warm, light breezes to mostly calm air
Cork: Cloudy, mostly dry, warm, serene air, chance of overnight fog patches
Galway: Cloudy, cooler, occasional showers, light breezes

RAIN POTENTIAL

FROST/SNOW

AUGUST 7 WEDNESDAY

A low pressure trough associated with an intense depression to the northwest of Ireland maintains a light, variable to southerly flow over the country. Hot, humid, and cloudy throughout with outbreaks of rain in most districts.

Belfast: Cloudy, rain, hot and humid air, chance of overnight fog patches
Dublin: Cloudy, light showers, hot and humid air, gentle breezes at times
Cork: Light breezes, cloudy, passing showers, chance of overnight misty patches
Galway: Freshening breezes, heavy cloud, mild, dry with increasing threats of rain

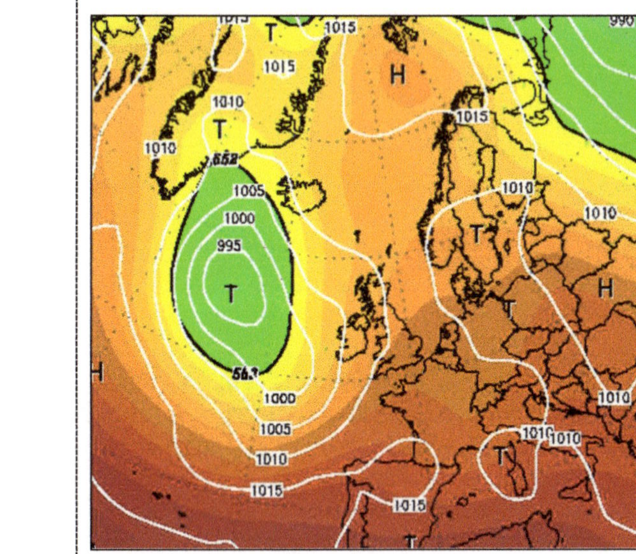

RAIN POTENTIAL

FROST/SNOW

AUGUST 8 THURSDAY

A large depression to the west of Ireland moves onto the country bringing strengthening south to southeast breezes, possibly reaching gale force for a time around the western coasts. Some patches of overnight fog or mists in inland parts at first, turning to persistent drizzle and scattered rain and spreading throughout the country when the winds strengthen, and continue scattered shower activity throughout the day. Some sunny breakthroughs likely between the bands of showers.

Belfast: Partly cloudy, brief showers, warm, humid, freshening breezes later in the day
Dublin: Partly cloudy, light showers, warm and humid, moderate southerly breezes
Cork: Partly cloudy, occasional showers, freshening breezes
Galway: Overnight showers, partly cloudy, mild, windy spells

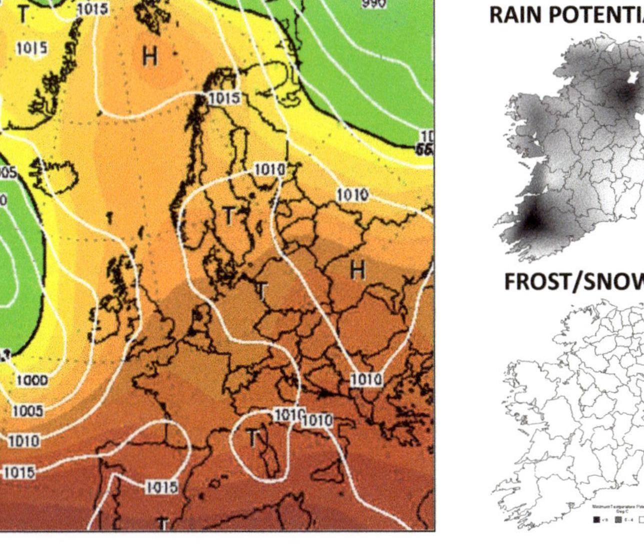

RAIN POTENTIAL

FROST/SNOW

AUGUST **9** FRIDAY

A depression off the western coasts of Ireland continues to direct moderate to fresh south to southwest wind over the country, and with an accompanying mix of sunny spells and showers in all districts. Widespread scattered thunderstorms likely, especially through central parts. Spells of sunshine will be longer in the east and south, while less frequent in the west and north.

Belfast: Unsettled and static air, mix of showers and sunny spells, blustery
Dublin: Sunny, warm, humid, chance of showers and thunderstorms, windy at times
Cork: Sunny, windy, occasional showers
Galway: Partly cloudy, overnight rain, possible thunderstorms, mild, blustery at times

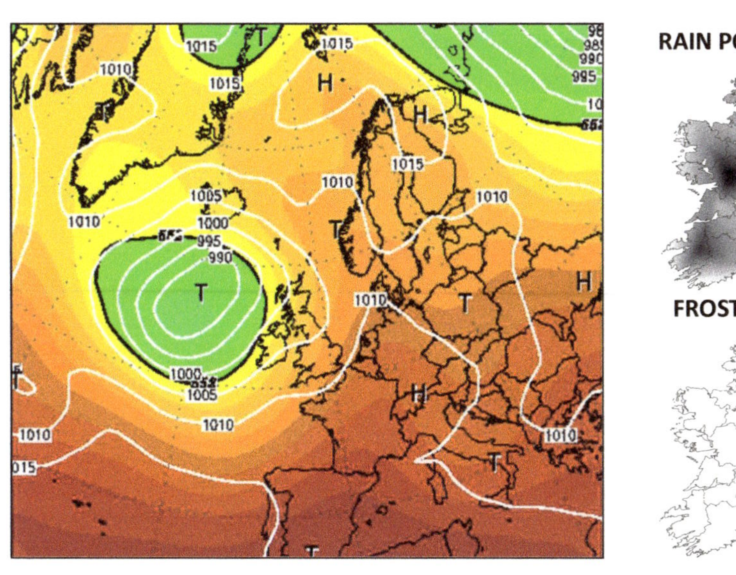

RAIN POTENTIAL

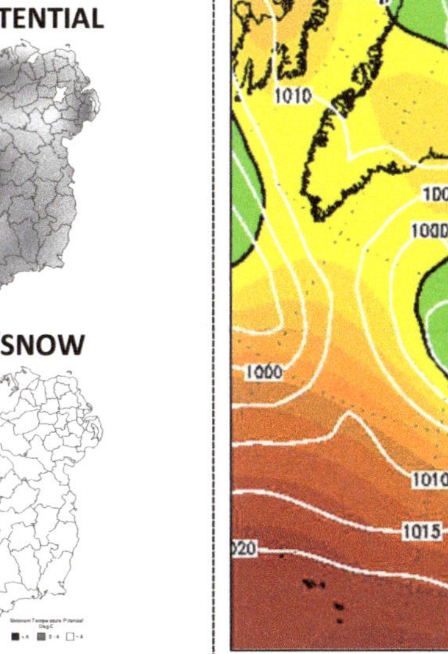

FROST/SNOW

AUGUST **10** SATURDAY

A large depression stalls off the western coasts of Ireland bringing a stormy air throughout the country, with strong southerly winds prevailing at first, but turning westerly later in the day. Widespread showers throughout the country, sometimes heavy, and occasionally accompanying thunderstorm activity, particularly in the east and northeastern parts. Cooler.

Belfast: Stormy air, cloudy, heavy rain, blustery, thunderstorms likely
Dublin: Overcast, cooler, windy at times, showers of light rain, possible thunderstorms
Cork: Windy and cloudy, scattered showers
Galway: Cloudy, scattered showers, mild, blustery at times, chance of thunderstorms

RAIN POTENTIAL

FROST/SNOW

AUGUST **11** SUNDAY

The depression that has brought stormy conditions over the previous couple of days begins to weaken while shifting to be centred just north of the country. Blustery winds prevail throughout, mostly in a northwest flow, decreasing later in the day. Overnight showers in most parts, clearing to a mostly dry and cloudy day in the southern half of the country, while showers and widespread cloud continue to prevail over most of the northern half until late in the day.

Belfast: Milder, mostly sunny, isolated showers, strong winds with occasional gusts
Dublin: Unsettled windy spells, sunny, mostly dry, warm and humid
Cork: Mostly cloudy, isolated showers, windy spells
Galway: Unsettled winds, partly cloudy, dry, slightly warmer

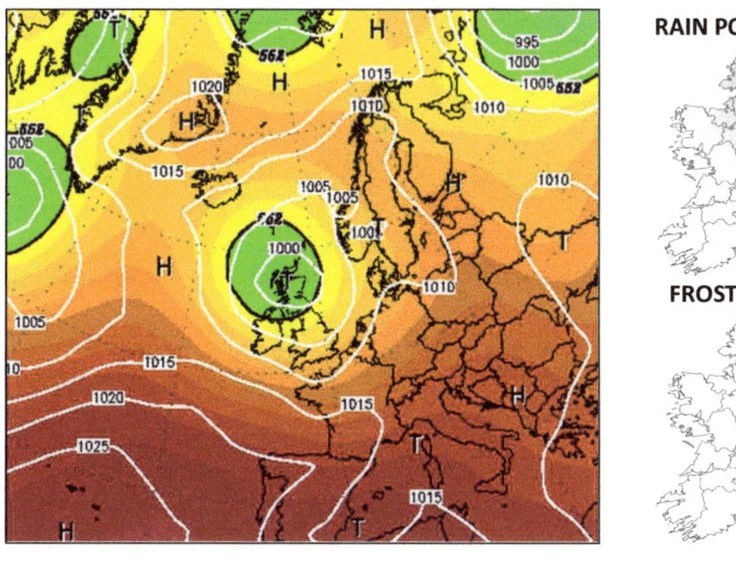

RAIN POTENTIAL

FROST/SNOW

AUGUST **12** MONDAY

A complex of high and low pressure centres prevail around the region, with a weak high pressure ridge between two low pressure centres moving over Ireland. Fresh northwest winds prevail at first, easing with a northerly change later in the day. Mostly dry throughout, with long sunny spells in the north and east, while cloudier in the south and west. Chance of lingering spotty showers mostly in the north and east.

Belfast: Unsettled and breezy air, sunny, warmer, dry
Dublin: Sunny, dry, warm, windy spells
Cork: Partly cloudy, mild, dry, moderate breezes
Galway: Mostly cloudy, dry and mild, windy at times

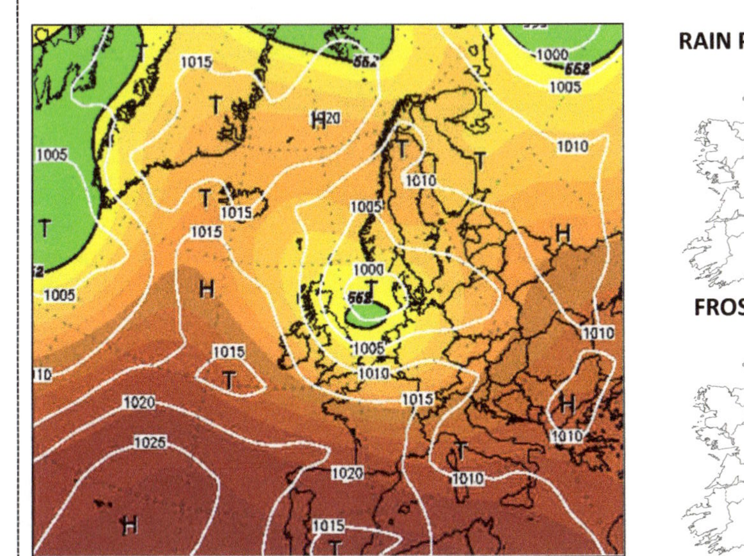

RAIN POTENTIAL

FROST/SNOW

AUGUST 13 TUESDAY

A large high pressure system situated to the southwest of Ireland directs a light to moderate northerly flow over the country. Mostly dry, warm and pleasant air throughout, with long sunny spells in the east and south and increased scattered cloud in the north and west. Chance of some brief showers along the northwest coasts.

Belfast: Partly cloudy, dry, warm, windy spells
Dublin: Warm, dry, sunny, windy spells
Cork: Warm, sunny, dry, fresh breezes at times
Galway: Cloudy, mild day cool night, dry, fresh breezes at times

RAIN POTENTIAL

FROST/SNOW

AUGUST 14 WEDNESDAY

A large high pressure system to the southwest of Ireland continues to direct a light to moderate north to northwest flow over the country. Dull and gloomy almost everywhere with chance of overnight patches of fog or mist, later turning to spotty showers in northern and western parts, while remaining mostly dry in the east and south.

Belfast: Overcast, warm and humid, brief showers, unsettled winds at times
Dublin: Overcast, isolated showers, warm, fresh breezes at times
Cork: Partly cloudy, warmer, dry, fresh breezes at times
Galway: Warm, overcast, passing drizzly showers, onshore fresh breezes

RAIN POTENTIAL

FROST/SNOW

AUGUST **15** THURSDAY

High pressure to the southwest of Ireland ridges over the country, while a depression to the northwest of Ireland pushes closer, with a light southwest flow prevailing throughout. Cloudy, hot, and humid everywhere. Mostly dry in the far southeast and along the eastern coastal fringes with patches of overnight fog likely, while drizzly showers prevail elsewhere, briefly clearing for a time in the afternoon.

Belfast: Cloudy, light showers, warm, windy at times
Dublin: Hot, cloudy, humid and dry, light breezes
Cork: Hot, mostly sunny and dry, light breezes
Galway: Warm, humid, cloudy, scattered showers, fresh southwest breezes

RAIN POTENTIAL

FROST/SNOW

AUGUST **16** FRIDAY

A large high moves over the country with northwest breezes prevailing, freshening ahead of the passage of a wet front. Widespread cloud everywhere with some overnight pockets of fog mostly in northern parts. A wet front crosses the west at first and spreads showers across the country in a westerly flow to reach the eastern and southern coasts later in the day, with rain clearing everywhere by the end of the day.

Belfast: Cloudy, brief showers, warm, windy spells, chance of overnight fog patches
Dublin: Overcast, slightly cooler, light showers, gentle breezes to mostly calm air
Cork: Overcast, drizzly showers, cooler, humid air, light breezes
Galway: Cloudy and dry, slightly cooler, light breezes

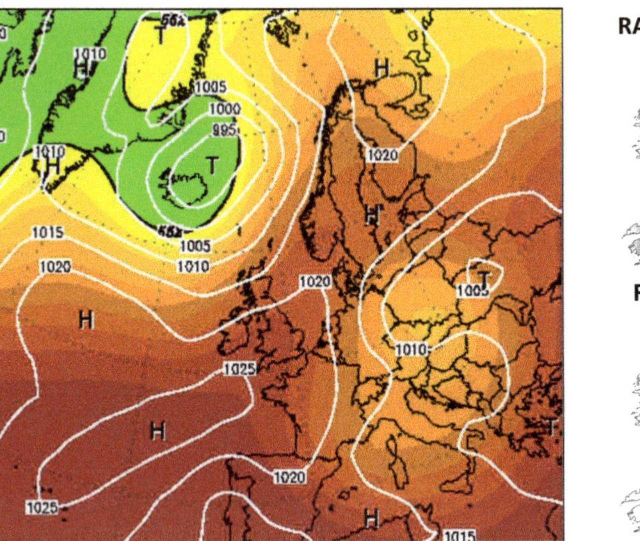

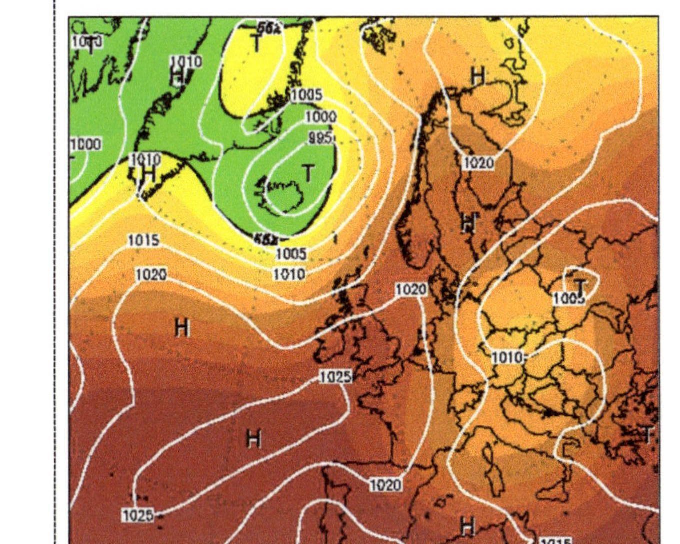

RAIN POTENTIAL

FROST/SNOW

AUGUST **17** SATURDAY

A large high ridges over the country with light mostly northerly breezes prevailing. Some brief overnight showers likely mostly in the west. Mostly dry everywhere throughout the day, with more frequent cloud in the north and very sunny in the south.

Belfast: Partly cloudy, cooler, dry, light shifting breezes
Dublin: Light shifting breezes, partly cloudy, dry, mild day cool night
Cork: Very sunny, warm and dry, gentle shifting breezes
Galway: Partly cloudy, dry, mild day cool night, light shifting breezes

RAIN POTENTIAL

FROST/SNOW

AUGUST **18** SUNDAY

A large anticyclonic system sits over Ireland with a mostly gentle westerly air prevailing. Some patches of overnight fog in many districts. A weak front crosses the country in southeasterly flow, bringing a mix of sunny spells and showers in most districts, cloudiest in the northwest, and sunniest in the south.

Belfast: Cloudy, light showers, mild day cool night, chance of fog patches
Dublin: Warm day cold night, mix of sunny spells and showers,possible fog patches
Cork: Sunny, warmer, dry, light shifting breezes, chance of shallow fog overnight
Galway: Cloudy, calm, mild day cool night, brief showers, chance of misty fog patches

RAIN POTENTIAL

FROST/SNOW

AUGUST **19** MONDAY

A depression to the north of Ireland strengthens winds along the high pressure ridge, bringing moderate to fresh northwest winds in most districts, with stronger gusts at times in the north. Some light showers in the northern half of the country, while mostly dry and sunny elsewhere.

Belfast: Partly cloudy, cooler, brief showers, unsettled winds
Dublin: Partly cloudy, mostly dry with odd threat of rain, milder, breezy at times
Cork: Sunny, cooler, dry, fresh breezes at times
Galway: Partly cloudy, cool and dry, moderate breezes

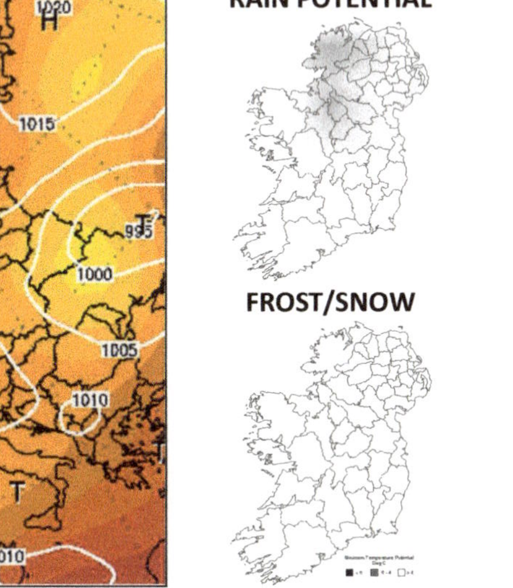

RAIN POTENTIAL

FROST/SNOW

AUGUST **20** TUESDAY

The high pressure ridge pushes back across the country bringing northwest winds throughout, mostly moderate to fresh but blustery at times with the chance of gale gusts around the coasts. Mostly cloudy with passing showers in the north and east, while mostly sunny and dry in the west and south.

Belfast: Cloudy, scattered showers, breezy spells
Dublin: Overcast, windy, isolated showers, cooler
Cork: Sunny, warm, dry, moderate breezes
Galway: Mostly sunny, dry, warmer, fresh breezes

RAIN POTENTIAL

FROST/SNOW

AUGUST 21 WEDNESDAY

A large anticyclonic system continues to direct a light to moderate northerly flow over the country. Widespread cloud prevails everywhere, with the occasional threat of rain here and there, despite a mostly dry day throughout. Chance of some overnight light drizzle in the north and east, clearing early in the day, and returning in the evening in the north.

Belfast: Cloudy, mild, dry, occasional breezy spells, misty patches possible
Dublin: Cloudy, light showers, windy at times
Cork: Overcast and cooler, dry with odd threat of rain, some windy spells
Galway: Overcast and cooler, dry with odd threat of rain, moderate breezes

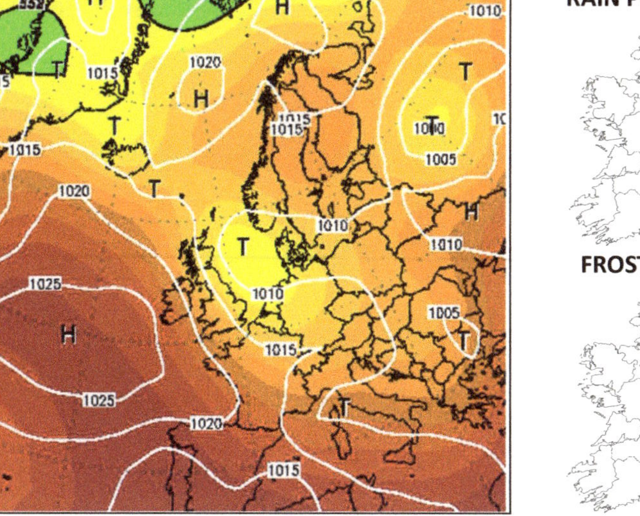

RAIN POTENTIAL

FROST/SNOW

AUGUST 22 THURSDAY

High pressure continues to direct light to moderate northerly breezes over the country, with fresher spells possible in northern parts. Mostly dry and sunny over the southern half of the country, while mostly cloudy and dry over the northern half, although there is a chance of some isolated brief drizzle spells here and there, except in the southeast. Cooler than of recent.

Belfast: Light northerly breezes, dry, mostly cloudy
Dublin: Cloudy, cooler, dry, fresh breezes at times
Cork: Mostly sunny and dry, mild day cool night, moderate breezes
Galway: Partly cloudy, dry, mild day cool night, moderate breezes

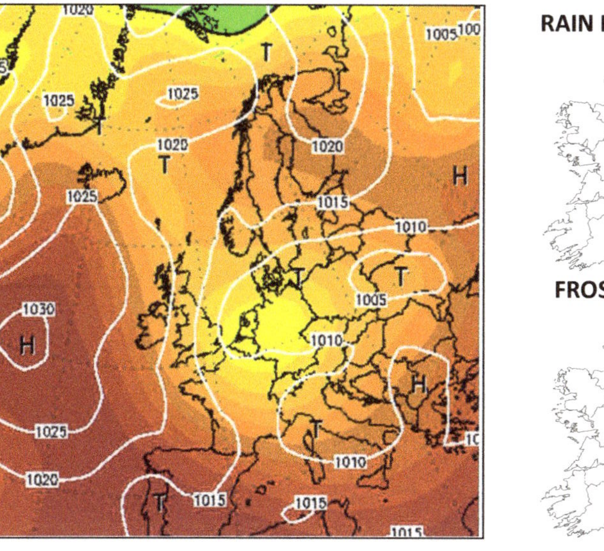

RAIN POTENTIAL

FROST/SNOW

AUGUST **23** FRIDAY

High pressure to the west, low pressure to the east and a moderately northerly flow prevails along the ridge between the two systems, with winds fresher at times. Mostly dry and sunny throughout, except for increased scattered cloud and the chance of some isolated showers of little significance, here and there in the west and north.

Belfast: Partly cloudy, cooler, dry, fresh northerly breezes at times
Dublin: Mostly sunny and dry, mild day cool night, windy spells
Cork: Dry, sunny, windy spells
Galway: Partly cloudy, dry, mild, freshening breezy spells with occasional gusts likely

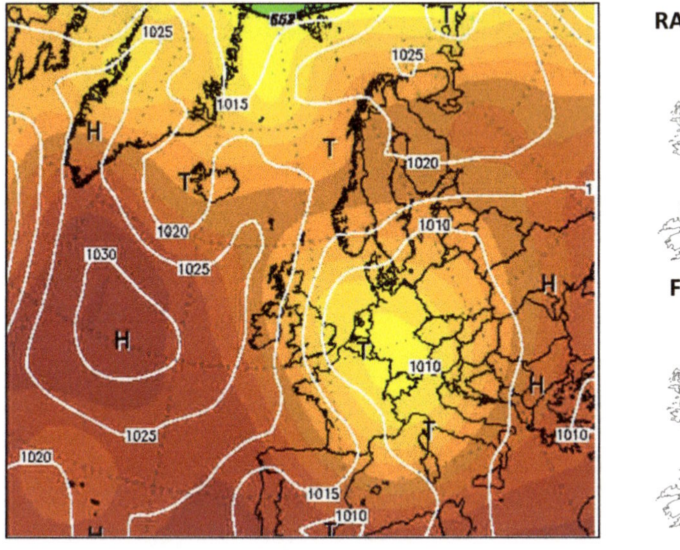

RAIN POTENTIAL

FROST/SNOW

AUGUST **24** SATURDAY

A large anticyclonic system continues to linger in the Atlantic Ocean to the west of Ireland, while low pressure lingers over Europe to the east, with a cooling, light, northerly flow prevailing throughout the country. Mostly cloudy in the north and west, with increasing threats of rain in western districts, and some spotty showers likely in the northwest. Mostly dry and sunny elsewhere, with particularly long sunshine hours in southern parts.

Belfast: Cloudy, coolish, moderate breezes, light drizzle patches likely
Dublin: Partly cloudy, dry, mild day cool night, moderate northerly breezes
Cork: Very sunny, dry and mild, light breezes
Galway: Cloudy and dry, increasing threats of rain, slightly warmer, gentle breezes

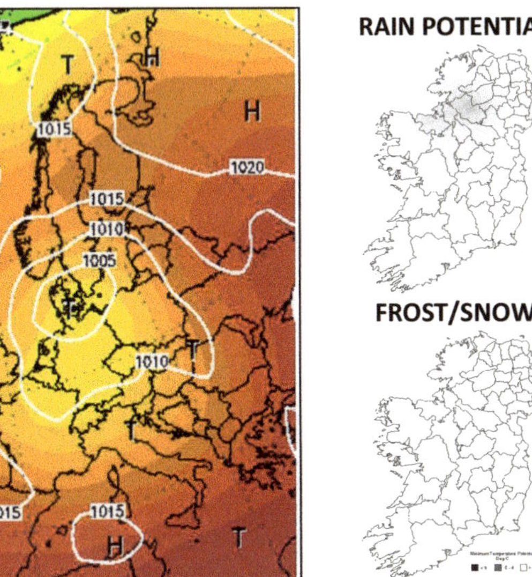

RAIN POTENTIAL

FROST/SNOW

AUGUST 25 SUNDAY

The low pressure than has lingered over Europe the past few days unites with another depression to the northwest of the country and together causes the high pressure ridge to move off the country and into the Atlantic Ocean. Widespread cloud gathers and brings a dull and gloomy air throughout, with light to moderate westerly winds generally prevailing. An unsettled day throughout, with outbreaks of rain, drizzle in most districts. Chance of some fog patches overnight in the north.

Belfast: Overcast, scattered showers, windy spells
Dublin: Overcast, cooler, rain, windy at times
Cork: Overcast, outbreaks of light rain, cooler, windy spells
Galway: Cloudy, cooler, rain, unsettled windy spells

RAIN POTENTIAL

FROST/SNOW

AUGUST 26 MONDAY

A weak depression crosses the country bringing unsettled and unstable winds, generally in a northeasterly flow. Dull and gloomy throughout, with outbreaks of drizzle and rain almost everywhere, mostly in the morning and evening with drier spells in the middle of the day. Chance of thunderstorms at times, particularly in the south of the country. Some brief sunny breakthroughs likely in the west and south.

Belfast: Unsettled and breezy, cloudy, mostly dry, cooler, odd threat of rain
Dublin: Unsettled windy spells, dull and gloomy, spotty showers and drizzle patches
Cork: Unsettled windy spells, cloudy, light showers
Galway: Unsettled and blustery at time, cloudy, brief showers

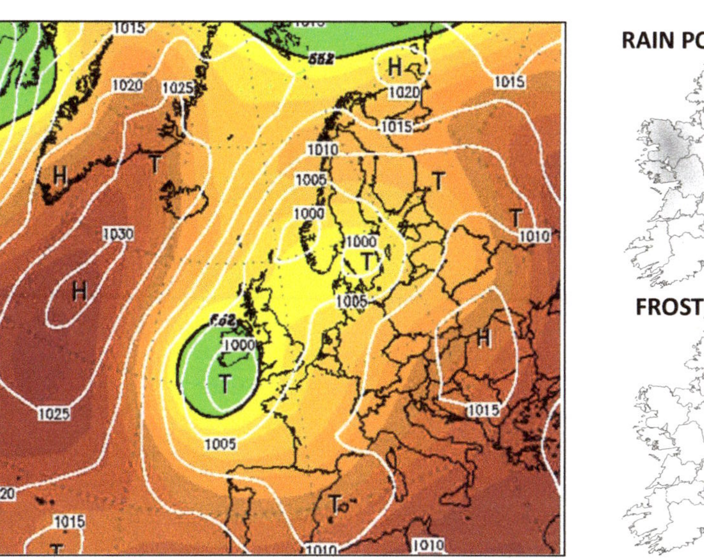

RAIN POTENTIAL

FROST/SNOW

AUGUST 27 TUESDAY

A weak depression continues to linger over Ireland with light to moderate northeasterly winds prevailing, although gustier for a time along the coastal fringes. Overnight outbreaks of rain or drizzle in many districts, easing during the day to mostly dry in some western districts and good sunny spells at times in the north and west. Even rain likely to return accompanied by thunderstorms, particularly in the southeast.

Belfast: Warm, mix of showers and sunny spells, windy at times
Dublin: Partly cloudy, isolated showers, fluctuating winds
Cork: Changeable, some sunny spells, rain later in the day, windy with occasional gusts
Galway: Cloudy, windy, dry

RAIN POTENTIAL

FROST/SNOW

AUGUST 28 WEDNESDAY

A weak depression continues to linger over the country bringing widespread cloud throughout and generally moderate north to northwesterly breezes, but occasionally strong and gusty with the showers. Mostly dry along the western coastal fringes, with widespread rain everywhere else. Afternoon thunderstorms likely, and accompanied by some heavy falls, particularly in the east and south.

Belfast: Cloudy, rain, warm, humid, changeable winds, chance of overnight fog
Dublin: Cloudy, scattered showers, fluctuating winds, chance of misty patches
Cork: Stormy air, cloudy, heavy rain, thunderstorms likely, blustery, possible gales
Galway: Cloudy, light showers, unsettled and changeable winds

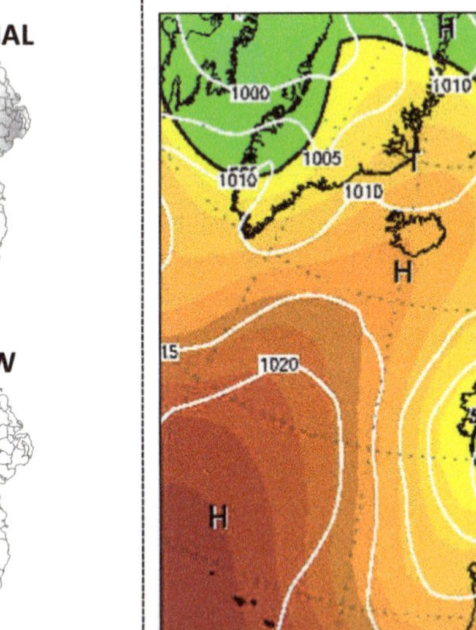

RAIN POTENTIAL

FROST/SNOW

AUGUST 29 THURSDAY

Low pressure continues to bring unstable conditions throughout the country with moderate north to northwest breezes prevailing. Mostly overcast throughout, with widespread rain and scattered thunderstorms likely overnight, clearing in the west during the day but persisting on and off throughout the day in the east. Chance of some sunny breakthroughs in the extreme north and far south briefly.

Belfast: Overcast, warm, drizzly showers, humid, chance of fog patches
Dublin: Static air, overcast, heavy rain, changeable winds
Cork: Cloudy, dry, windy spells, mostly dry, possible drizzle patches for a time
Galway: Overcast, dry with passing threats of rain, mild, unsettled breezes at times

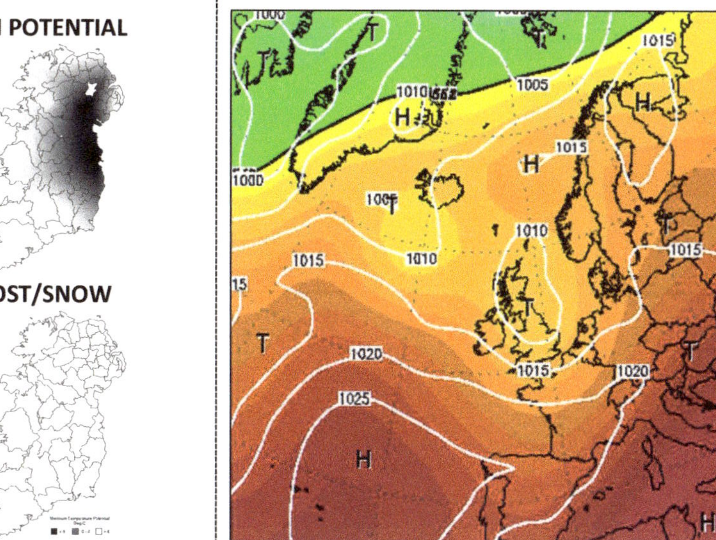

RAIN POTENTIAL

FROST/SNOW

AUGUST 30 FRIDAY

The depression that has lingered over Ireland the past few days weakens as a large high to the south of the country advances northwards, bringing a west to southwest change. Winds generally remain light to moderate. A changeable day with a mix of sunny spells and light rain or drizzle outbreaks in most districts. Sunniest in the northwest and southeast.

Belfast: Warm, humid, occasional sunny spells, light showers
Dublin: Partly cloudy, warmer, dry, shifting windy spells
Cork: Warmer day colder night, cloudy, dry, fluctuating windy spells
Galway: Dry and cloudy, mild day cool night, fluctuating winds with occasional gusts

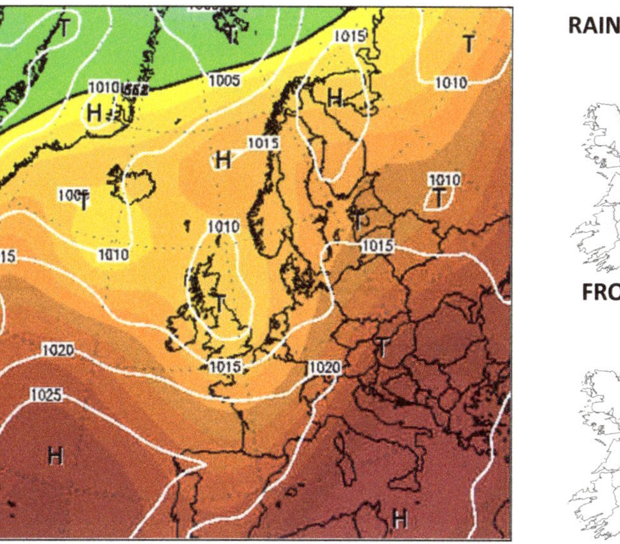

RAIN POTENTIAL

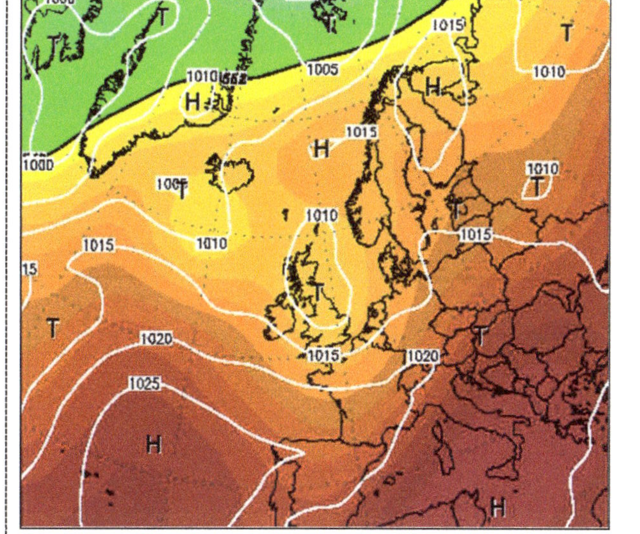

FROST/SNOW

AUGUST 31 SATURDAY

The low pressure trough slips northwards off the country while high pressure ridges to the south. Mostly dry and sunny throughout, except for the chance of the occasional spotty showers in the northeast and southwest. Light winds prevail in a northwest to variable flow.

Belfast: Partly cloudy, dry, warm, light breezes
Dublin: Sunny, warm, mostly dry with odd threat of rain, moderate breezes
Cork: Cooler, mostly sunny and dry, windy spells
Galway: Breezy spells, partly cloudy, mild, dry

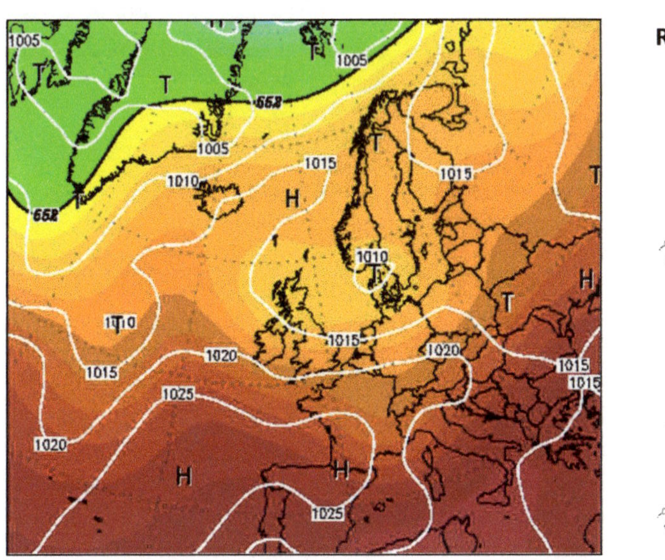

SEPTEMBER

PHASES OF THE MOON

3rd	New Moon
5th	Apogee
5th	Crossing Equator
11th	First Quarter
12th	Southern Declination
18th	Full Moon
18th	Perigee #3
19th	Crossing Equator
24th	Third Quarter
24th	Northern Declination

1 Sun	2 Mon	3 Tue	4 Wed	5 Thur	6 Fri	7 Sat	8 Sun	9 Mon	10 Tue	11 Wed	12 Thur	13 Fri	14 Sat	15 Sun	16 Mon	17 Tue	18 Wed	19 Thur	20 Fri	21 Sat	22 Sun	23 Mon	24 Tue	25 Wed	26 Thur	27 Fri	28 Sat	29 Sun	30 Mon
		N		A XhS						1Q	V						F P3	XhN					3Q ^						

MONTHLY SUMMARY

1st-3rd Cool overnight temperatures. Chance of some light ground frosts in sheltered parts, mostly in the west.
1st-10th Atlantic depressions cross the country bringing a period of widespread unsettled weather in a moderate southwest to westerly flow. A mix of sunny spells and showers prevail.
2nd Thunderstorms possible.
2nd-3rd Mostly sunny spell.
6th-7th Chance of thunderstorms.
9th-10th Chance of scattered patches of fog.
10th A thunderstorm band with heavy rain crosses the northern half of the country.

Over the next four weeks Atlantic depressions close to the west coasts will be the dominant influence on the weather, bringing a mix of plenty of rain and wind, but also mild temperatures as low pressure mostly prevails, especially during the second half of this outlook period, with anticyclonic conditions mostly during the second week. Rainfall will be above normal throughout the country, especially in some parts of the west and south even though the number of days when some rain will be seen will be around the average. Some parts of Munster and Connaught may record more than twice as much rain as is typical for this time of year. Sunshine hours will be slightly higher than average almost everywhere, with a couple of very sunny spells expected during the second and third weeks. Only the southeast can expect slightly less sunshine hours than is the norm for this outlook period. Temperatures are expected to be fairly consistently above normal throughout this outlook period, with some most districts around 2°C above their norm, and in some eastern parts possibly up to 3°C above the norm for this time of year. Despite the above average maximum and minimum temperatures during this outlook period, no extremes are expected. On some days however, temperatures may be higher than those of the previous four weeks. During this outlook period there is likely to be around 13-19 rain days in most districts, but up to 20-22 rain days in the west and south, 8 days when thunder is heard, hail on 2 days, fog on 5 days, and gale force winds on 5 days.

11th-15th A series of fronts cross the country in a strong southwesterly flow, bringing high humidity, lines of showers and some outbreaks of heavy rain at times. Temperatures will be much higher than normal; in some places by up to 5°C above the norm.

12th Very heavy rain likely.
11th Chance of thunderstorms.
15th Heavy rain in the west and north. Unusually warm.
16th-20th A large anticyclonic system crosses the country in an easterly track. Very sunny, warm, light winds, and mostly spell despite some brief showers along the Atlantic coasts at first. Temperatures during this period may drop to a little below the average for this time of year.
17th-18th Very sunny spell with many districts seeing between 10-12 hours of sunshine each day.
20th-22nd Unusually warm spell with the chance of fog in some parts.
21st Chance of thunderstorms.
21st-31st A series of Atlantic depressions affect Ireland bringing unsettled conditions throughout, especially over the western half of the country, with unsettled conditions continuing into the start of next month.
21st Very heavy rain in some places.
23rd Chance of heavy rain.
24th-27th Brief mostly dry and very sunny spell. Cooler than normal with the chance of isolated pockets of ground frosts.
25th Very sunny spell.
28th Chance of thunderstorms and associated hail showers.
28th-30th Cooler. Very strong winds with chance of gale gusts at times, especially in the northwest.
30th Former tropical cyclone nears the southern coasts of Ireland with wet fronts extending heavy rain across the country, with very heavy falls over the western half of the country.

IRELAND RAINFALL ESTIMATES

September 2024

d=mainly dry, n= nonrecordable, l=light shrs, s=significant shrs, r=rain, h=hvy falls

		ULSTER PROVINCE					CONNAUGHT PROVINCE						LEINSTER PROVINCE				MUNSTER PROVINCE						
		Alderg	Hillsbo	Armag	Loughl	Malin H	Sligo	Ardtari	Belmul	Drums	Galway	Roscoi	Markre	Edende	Kilkenn	Dublin	Shannc	Newpo	Tralee	Killarne	Cork	All	Moon
1st	Sun	l	l	d	l	n	n	l	s	l	l	s	l	l	l	d	s	l	s	s	l	27	
2nd	Mon	d	d	n	d	l	d	d	d	s	n	l	d	n	l	s	s	l	s	l	s	23	
3rd	Tue	l	d	s	d	d	l	d	d	l	l	l	l	l	l	d	d	l	d	d	d	23	N
4th	Wed	l	s	d	s	s	r	s	l	s	d	l	r	s	l	s	s	s	l	l	l	93	XhS
5th	Thur	l	l	l	l	l	l	l	l	l	l	l	s	n	d	s	l	n	d	d	d	28	A
6th	Fri	l	l	n	s	l	s	s	n	d	l	s	r	l	l	l	l	l	l	d	d	51	
7th	Sat	s	l	s	s	s	s	s	s	l	l	l	s	s	d	s	n	s	l	d	n	69	
8th	Sun	l	l	l	d	l	l	d	l	d	l	l	l	l	d	l	d	l	d	l	d	18	
9th	Mon	s	r	l	r	s	h	r	s	h	h	r	r	l	l	s	s	h	r	l	l	204	
10th	Tue	s	n	s	s	s	s	s	l	s	s	l	l	l	l	d	l	l	d	s	r	77	
11th	Wed	r	h	l	s	r	s	s	d	r	h	r	s	r	h	l	l	s	l	h	l	201	1Q
12th	Thu	s	s	r	s	l	l	s	r	l	s	s	l	r	n	h	r	s	h	l	h	166	V
13th	Fri	l	n	l	d	s	n	d	l	n	d	n	d	l	d	l	l	n	l	n	n	20	
14th	Sat	s	s	l	s	s	s	s	l	r	h	r	s	s	l	s	r	h	h	l	l	147	
15th	Sun	s	l	r	l	h	s	l	h	s	r	s	s	s	n	d	l	h	l	s	l	146	
16th	Mon	s	d	s	d	l	l	d	d	d	d	d	d	d	d	l	l	n	l	d	s	15	
17th	Tue	d	d	d	d	d	d	d	d	d	d	d	n	d	d	d	d	d	d	d	d	1	
18th	Wed	d	d	d	d	d	d	d	d	d	d	d	d	d	d	d	d	d	d	d	d	0	F P3 XhN
19th	Thur	d	d	d	s	d	l	s	d	d	d	d	d	d	d	d	d	n	l	d	d	5	
20th	Fri	n	l	d	l	d	l	l	l	s	s	l	l	d	d	d	d	s	s	l	d	31	
21st	Sat	s	r	d	s	l	s	s	s	s	l	s	l	s	s	r	l	l	s	r	h	114	
22nd	Sun	l	l	s	l	l	l	l	d	d	l	s	d	l	l	n	l	l	r	l	l	40	
23rd	Mon	s	r	d	l	d	l	l	l	l	d	l	d	l	l	s	s	l	r	l	r	68	
24th	Tue	s	d	r	d	d	d	d	d	d	d	d	d	d	d	d	d	d	l	n	d	24	3Q ^
25th	Wed	d	d	d	d	d	d	d	n	d	l	d	d	d	d	d	d	d	d	d	d	1	
26th	Thu	s	s	d	d	d	s	d	l	l	s	l	d	l	n	d	l	l	r	s	s	55	
27th	Fri	n	d	l	l	n	s	l	l	l	r	l	l	d	d	n	d	s	l	r	d	48	
28th	Sat	l	s	d	l	r	s	l	s	r	s	s	r	r	l	s	s	s	s	s	s	142	
29th	Sun	l	s	l	r	s	r	r	s	r	h	h	s	l	l	d	l	h	s	h	d	189	
30th	Mon	h	h	l	r	r	h	r	h	h	h	h	h	h	h	l	h	s	h	h	h	413	
	Estimate:	112	140	83	104	112	128	104	102	149	195	141	115	100	73	69	87	148	192	169	116	2437	
	Average:	76	81	68	99	96	104	106	102	79	100	105	104	78	70	60	76	132	128	87	95	1846	
	Trend:	wtr	wtr	wtr	av	wtr	wtr	av	av	wtr	wtr	wtr	wtr	wtr	av	av	wtr	wtr	wtr	wtr	wtr	wtr	

IRELAND SUNSHINE ESTIMATES

September 2024

F=fine (8-12 hours of sunshine), pc= partly cloudy (4-7 hours), c=cloudy (1-3 hours), o=overcast (0 hours)

		ULSTER PROVINCE					CONNAUGHT PROVINCE						LEINSTER PROVINCE				MUNSTER PROVINCE						
		Alderg	Hillsbo	Armag	Loughl	Malin H	Sligo	Ardtar	Belmul	Drums	Galway	Roscoi	Markre	Edende	Kilkenn	Dublin	Shannc	Newpo	Tralee	Killarne	Cork	All	Moon
1st	Sun	o	o	o	o	o	o	o	o	c	o	o	o	c	c	c	c	c	pc	pc	pc	28	
2nd	Mon	pc	pc	pc	pc	pc	pc	pc	pc	pc	F	pc	pc	pc	pc	c	F	pc	pc	pc	pc	119	
3rd	Tue	F	F	F	F	F	F	F	F	F	pc	pc	F	F	F	F	pc	F	F	F	F	185	N
4th	Wed	o	o	o	o	o	o	o	o	o	c	c	o	o	c	o	pc	c	c	c	c	25	XhS
5th	Thur	c	c	c	c	c	c	c	c	c	c	c	c	c	c	c	c	c	pc	pc	pc	57	A
6th	Fri	pc	pc	pc	pc	pc	pc	pc	pc	pc	pc	pc	pc	pc	pc	pc	c	pc	F	F	F	127	
7th	Sat	F	F	F	F	F	pc	F	pc	pc	pc	pc	F	pc	pc	F	pc	pc	pc	pc	pc	142	
8th	Sun	F	F	F	F	F	pc	F	pc	F	F	pc	F	F	F	F	F	F	pc	pc	pc	171	
9th	Mon	o	o	o	o	o	o	o	o	o	o	o	o	o	o	o	o	o	o	o	o	4	
10th	Tue	c	c	c	c	c	c	c	c	c	c	c	c	c	c	c	c	c	o	o	o	34	
11th	Wed	pc	pc	pc	pc	pc	pc	pc	pc	pc	pc	pc	pc	pc	pc	pc	F	pc	pc	pc	pc	125	1Q
12th	Thu	pc	pc	pc	pc	pc	pc	pc	pc	pc	c	pc	pc	pc	c	pc	c	c	c	c	c	80	V
13th	Fri	pc	pc	pc	pc	pc	pc	pc	pc	pc	pc	pc	pc	pc	F	pc	pc	pc	F	F	F	152	
14th	Sat	o	o	o	o	o	o	o	o	o	o	o	o	o	o	o	o	o	o	o	o	1	
15th	Sun	c	c	c	c	c	o	c	o	c	c	c	c	c	c	pc	pc	pc	o	o	o	55	
16th	Mon	pc	pc	pc	pc	pc	pc	pc	pc	pc	pc	pc	pc	pc	c	c	pc	pc	c	c	c	95	
17th	Tue	F	F	F	F	F	F	F	F	F	F	F	F	F	F	F	F	F	F	F	F	211	
18th	Wed	F	F	F	F	F	F	F	F	F	F	F	F	F	F	F	F	F	F	F	F	214	F P3 XhN
19th	Thur	pc	pc	pc	pc	pc	pc	pc	pc	pc	pc	pc	pc	pc	pc	pc	pc	pc	c	c	c	111	
20th	Fri	o	o	o	o	o	o	o	o	o	o	o	o	o	o	c	c	c	o	o	o	15	
21st	Sat	c	c	c	c	c	c	c	c	c	c	c	c	c	c	c	c	c	c	c	c	32	
22nd	Sun	c	c	c	c	c	c	c	c	c	c	c	c	c	pc	pc	pc	pc	c	c	c	73	
23rd	Mon	c	c	c	c	c	c	c	c	c	c	c	c	c	c	o	c	c	c	c	c	33	
24th	Tue	pc	pc	pc	pc	pc	pc	pc	pc	pc	F	F	pc	pc	pc	pc	F	F	pc	pc	pc	147	3Q ^
25th	Wed	F	F	F	F	F	F	F	F	F	F	F	F	F	F	F	F	F	F	F	F	205	
26th	Thu	pc	pc	pc	pc	pc	c	pc	c	pc	c	c	pc	pc	pc	F	o	c	o	o	o	85	
27th	Fri	pc	pc	pc	pc	pc	pc	pc	pc	pc	pc	pc	pc	pc	pc	pc	F	F	F	F	F	129	
28th	Sat	c	c	c	c	c	c	c	c	c	c	c	c	c	c	c	c	c	pc	pc	pc	57	
29th	Sun	pc	pc	pc	pc	pc	c	pc	c	pc	c	c	pc	pc	pc	pc	c	pc	pc	pc	pc	87	
30th	Mon	o	o	o	o	o	o	o	o	o	o	o	o	o	o	o	o	o	o	o	o	2	
Estimate hours:		142	142	142	142	142	127	142	127	144	136	131	142	144	147	157	144	146	137	137	137	2804	
Average hours:		118	118	111	111	105	99	101	110	96	111	120	111	119	115	129	111	95	109	108	124	2219	
Trend:		more	more	more	more	more	more	more	more	more	more	more	more	more	more	more	more	more	more	more	more	more	

SEPTEMBER 1 SUNDAY

A low pressure trough flows over the north of Ireland while a high pressure ridge covers the south, with a light southwest to westerly flow prevailing. Widespread cloud throughout, with some brief sunny breakthroughs mostly in the east and south. A band of rain crosses the country moving up from the southwest and leaving through the northwest, bringing light drizzle patches and showers into most districts. Chance of overnight fog patches in the northern and southern parts.

Belfast: Overcast, showers of light rain, mostly calm air, chance of misty patches
Dublin: Cloudy, mostly dry, gentle breezes, chance of misty patches
Cork: Mostly cloudy, warm, spotty showers, light breezes, chance of misty patches
Galway: Overcast, breezy spells, light showers and passing drizzle patches

RAIN POTENTIAL

FROST/SNOW

SEPTEMBER 2 MONDAY

A weak trough continues to bring patches of light rain or drizzle throughout most of the country, crossing the western coasts at first and spreading into the east as the day progresses. Light winds in a mostly northwesterly flow prevails. Some sunny spells develop behind the showers, particularly in the west and north. Chance of overnight patches of fog in some southern parts.

Belfast: Partly cloudy, dry, mild, occasional breezy spells
Dublin: Cloudy, warmer, brief showers overnight, gentle breezes
Cork: Cooler, partly cloudy, light showers, light breezes, chance of fog patches
Galway: Sunny, isolated showers overnight, fresh breezes at times

RAIN POTENTIAL

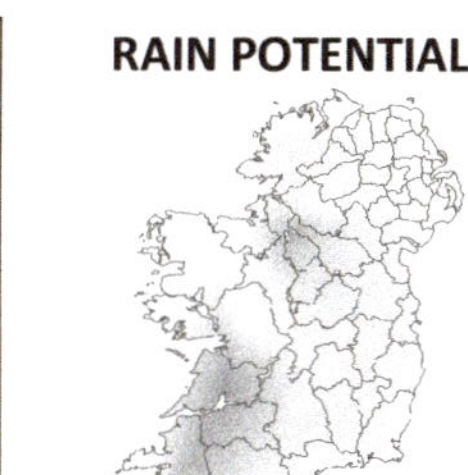

FROST/SNOW

SEPTEMBER 3 TUESDAY

Low pressure trough continues to prevail over the country directing a light, west to southwest flow. A changeable day with good sunny spells throughout and mostly dry along the western coasts and southern parts. Light showers likely elsewhere, and the chance of thunderstorms in the northwest of the country.

Belfast: Sunny, dry, warm day cool night, gentle shifting breezes
Dublin: Sunny, dry, moderate breezes, cool overnight
Cork: Sunny, dry, light shifting breezes
Galway: Mostly sunny, fluctuating breezes, some light showers overnight

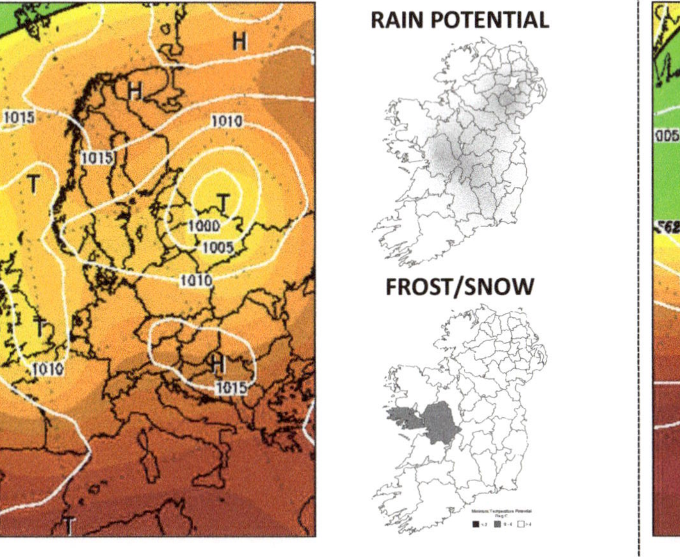

RAIN POTENTIAL

FROST/SNOW

SEPTEMBER 4 WEDNESDAY

The low pressure trough maintains a light to moderate flow over the country, southeasterly for a time but backing northwesterly later in the day. Mostly cloudy in the west and south, and dull and gloomy in the north and east. Occasional light showers throughout, heavier in the northwest and more isolated in the far south. Chance of misty patches or fog briefly in some northern and southern districts.

Belfast: Overcast, light shifting breezes, some showers and outbreaks of drizzle patches
Dublin: Overcast, cooler, fluctuating breezes, light showers
Cork: Cloudy, warmer, light showers, light shifting breezes, chance of fog patches
Galway: Cloudy, dry, mild, light breezes

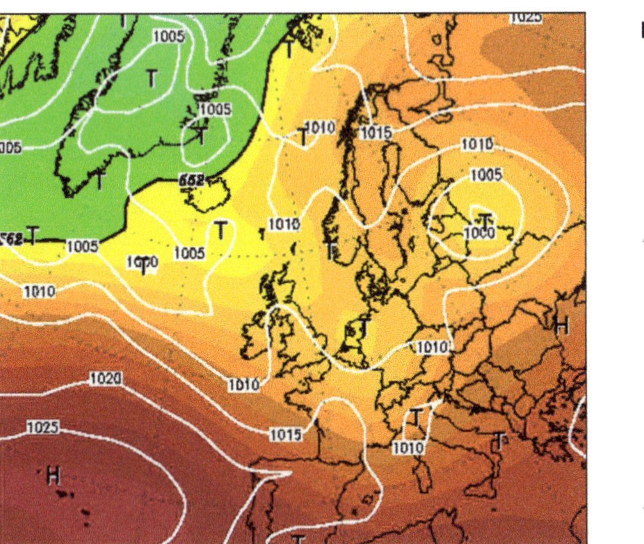

SEPTEMBER 5 THURSDAY

Low pressure continues to bring light to moderate breezes throughout the country, generally in a westerly flow. A changeable day with a mix of brief sunny spells and occasional showers, particularly in the afternoon and evening. Mostly dry around the southern and eastern coastal fringes.

Belfast: Cloudy, cooler, spotty showers, light breezes
Dublin: Cloudy, spotty showers, slightly warmer, light breezes
Cork: Dry, mild, some sunny spells, gentle breezes
Galway: Cloudy, isolated showers, moderate breezes

RAIN POTENTIAL

FROST/SNOW

SEPTEMBER 6 FRIDAY

Low pressure trough maintains a light to moderate westerly flow across the country. Another changeable day with a mix of sunny spells and light showers, with showers sometimes accompanied by thunderstorm activity, particularly in northern parts. Mostly sunny in the far south, with some precipitation elsewhere.

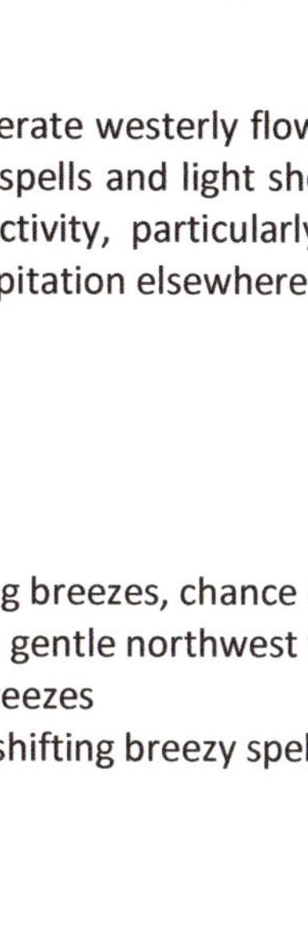

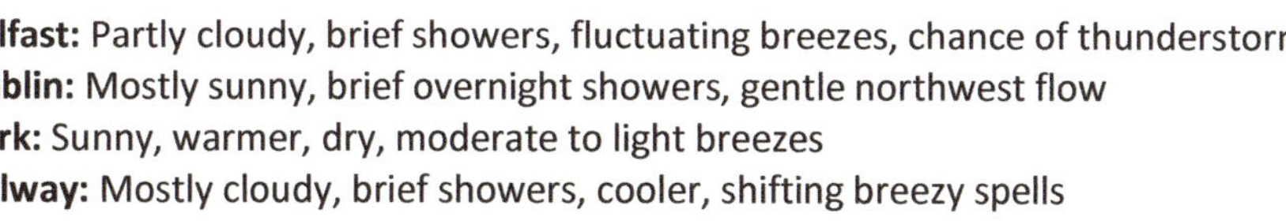

Belfast: Partly cloudy, brief showers, fluctuating breezes, chance of thunderstorms
Dublin: Mostly sunny, brief overnight showers, gentle northwest flow
Cork: Sunny, warmer, dry, moderate to light breezes
Galway: Mostly cloudy, brief showers, cooler, shifting breezy spells

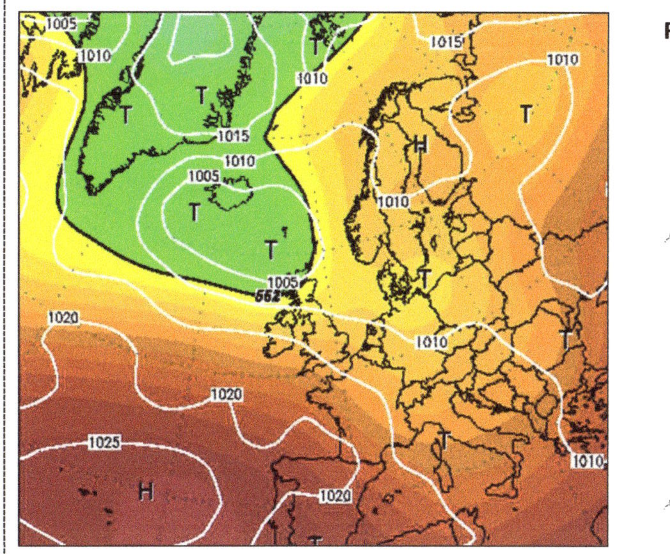

RAIN POTENTIAL

FROST/SNOW

SEPTEMBER **7** SATURDAY

A high pressure ridge slowly nears the southern coasts of Ireland while a low pressure trough continues to linger over the country bringing light to moderate west to northwest breezes, freshening at times to blustery. Scattered showers, some sunny spells, and the occasional thunderstorm prevail everywhere, particularly in the north, although mostly dry along the far southern coasts.

Belfast: Sunny, overnight showers, blustery at times, chance of thunderstorms
Dublin: Sunny, cooler, light showers, fluctuating windy spells
Cork: Partly cloudy, isolated showers, cooler, breezy spells
Galway: Partly cloudy, cooler, blustery at times, drizzly showers

RAIN POTENTIAL

FROST/SNOW

SEPTEMBER **8** SUNDAY

Low pressure continues to dominate over the country directing a moderate northwesterly flow. Mostly sunny throughout, despite a weak front bringing some scattered thunderstorms and brief showers mostly over the northern half of the country. Dry in the south.

Belfast: Sunny, mild day cool night, isolated overnight showers, windy, occasional gusts
Dublin: Sunny, light overnight showers, windy
Cork: Partly cloudy, mostly dry, increasing threats of rain, fluctuating breezes
Galway: Sunny, mild day cool night, light overnight showers, moderate breezes

RAIN POTENTIAL

FROST/SNOW

SEPTEMBER 9 MONDAY

While high pressure ridges to the south of Ireland, an intense depression to the west of the country directs a moderate southeasterly flow, changing to southwesterly later in the day. Overnight patches of fog likely in the north and west, with cloud thickening and becoming widespread throughout. Rain spreads from the west at first and moves eastward across the country, with heavy rain in many parts of the west, northwest, and far northeast, and lighter showers elsewhere.

Belfast: Unsettled windy spells, dull and gloomy air, rain
Dublin: Overcast, occasional showers, mild day cool night, windy at times
Cork: Overcast, light showers, breezy with occasional gusts
Galway: Overcast, heavy rain, cooler, unsettled and shifting winds

RAIN POTENTIAL

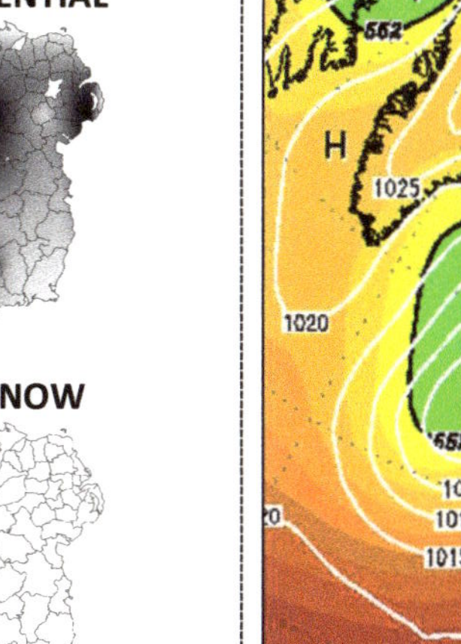

FROST/SNOW

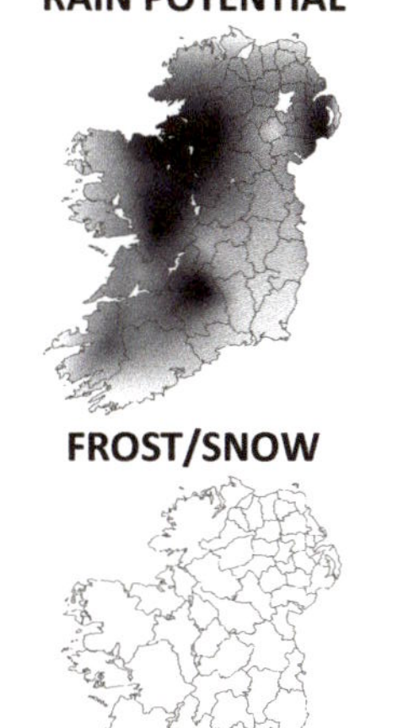

SEPTEMBER 10 TUESDAY

A large depression to the west of Ireland brings a warmer southwest flow, with an unsettled and stormy air across the country with strengthening southwest winds, and the chance of winds reaching gale force for a time around the southern coastal fringes. Widespread cloud and scattered outbreaks of rain throughout, although mostly dry with occasional sunny spells breaking through in the central eastern and western coastal fringes.

Belfast: Cloudy, blustery with chance of gale gusts, isolated showers
Dublin: Warmer, cloudy, mostly dry and humid air, freshening winds
Cork: Overcast, rain, stormy air, strong winds with possible gale gusts
Galway: Cloudy, milder, scattered showers, fresh winds with occasional gusts

RAIN POTENTIAL

FROST/SNOW

SEPTEMBER 11 WEDNESDAY

An unsettled and stormy air prevails over Ireland as a depression to the northwest of the country continues to direct and fresh southwesterly flow, backing southerly and decreasing later in the day. A wet front crosses the north of the country at first and proceeds to move down to the south by the end of the day. Widespread and frequent heavy rain in the north, through the midlands and eastern districts to occasional showers in the south. Mostly dry in the central western coastal fringes.

Belfast: Stormy air, changeable, warmer, heavy rain, some sunny breakthroughs
Dublin: Warm, scattered cloudy spells, brief showers, blustery at times
Cork: Light showers, occasional sunny spells, blustery
Galway: Heavy overnight rain, some sunny spells, strong and blustery winds at times

RAIN POTENTIAL

FROST/SNOW

SEPTEMBER 12 THURSDAY

A stormy air continues over Ireland as the depression to the northwest expands, with mostly south to southwest winds prevailing, strong and blustery and up to gale force at times. The passage of another wet front spreads rain from the southwest at first, followed by a short dry spell with some sunny breakthroughs, and with rain returning later in the day and becoming widespread. Rain likely to be very heavy at times in the south of the country as well as around the eastern coastal districts.

Belfast: Cooler, mix of showers and sunny spells, strong winds with gusts at times
Dublin: Unsettled and stormy air, heavy rain, some sunny spells, windy with gusts
Cork: Warmer, cloudy, heavy rain, strong gusty winds at times
Galway: Cloudy, showers, strong gusty winds

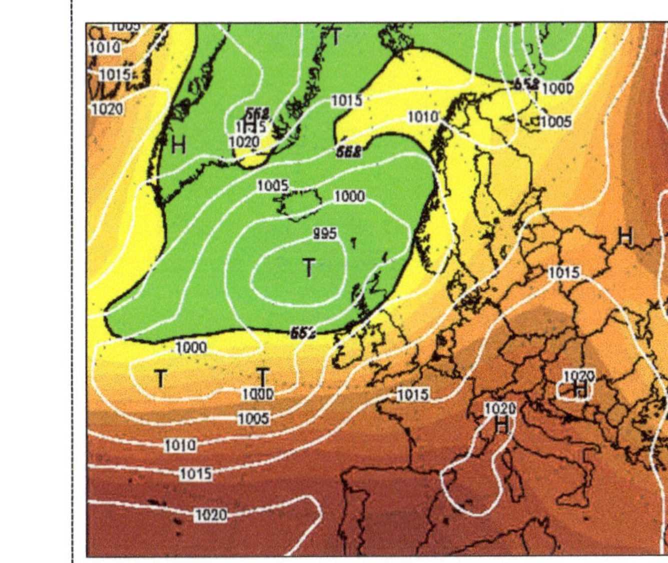

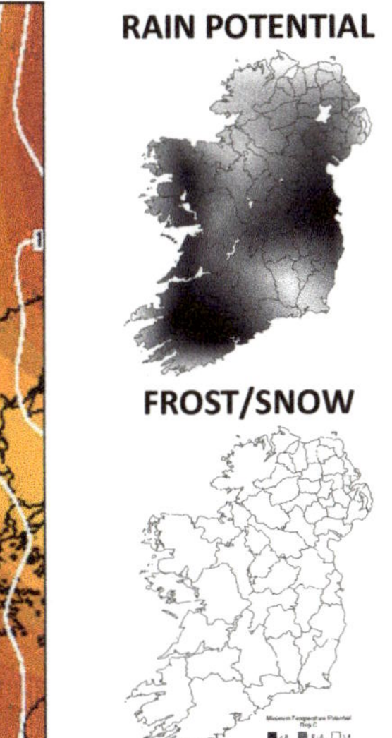

SEPTEMBER 13 FRIDAY

Low pressure continues to cover the country in a strong southwest to westerly flow at first, easing to a moderate to fresh west to northwest flow as the frontal trough weakens as it clears southeastward. Widespread drizzle patches and some rainy spells overnight are likely to clear by noon, bringing a mostly dry, sunny, and windy afternoon.

Belfast: Unsettled air, mostly sunny, isolated showers, blustery with occasional gusts
Dublin: Mostly sunny, brief showers, warm, strong shifting winds at times
Cork: Unsettled and blustery air, isolated showers, warm and sunny day
Galway: Partly cloudy, dry, gusty winds, possible gales at times

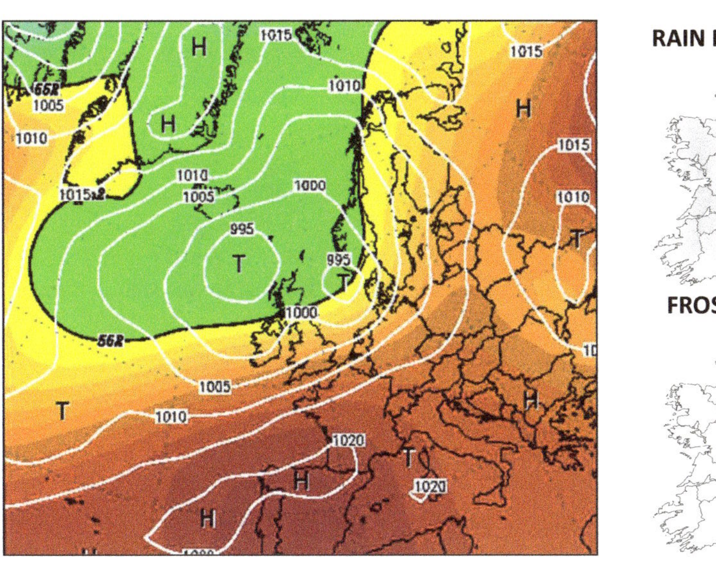

RAIN POTENTIAL

FROST/SNOW

SEPTEMBER 14 SATURDAY

The depression to the north of Ireland moves away to the east, while a weak high pressure ridge nears the southern coasts, with a lighter to moderate westerly flow prevailing. Despite being mostly dry, dull and gloomy everywhere at first, widespread showers will develop in most districts and heavy rain at times in the southwest. Chance of some mist or fog patches for a short time in many inland districts.

Belfast: Overcast, windy, scattered showers
Dublin: Overcast, cooler, windy spells, light showers
Cork: Overcast, cooler, patches of light drizzy showers, light breezes, chance of fog
Galway: Overcast and stormy air, heavy rain, blustery at times, chance of gales

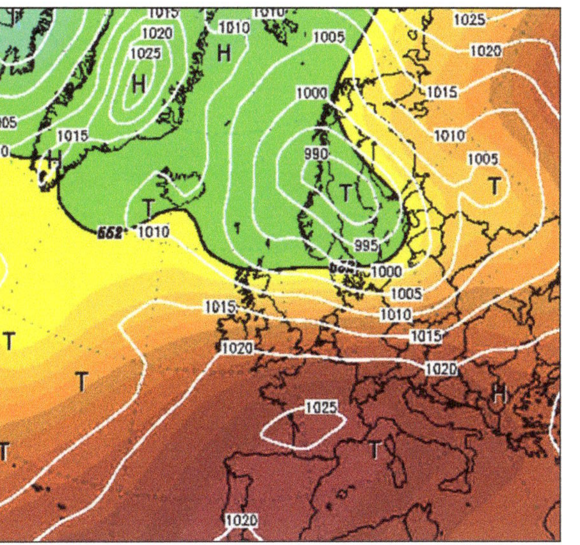

RAIN POTENTIAL

FROST/SNOW

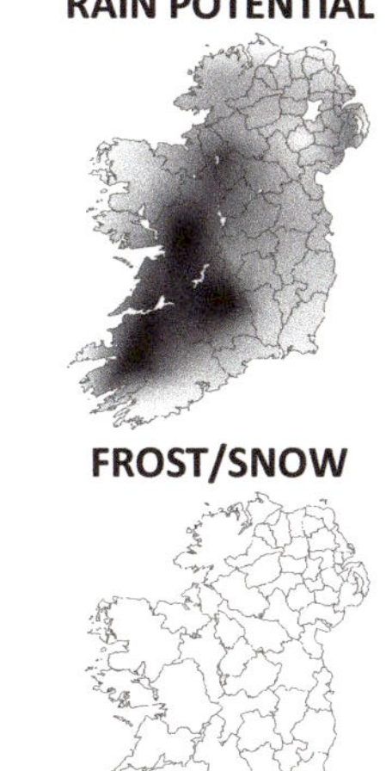

SEPTEMBER **15** SUNDAY

High pressure ridges to the east of Ireland while a low pressure trough sits to the west, with a fresh southwest backing southeast flow prevailing. Mostly dry with some sunny spells in the central eastern counties, while mostly cloudy and dry along the southern coasts. The passage of a wet front brings widespread cloud and rain elsewhere, with heavy rain in some parts in the northwest and west of the country. Chance of fog turning to drizzle in parts of the north and south.

Belfast: Mostly cloudy, moderate winds, passing showers
Dublin: Partly cloudy, dry, warmer, moderate breezes
Cork: Overcast, isolated showers, light shifting breezes
Galway: Cloudy, windy, rain

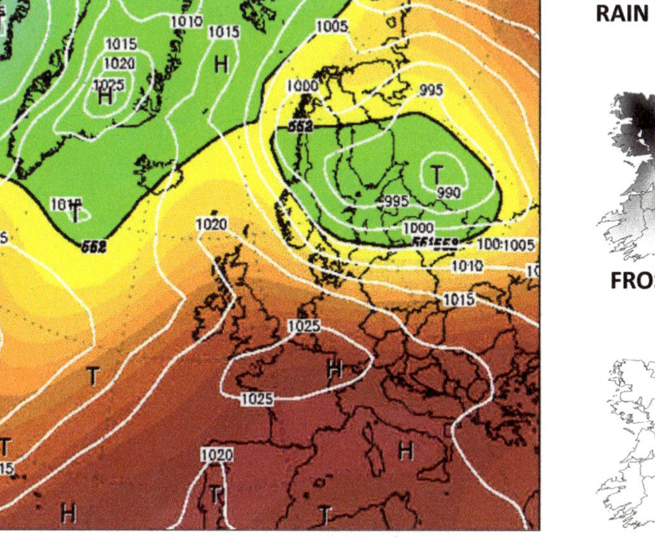

RAIN POTENTIAL

FROST/SNOW

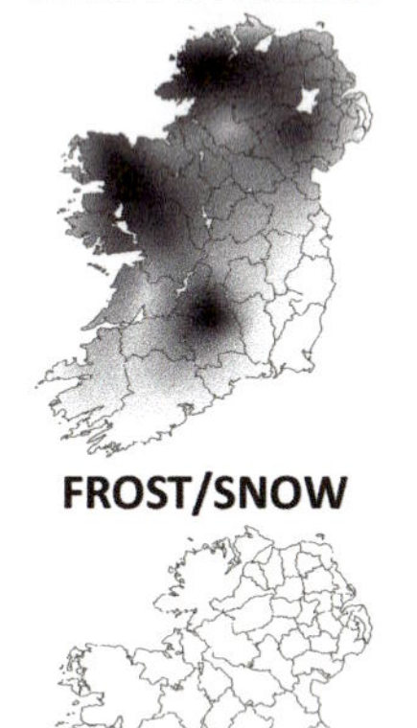

SEPTEMBER **16** MONDAY

A low pressure trough prevails over Ireland while a high pressure ridge nears the western coasts, with a light to moderate variable to mostly westerly flow prevailing. Overnight rain clears to isolated showers for a time, clearing to a mostly dry day with increasing sunny spells in most districts. A few light showers likely in some northern parts, along the eastern coasts and into the southeast. Chance of overnight patches of fog in southern parts.

Belfast: Warmer, dry, occasional sunny spells, fluctuating breezes
Dublin: Cloudy, light showers, cooler, shifting breezy spells
Cork: Changeable, mix of showers and occasional sunny spells, light shifting breezes
Galway: Partly cloudy, dry, shifting breezy spells

RAIN POTENTIAL

FROST/SNOW

SEPTEMBER **17** TUESDAY

A large anticyclonic system moves onto the country directing light to moderate northerly breezes throughout. Mostly dry and sunny almost everywhere, except for the chance of some isolated showers in the north, and possible patches of fog in the west.

Belfast: Sunny, dry, cooler, breezy
Dublin: Mild day cool night, sunny, dry, moderate breezes
Cork: Sunny, breezy spells, mostly dry and pleasant air
Galway: Sunny, dry, moderate breezes at times, chance of overnight fog patches

RAIN POTENTIAL

FROST/SNOW

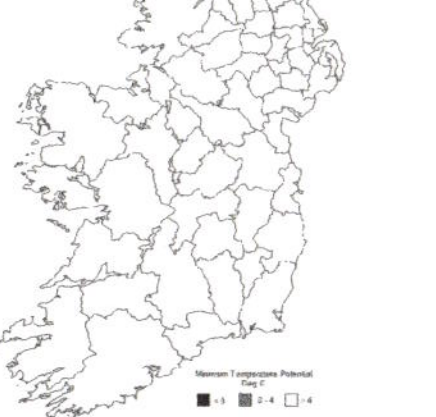

SEPTEMBER **18** WEDNESDAY

A large high crosses the country bringing light northeasterly breezes at first, shifting to southeasterly later. Dry and sunny everywhere, with cooler overnight temperatures that may see some patches of frost in inland parts in the north. Overnight fog likely in some norther, western, and southern districts.

Belfast: Sunny, dry, fluctuating breezes, chance of overnight fog patches
Dublin: Sunny, dry, mild day cool night, pleasant air with moderate breezes
Cork: Sunny, dry, gentle breezes, chance of overnight foggy patches
Galway: Dry, sunny, warm day cold night, mostly calm air, fog possible

RAIN POTENTIAL

FROST/SNOW

SEPTEMBER **19** THURSDAY

The large high lingering over Ireland continues to drift eastward, with light to moderate southeasterly breezes prevailing. Some overnight fog and misty patches likely in many districts, clearing to a warmer and generally sunny day throughout. Mostly dry everywhere, except for a few drizzly spells in the southwest.

Belfast: Partly cloudy, dry, warmer, shifting breezy spells
Dublin: Partly cloudy, dry, light breezes, chance of shallow fog patches
Cork: Mostly cloudy, dry, mild, moderate seas breezes, chance of overnight fog patches
Galway: Partly cloudy, warm and dry, cool overnight, gentle breezes

RAIN POTENTIAL

FROST/SNOW

SEPTEMBER **20** FRIDAY

The high pressure ridge moves away to the east of Ireland while a low pressure trough ahead of a weak depression to the west of the country moves into the west. Winds generally light in a southerly flow, fresher at times along the western coasts. A changeable day with widespread cloud everywhere and showers on and off over the western half of the country, while remaining mostly dry along the eastern and southern coastal districts.

Belfast: Overcast, drizzly showers, unsettled winds
Dublin: Mostly dry and cloudy, warm breezy spells
Cork: Overcast, mostly dry with passing threats of rain, moderate southerly flow
Galway: Overcast, light passing showers, warm southerly breezes

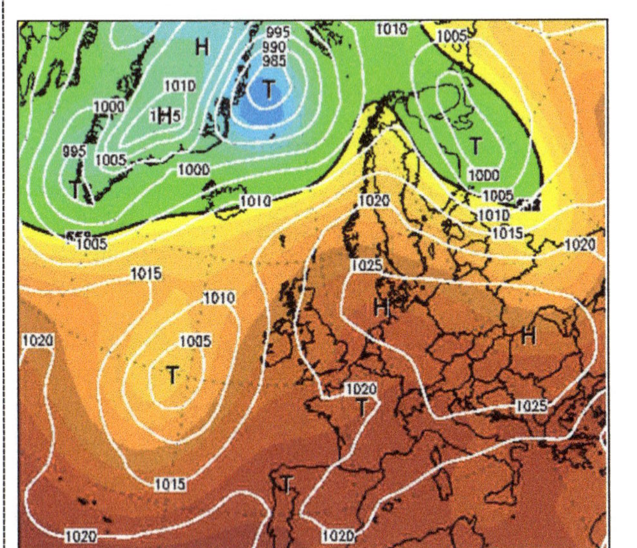

SEPTEMBER **21** SATURDAY

Low pressure and a moderate to freshening southerly flow prevails. A series of cold fronts cross the country bringing outbreaks of rain and drizzle in the west and south at first, spreading into the east and north as the day progresses, with some periods of rain heavy at times particularly around the southeast and northeast coastal areas. Some brief sunny outbreaks likely between the showers.

Belfast: Cloudy, warmer, rain, blustery for a time
Dublin: Cloudy, rain, windy spells
Cork: Cloudy, rain, blustery at times
Galway: Warmer, cloudy, moderate breezes, light showers

RAIN POTENTIAL

FROST/SNOW

SEPTEMBER **22** SUNDAY

A cluster of low pressure systems to the northwest of Ireland gather, while high pressure ridges to the far southwest, with a low pressure trough covering the country and directing a mostly calm and serene air in a southerly flow over the country. Mist and fog patches in many districts overnight, later turning to outbreaks of spotty showers in most districts, then clearing by later in the day with increasing sunny spells developing before cloud gathers and rain threatens again.

Belfast: Cooler, mostly cloudy, brief showers, gentle breezes, chance of misty patches
Dublin: Partly cloudy, warmer, isolated showers, mostly calm air, chance of fog
Cork: Mostly cloudy, brief spotty showers, gentle breezes, odd misty patches likely
Galway: Mostly cloudy, cooler, occasional showers, serene air, chance of misty patches

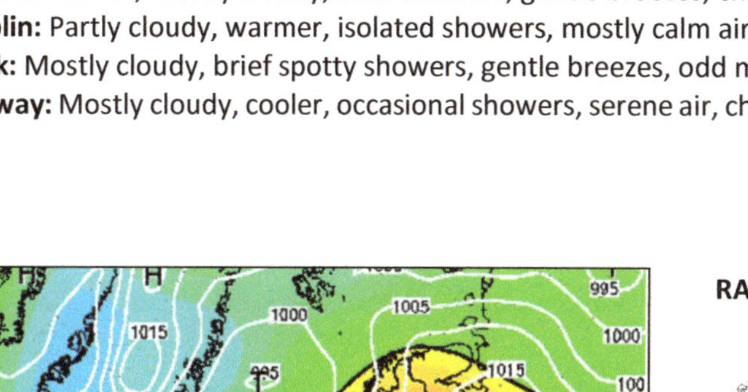

RAIN POTENTIAL

FROST/SNOW

SEPTEMBER **23** MONDAY

A deep depression moves closer to the west of Ireland bringing fresh to strong south to southwest winds throughout, possibly reaching gale force for a time particularly along the western coasts. Rain crosses into the west of the country overnight with stormy conditions spreading everywhere during the day. Heavy rain likely at times in the south and northeast while precipitation is more scattered and lighter elsewhere. Some sunny spells later in the day.

Belfast: Cloudy, warmer, wet and windy, chance of gales for a time
Dublin: Overcast, cooler, strengthening winds, passing showers
Cork: Cloudy, rain, stormy air, chance of gale winds
Galway: Cloudy, cooler, dry, strong gusty winds

RAIN POTENTIAL

FROST/SNOW

SEPTEMBER **24** TUESDAY

A depression to the northwest of Ireland contracts northwards allowing a high pressure ridge to approach from the southeast of the country, with winds easing to light, mostly in a west to southwest flow. Rain clears the northeast of the country overnight bringing a mostly dry, sunny day throughout, with some fog patches likely mostly in the south.

Belfast: Cooler, mostly sunny, dry, fluctuating winds
Dublin: Partly cloudy, dry, mild day cool night, windy spells
Cork: Mostly sunny and dry, winds easing to light, chance of shallow overnight fog
Galway: Sunny, dry, mild day cold night, moderate breezes at times

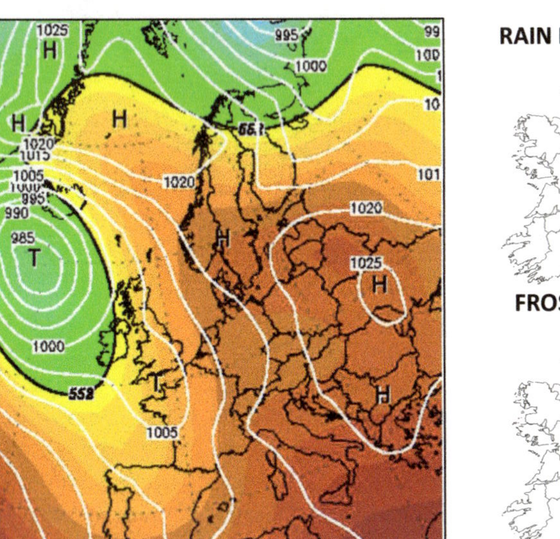

RAIN POTENTIAL

FROST/SNOW

SEPTEMBER **25** WEDNESDAY

A high pressure ridge continues to slowly advance towards Ireland directing a light southerly flow throughout, with a fresher breeze developing later along the northern coasts. Mostly dry, sunny, and mild day everywhere, with chance of fog patches mostly in northern parts. Chance of patches of mild frost in some inland districts.

Belfast: Sunny, dry, shifting breezy spells, chance of overnight fog patches
Dublin: Sunny, dry, mild day cool night, light shifting breezes
Cork: Dry, sunny, mild day cool night, gentle breezes
Galway: Sunny, mild day cold night, isolated showers, light breezes

RAIN POTENTIAL

FROST/SNOW

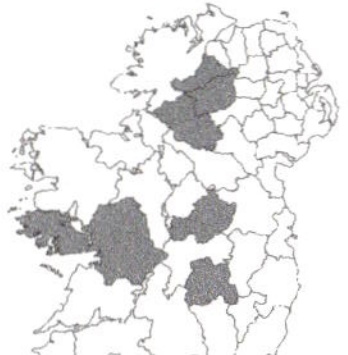

SEPTEMBER **26** THURSDAY

A large depression moves back into place to the west of Ireland returning fresh to strong southerly winds everywhere, very strong and blustery at times over the western half of the country. Mostly dry overnight before a rain band crosses the southwestern coasts bringing drizzle through the south at first, spreading later into the west and up into the north, with some heavier falls at times mostly in the southwest. Meanwhile, it is likely to remain a mostly dry and sunny day in the east.

Belfast: Changeable, mix of showers and sunny spells, blustery, chance of gale winds
Dublin: Mostly dry and sunny, mild day cold night, strong winds at times
Cork: Overcast, fresh onshore breezes, showers
Galway: Unsettled and breezy at times, cloudy, milder, scattered showers

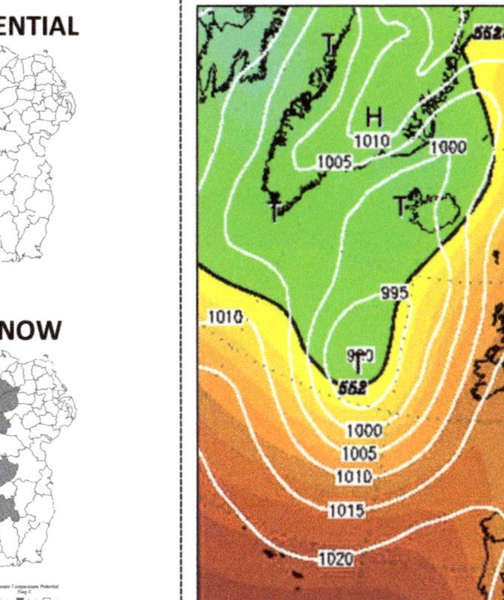

RAIN POTENTIAL

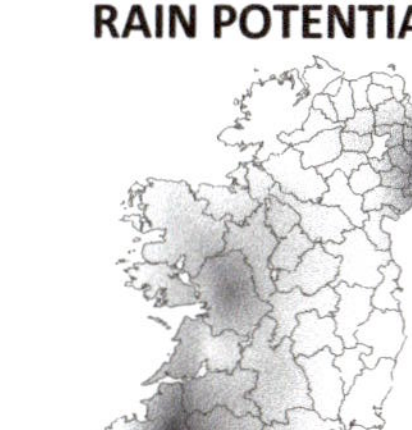

FROST/SNOW

SEPTEMBER **27** FRIDAY

Low pressure prevails over the country in a light to moderate westerly flow before backing southerly later in the day. Unsettled and changeable throughout. Mostly dry and sunny in the east, with a mix of sunny spells and showers elsewhere. Chance of fog patches in some southern districts overnight.

Belfast: Partly cloudy, warmer, dry, breezy
Dublin: Unsettled and changeable air, partly cloudy, isolated showers
Cork: Sunny, warmer, dry, shifting breezes, chance of overnight shallow fog patches
Galway: Rain, moderate winds, good sunny spells

RAIN POTENTIAL

FROST/SNOW

SEPTEMBER **28** SATURDAY

Low pressure prevails throughout the country with moderate to fresh southwest winds prevailing, but more blustery at times along the western coasts. A changeable day, with a mix of sunny outbreaks and scattered showers everywhere. Some outbreaks of heavy rain accompanying thunderstorm activity likely, moving across the country from the west at first and into the east later in the day.

Belfast: Cloudy, scattered showers, cooler, strengthening winds later in the day
Dublin: Cloudy, cooler, scattered showers, moderate breezes strengthening later
Cork: Changeable, breezy at times, rain, some sunny spells
Galway: Cloudy, rain, cooler, fresh breezes with occasional gusts

RAIN POTENTIAL

FROST/SNOW

SEPTEMBER 29 SUNDAY

A stormy air prevails as a large depression to the northwest of Ireland brings strong and blustery west to southwest winds, with winds possibly reaching gale force at times, particularly in the north and west. Mostly dry with occasional sunny spells in the east, while a frontal trough associated with the depression to the west brings widespread cloud and heavy rain elsewhere, with very heavy downpours likely in the west. Chance of thunderstorms.

Belfast: Stormy air, strong gusty winds, cooler, mix of showers and sunny outbreaks
Dublin: Partly cloudy and dry, strong gusty winds, chance of gales
Cork: Partly cloudy, mostly dry with increasing threats of rain, strong gusty winds
Galway: Stormy air, cloudy, heavy rain

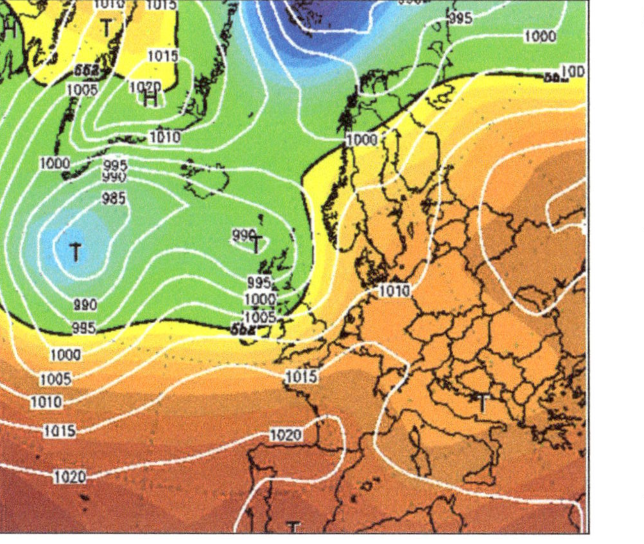

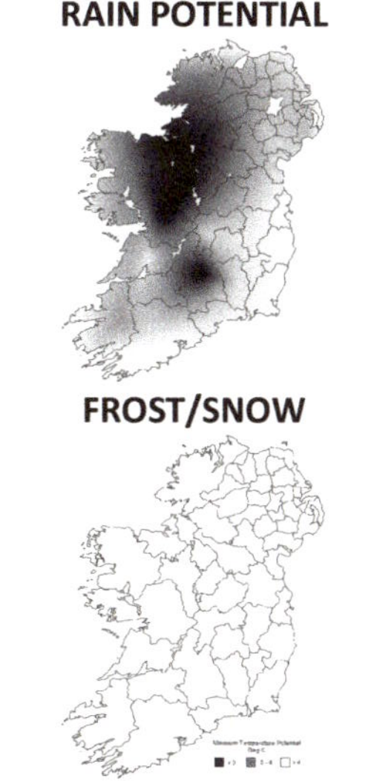

SEPTEMBER 30 MONDAY

The depression to the west of Ireland intensifies into a deep storm off the western coasts. Very strong south to southwest winds prevail, with gale force gusts likely. Despite a mostly dry start to the day in the east, storm conditions spread throughout the country with persistent drizzle and showers in the east, and heavy rain elsewhere, sometimes accompanied by thunderstorm activity, particularly in the west and south.

Belfast: Stormy, overcast, heavy rain, strong forceful winds and gales likely
Dublin: Overcast, warmer, light squally showers, blustery winds to gale force at times
Cork: Stormy, overcast, heavy rain, chance of gale winds and scattered thunderstorms
Galway: Stormy, overcast, cooler, heavy rain, chance of thunderstorms and gale winds

OCTOBER

PHASES OF THE MOON

2nd	New Moon
2nd	Apogee
2nd	Crossing Equator
9th	Southern Declination
10th	First Quarter
17th	Full Moon
17th	Perigee #2
17th	Crossing Equator
22nd	Northern Declination
24th	Third Quarter
29th	Apogee
30th	Crossing Equator

1	2	3	4	5	6	7	8	9	10	11	12	13	14	15	16	17	18	19	20	21	22	23	24	25	26	27	28	29	30	31
Tue	Wed	Thur	Fri	Sat	Sun	Mon	Tue	Wed	Thur	Fri	Sat	Sun	Mon	Tue	Wed	Thur	Fri	Sat	Sun	Mon	Tue	Wed	Thur	Fri	Sat	Sun	Mon	Tue	Wed	Thur
	N A XhS							V	1Q							F P2 XhN					^		3Q					A	XhS	

MONTHLY SUMMARY

1st-3rd A decaying hurricane may cross the country bringing stormy conditions with bands of heavy rain into the western half of the country, with particularly heavy rain into parts of the west and southwest. Gusty south to southwest winds prevail and up to gale force at times, especially in the northwest, which may cause localised damage in a number of districts as well as the chance of damage to the power grid. Heavy rain may bring some localised flooding during this period. Generally mild temperatures everywhere.
1st-10th A series of Atlantic depressions affect Ireland bringing unsettled conditions throughout, especially over the western half of the country.
1st Very heavy rain in the northwest of the country and in some parts of the Munster Province. Chance of localised flooding in the Belmullet, Shannon and Killarney district.
2nd-4th Thunderstorms and hail showers possible, especially in the southeast and northwest. Strong and unsettled southeasterly winds may develop into funnel formations in the northwest of Ireland.
2nd-8th Unusually milder maximum temperatures than is typical for this time of year.
4th-6th Fog patches possible.
6th-8th Very dull and gloomy spell.
9th-10th Outbreaks of heavy rain may be accompanied by thunderstorm activity.

Over the next four weeks, generally it will be very mild and sunny, with most rainfall seen in the south and east, with south to southwest winds predominating. Low pressure prevails at the start of this outlook period, and again over the third week, with high pressure systems being the dominant weather features in the second and fourth weeks. Rainfall will be close to normal in the northwest, while up to 50% wetter in some districts in the east and south. Except for a mainly dry spell over the second week, some rain activity can be anticipated on most other days during this outlook period. Sun hours will be above normal overall by the end of this outlook period, especially in the north and northwest, with the second week being very sunny, and despite the third and fourth weeks being relatively dull. Temperatures will be up to 1-2°C warmer than is normal for this time of year, mostly due to a south to southwest flow over much of this outlook period, with temperatures being fairly consistently high throughout the country. Only a few frost days are expected, and these only in inland districts. Winds will be close to normal for this time of year with the strongest gusts expected in the last couple of days. During this outlook period there is likely to be around 18-23 rain days (mostly light) in the west and northern districts, and around 12-17 days elsewhere.

Thunderstorms may be heard on about 8 days, fogs on at least 10-12 days, gales on 1 day, and light frost pockets only in inland parts on 3 days, but no hail or snow expected.

11th-21st Low pressure dominates as depressions brush the west of Ireland on a regular basis during this period. Rain or showers likely in most districts almost daily. Warmer than normal with light winds at first, but reaching gale force towards for a spell in the middle of this period.
13th Very sunny in the southeast of the country.
15th Widespread heavy rain.
16th Strong winds across the country, possibly reaching gale force for a time.
19th-24th Very sunny spell.
20th Cooler spell. Chance of isolated pockets of frosts in inland districts.
20th-21st Widespread heavy rain.
22nd-23rd Unusually mild temperatures.
22nd-26th A large anticyclonic system ridges to the east of Ireland bringing a mostly dry spell under a light to moderate southerly flow, backing southeasterly later in the period. Mostly sunny throughout until the last couple of days when cloud is likely to increase resulting in widespread patches of mist or fog. Cooler than of recent, but still likely to be warmer than is typical for this time of year.
24th Widespread mist or fog patches likely.
25th-30th Very dull and gloomy spell with widespread overcast skies.
27th Widespread heavy rain.
31st Low pressure dominates the country.

IRELAND RAINFALL ESTIMATES
October 2024

d=mainly dry, n= nonrecordable, l=light shrs, s=significant shrs, r=rain, h=hvy falls

		ULSTER PROVINCE					CONNAUGHT PROVINCE						LEINSTER PROVINCE				MUNSTER PROVINCE						
		Alderg	Hillsbo	Armag	Loughl	Malin H	Sligo	Ardtar	Belmul	Drums	Galway	Rosco	Markre	Edende	Kilken	Dublin	Shann	Newpo	Tralee	Killarne	Cork	All	Moon
1st	Tue	r	s	h	h	r	h	h	h	r	h	s	h	l	r	l	h	h	h	h	h	411	
2nd	Wed	l	s	d	s	d	s	s	s	s	l	s	s	r	h	l	d	s	l	s	s	95	N XhS A
3rd	Thur	s	r	d	s	s	r	s	n	h	h	h	r	r	l	s	s	h	l	s	r	180	
4th	Fri	l	d	l	l	s	l	l	r	s	l	n	l	l	d	s	l	s	s	d	n	62	
5th	Sat	d	d	n	s	l	s	s	d	d	l	d	d	d	d	n	d	l	d	d	d	13	
6th	Sun	l	l	d	s	s	s	s	r	s	s	l	s	n	l	n	l	r	l	h	d	103	
7th	Mon	s	s	l	l	l	s	l	r	r	s	s	d	l	s	l	s	s	h	s	r	128	
8th	Tue	s	n	l	d	l	l	d	s	l	l	s	s	l	l	d	l	s	l	s	s	34	
9th	Wed	d	n	d	s	d	l	s	s	l	s	l	s	n	s	s	l	s	s	h	l	79	V
10th	Thu	r	h	d	l	s	l	l	s	s	s	r	s	h	h	r	h	l	h	r	h	194	1Q
11th	Fri	r	l	h	r	l	s	r	s	l	l	l	s	l	s	s	l	l	l	n	r	128	
12th	Sat	l	l	l	l	s	s	l	s	l	d	l	s	l	d	d	n	l	l	l	n	25	
13th	Sun	n	n	n	d	d	d	d	n	d	n	d	d	d	d	d	d	d	l	l	d	5	
14th	Mon	s	r	d	h	l	r	h	l	s	h	r	r	s	s	d	l	h	l	r	l	187	
15th	Tue	s	l	s	h	r	r	h	h	r	r	s	s	s	s	r	h	h	h	r	r	238	XhN
16th	Wed	l	l	s	s	r	s	s	s	l	d	s	s	l	n	s	l	s	s	s	s	84	
17th	Thur	l	d	l	l	s	s	l	l	d	d	d	s	d	n	d	d	d	d	l	d	19	F P2
18th	Fri	s	r	d	l	d	l	l	s	l	l	n	l	s	r	r	l	s	s	l	h	100	
19th	Sat	l	d	l	d	l	d	d	l	d	d	d	d	d	d	l	d	d	d	d	l	8	
20th	Sun	s	r	d	r	l	r	r	l	r	r	s	r	r	h	s	s	r	r	h	h	217	
21st	Mon	h	l	h	l	h	s	l	h	s	s	s	l	d	l	r	s	l	s	r	l	160	^
22nd	Tue	l	d	l	d	d	d	d	n	l	d	d	d	d	n	d	n	d	d	n	s	5	
23rd	Wed	d	d	d	d	d	d	d	d	d	d	d	d	d	d	d	d	d	d	d	d	0	
24th	Thu	d	d	d	d	d	d	d	d	d	d	d	d	d	d	d	d	d	d	d	d	0	3Q
25th	Fri	d	d	d	d	d	d	d	d	d	d	d	d	d	d	d	d	d	d	d	d	0	
26th	Sat	d	d	d	l	d	d	l	d	l	d	d	d	d	d	d	d	d	d	l	d	6	
27th	Sun	s	r	d	s	s	r	s	r	r	r	h	h	r	h	s	r	r	s	s	r	201	
28th	Mon	l	l	s	d	d	n	d	l	d	l	n	d	s	l	d	d	d	h	h	s	57	
29th	Tue	s	r	n	d	l	l	d	d	s	s	l	l	s	r	s	l	d	s	r	r	78	A XhS
30th	Wed	s	l	s	s	d	s	s	s	l	d	l	s	l	l	d	n	l	l	l	l	52	XhS
31st	Thur	s	s	l	l	l	l	l	l	n	l	d	s	s	r	l	l	r	r	s	h	105	
	Estimate:	121	114	125	126	103	144	126	200	123	152	119	131	119	174	108	135	183	195	256	219	2972	
	Average:	90	100	86	129	121	134	136	146	112	129	140	134	104	93	79	105	176	183	133	138	2467	
	Trend:	wtr	wtr	wtr	av	drr	av	av	wtr	wtr	wtr	drr	av	wtr	wtr	wtr	wtr	av	wtr	wtr	wtr	wtr	

IRELAND SUNSHINE ESTIMATES
October 2024\

F=fine (8-12 hours of sunshine), pc= partly cloudy (4-7 hours), c=cloudy (1-3 hours), o=overcast (0 hours)

		ULSTER PROVINCE					CONNAUGHT PROVINCE						LEINSTER PROVINCE				MUNSTER PROVINCE						
		Alderg	Hillsbo	Armag	Loughl	Malin I	Sligo	Ardtari	Belmul	Drums	Galway	Roscoi	Markre	Edende	Kilkeni	Dublin	Shannc	Newpo	Tralee	Killarne	Cork	All	Moon
1st	Tue	o	o	o	o	o	o	o	o	o	o	o	o	o	o	o	o	o	o	o	o	2	
2nd	Wed	pc	pc	pc	pc	pc	F	pc	F	pc	F	F	pc	pc	pc	c	F	pc	F	F	F	133	N XhS A
3rd	Thur	c	c	c	c	c	c	c	c	c	c	c	c	c	c	c	c	c	c	c	c	32	
4th	Fri	c	c	c	c	c	c	c	c	pc	pc	pc	c	pc	pc	pc	pc	pc	pc	pc	pc	85	
5th	Sat	pc	pc	pc	pc	pc	pc	pc	pc	pc	pc	pc	pc	pc	pc	c	F	pc	F	F	F	114	
6th	Sun	c	c	c	c	c	o	c	o	c	o	o	c	c	c	pc	o	c	c	c	c	33	
7th	Mon	o	o	o	o	o	o	o	o	o	o	o	o	o	o	o	o	o	o	o	o	3	
8th	Tue	c	c	c	c	c	c	c	c	c	c	c	c	c	c	c	o	o	o	o	o	29	
9th	Wed	pc	pc	pc	pc	pc	F	pc	F	pc	F	F	pc	pc	pc	pc	F	F	F	F	F	163	V
10th	Thu	o	o	o	o	o	o	o	o	o	o	o	o	o	o	o	o	o	o	o	o	4	1Q
11th	Fri	c	c	c	c	c	pc	c	pc	c	pc	pc	c	c	c	c	c	c	c	c	c	65	
12th	Sat	pc	pc	pc	pc	pc	pc	pc	pc	pc	pc	pc	pc	pc	pc	pc	pc	pc	F	F	F	150	
13th	Sun	F	F	F	F	F	pc	F	pc	F	pc	pc	F	F	F	F	pc	F	pc	pc	pc	164	
14th	Mon	pc	pc	pc	pc	pc	pc	pc	pc	pc	pc	pc	pc	pc	pc	pc	pc	pc	pc	pc	pc	101	
15th	Tue	c	c	c	c	c	c	c	c	o	o	o	c	o	o	c	o	o	o	o	o	15	XhN
16th	Wed	pc	pc	pc	pc	pc	c	pc	c	pc	c	c	pc	pc	pc	pc	c	pc	pc	pc	pc	84	
17th	Thur	c	c	c	c	c	o	c	o	c	c	o	c	c	c	c	c	c	c	c	c	45	F P2
18th	Fri	o	o	o	o	o	o	o	o	o	o	o	o	o	o	o	o	o	o	o	o	2	
19th	Sat	F	F	F	F	F	pc	F	pc	F	F	F	F	F	F	F	F	F	F	F	F	179	
20th	Sun	pc	pc	pc	pc	pc	pc	pc	pc	pc	pc	pc	pc	pc	pc	pc	pc	pc	c	c	c	116	
21st	Mon	pc	pc	pc	pc	pc	pc	pc	pc	pc	pc	pc	pc	pc	pc	pc	F	pc	F	F	F	136	^
22nd	Tue	pc	pc	pc	pc	pc	pc	pc	pc	pc	pc	pc	pc	pc	pc	pc	pc	pc	pc	pc	pc	118	
23rd	Wed	c	c	c	c	c	c	c	c	c	c	c	c	c	c	c	pc	c	o	o	o	45	
24th	Thu	pc	pc	pc	pc	pc	pc	pc	pc	pc	pc	pc	pc	pc	c	pc	c	c	o	o	o	94	3Q
25th	Fri	c	c	c	c	c	c	c	c	o	c	c	c	o	o	o	c	c	o	o	o	22	
26th	Sat	o	o	o	o	o	o	o	o	o	o	o	o	o	o	o	o	o	o	o	o	5	
27th	Sun	o	o	o	o	o	o	o	o	o	o	o	o	o	o	o	o	o	c	c	c	5	
28th	Mon	c	c	c	c	c	o	c	o	c	o	o	c	c	c	c	c	c	o	o	o	24	
29th	Tue	o	o	o	o	o	o	o	o	o	o	o	o	o	o	o	o	o	o	o	o	0	A XhS
30th	Wed	c	c	c	c	c	o	c	o	c	o	o	c	c	c	c	c	c	pc	pc	pc	43	XhS
31st	Thur	c	c	c	c	c	c	c	c	c	c	c	c	c	c	pc	c	c	c	c	c	58	
Estimate hours:		104	104	104	104	104	100	104	100	104	105	103	104	104	104	108	110	107	100	100	100	2068	
Average hours:		91	91	87	82	74	74	76	82	75	85	88	85	86	84	96	82	71	80	78	87	1653	
Trend:		more	more	more	more	more	more	more	more	more	more	more	more	more	more	more	more	more	more	more	more	more	

OCTOBER **1** TUESDAY

An intense depression to the immediate west of Ireland directs a stormy air across the country, with a general southerly flow of winds to gale force at times. A series of very active rain belts slowly cross the country from the west at first and spreading eastward, bringing widespread light rain in the east and very heavy rain elsewhere, especially in western parts.

Belfast: Stormy air, overcast, light squally showers, strong gusty winds, possible gales
Dublin: Widespread cloud, dull and gloomy, light showers, gusty winds
Cork: Storm, overcast, heavy rain, blustery with gale winds likely
Galway: Overcast, heavy rain, strong gusty winds with chance of gale gusts

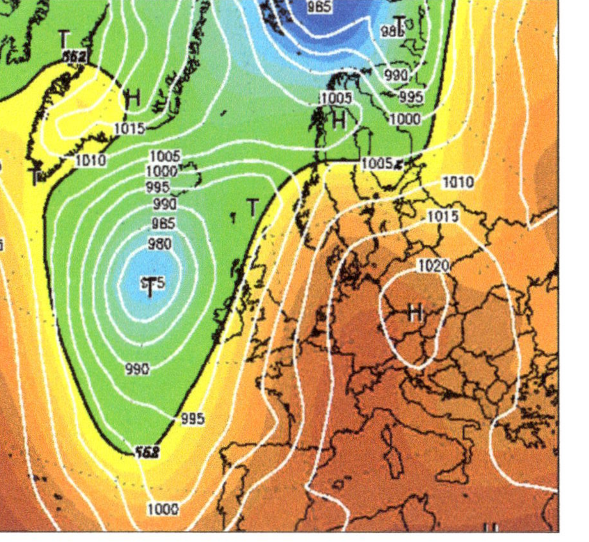

RAIN POTENTIAL

FROST/SNOW

OCTOBER **2** WEDNESDAY

An intense depression slowly moves into the west of the country, with unsettled and stormy south to southeast winds prevailing, with gale gusts in many places. A band of thunderstorms moves up through the country bringing heavy rain at times into the east, with a mix of sunny spells and lighter rain elsewhere.

Belfast: Mix of sunny spells and showers, strong blustery winds, warmer, gales likely
Dublin: Cloudy, light showers, cooler, strong winds with gale gusts likely
Cork: Sunny, mild day cool night, overnight showers, strong gusty winds, possible gales
Galway: Sunny, warmer day cool night, brief squally showers, strong gusty winds

RAIN POTENTIAL

FROST/SNOW

OCTOBER 3 THURSDAY

Low pressure to the west of Ireland continues to maintain a very unstable airflow over the country, despite winds easing to a moderate to light southeast flow late in the day. Further bands of thunderstorms and showers cross the country, with heavy rain in many central western districts, and scattered lighter rain, often also accompanied by thunderstorms, elsewhere.

Belfast: Cloudy, rain, blustery at times
Dublin: Cloudy, scattered showers, blustery at times, chance of thunderstorms
Cork: Cloudy, cooler, rain, blustery
Galway: Cloudy, windy, heavy rain, chance of overnight thunderstorms

RAIN POTENTIAL

FROST/SNOW

OCTOBER 4 FRIDAY

A decaying depression maintains a slow passage over the country in a lighter, easterly flow. A changeable day with further bands of thunderstorms and rain in most districts, interspersed with occasional sunny spells, although most dry along the southeast coastal fringes. Chance of evening mist patches developing in some northern and far southern areas.

Belfast: Mostly cloudy and dry, lighter breezes, chance of shallow fog overnight
Dublin: Rain, some sunny spells, light breezes, chance of overnight thunderstorms
Cork: Partly cloudy, warmer, isolated showers, chance of overnight fog patches
Galway: Partly cloudy, isolated showers, chance of overnight thunderstorms

RAIN POTENTIAL

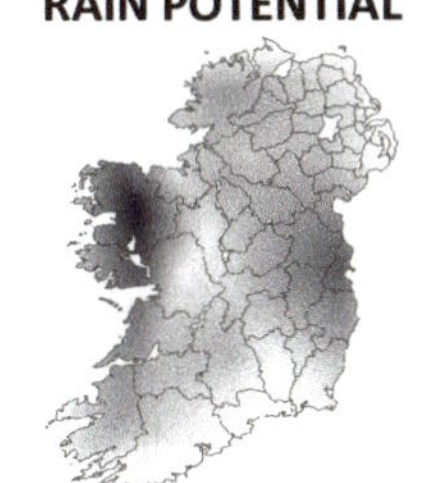

FROST/SNOW

OCTOBER **5** SATURDAY

The depression of the past few days decays into a low pressure trough lingering over Ireland in a light to mostly calm flow throughout. Scattered fogs in most districts, clearing to a mostly sunny and dry day everywhere, except for more frequent cloud and the chance of some isolated showers in the north and northwest of the country.

Belfast: Mostly calm, dry, and mild, chance of fog patches, some sunny spells
Dublin: Mostly cloudy, isolated showers, gentle breezes, chance of fog
Cork: Sunny, mostly dry and mild, light breezes to calm, fog patches likely
Galway: Mostly sunny, isolated overnight showers, cooler, light breezes, possible fog

RAIN POTENTIAL

FROST/SNOW

OCTOBER **6** SUNDAY

Low pressure continues to prevail over Ireland with a mostly light southerly flow prevailing, fresher for a brief time around the southwest. Mostly dry and cloudy with patches of fog in most eastern coastal areas as well as in the southeast of the country, while mostly cloudy with patches of overnight fog and outbreaks of showers elsewhere.

Belfast: Warmer day cooler night, cloudy, light showers, gentle breezes, chance of fog
Dublin: Partly cloudy, isolated showers, gentle breezes, fog patches possible
Cork: Cloudy, light breezes to mostly calm air, dry, chance of mists and fog patches
Galway: Overcast, scattered showers, light breezes, chance of overnight fog patches

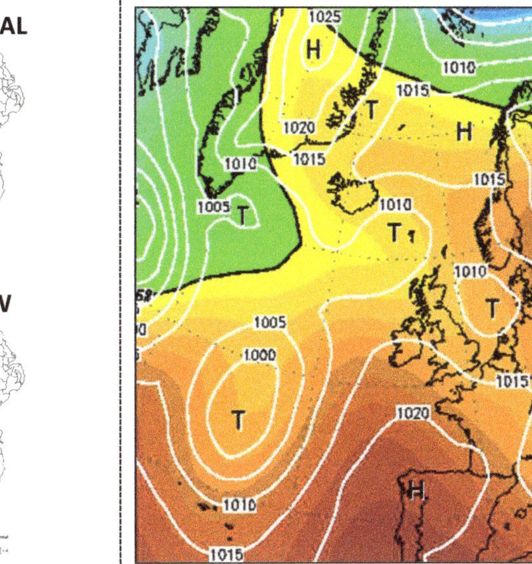

RAIN POTENTIAL

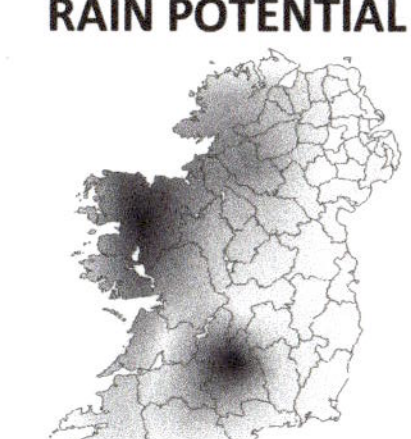

FROST/SNOW

OCTOBER **7** MONDAY

A depression to the southwest of Ireland extends a strong and sometimes gusty southeast flow over the country, with winds occasionally reaching gale force at times, particularly in the west. Wet fronts cross the country moving in from the southwest and spreading into all districts as the day progresses, with heavier falls in the southwest and west, and lighter showers elsewhere.

Belfast: Cooler, overcast, scattered showers, strengthening winds with gusts
Dublin: Changeable, widespread cloud, light showers, warmer, strengthening winds
Cork: Overcast, rain, windy
Galway: Blustery winds to gales, overcast, squally showers

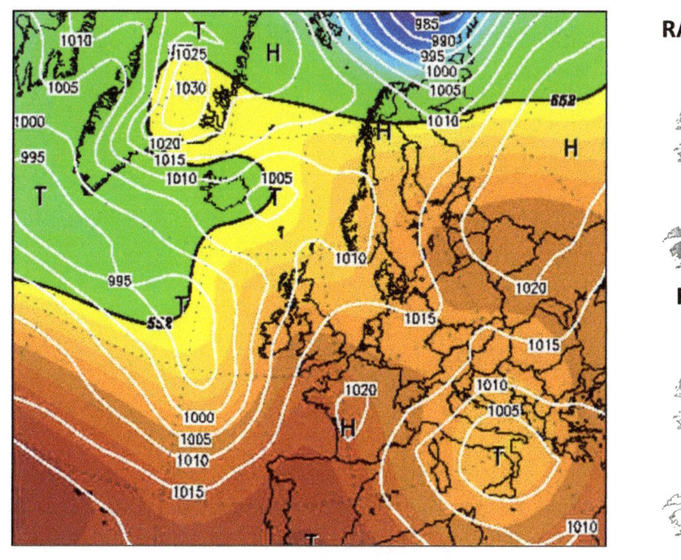

RAIN POTENTIAL

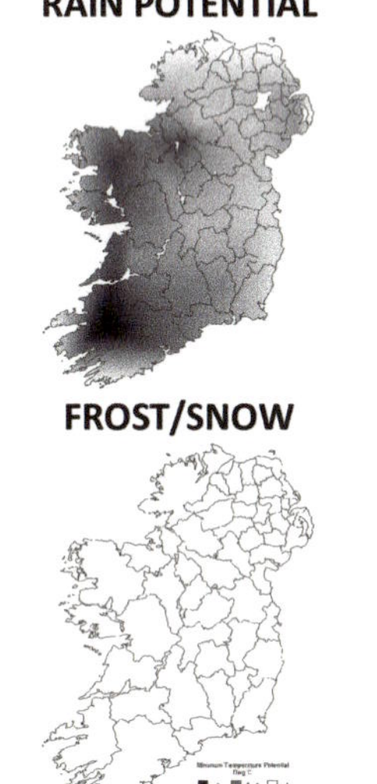

FROST/SNOW

OCTOBER **8** TUESDAY

A large depression drifts to the northwest of Ireland directing fresh to strong westerly winds over the northern half of the country, while fresh to moderate in the southern half. Some light showers in most districts, more frequent in the west and north, while isolated in the east and south.

Belfast: Cloudy, isolated showers, strong gusty winds, chance of gales
Dublin: Cloudy, cooler, mostly dry, passing threats of rain, strong gusty winds at times
Cork: Overcast, windy, brief light showers
Galway: Strong blustery winds, widespread cloud, brief showers, cooler

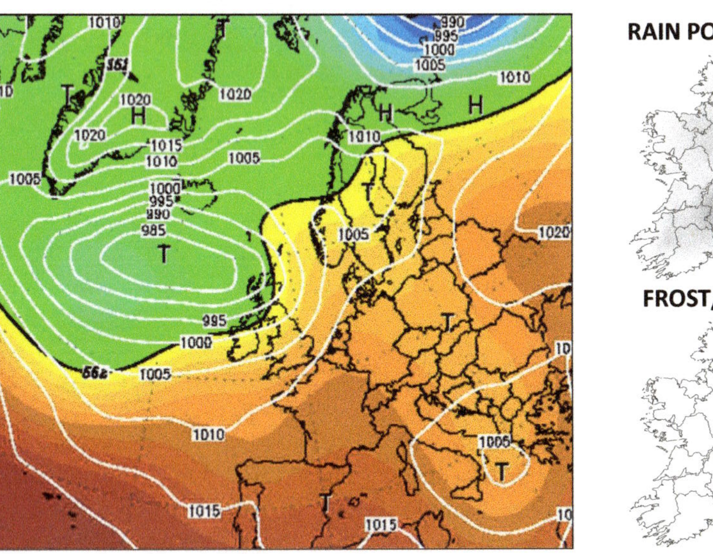

RAIN POTENTIAL

FROST/SNOW

OCTOBER 9 WEDNESDAY

The depression to the west of Ireland strengthens as it drifts deeper into the Atlantic Ocean, with winds generally light to moderate in a south to southwest flow. A series of frontal troughs brings overnight showers in most districts, before clearing to a mostly dry and sunny day. Further showers return in the evening, particularly in the west, and possibly accompanied by thunderstorm activity.

Belfast: Mostly sunny, isolated showers, windy at times, chance of thunderstorms
Dublin: Mostly sunny, light showers, windy spells
Cork: Sunny, light showers, mild day cool night, windy
Galway: Mix of sunny spells and rain, possible thunderstorms, windy

RAIN POTENTIAL

FROST/SNOW

OCTOBER 10 THURSDAY

The depression to the west of Ireland drifts closer to the country once more, directing a fresh to strong south to southeasterly flow, with windy gusts possibly up to gale force at times. A stormy air prevails throughout the country, as a series of showery troughs move up from the southern coasts to the northern coasts with rain widespread and likely to be particularly heavy in the southern half of the country, and thunderstorms likely in the north.

Belfast: Stormy, overcast, heavy rain, blustery, possible gales, chance of thunderstorms
Dublin: Overcast, rain, blustery at times
Cork: Overcast, heavy rain, windy with onshore gusts likely
Galway: Overcast, rain, strengthening winds with chance of gale gusts

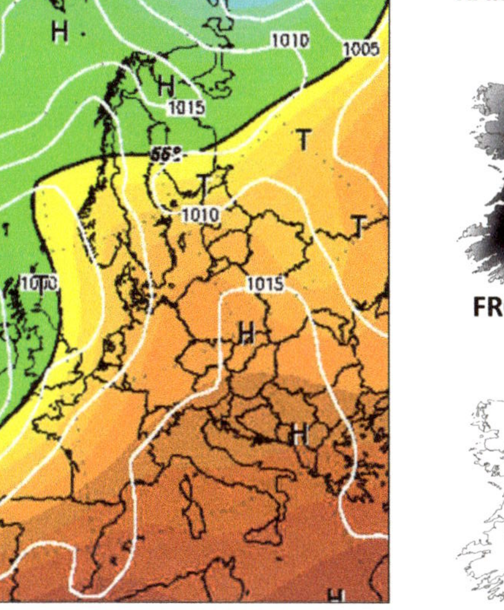

RAIN POTENTIAL

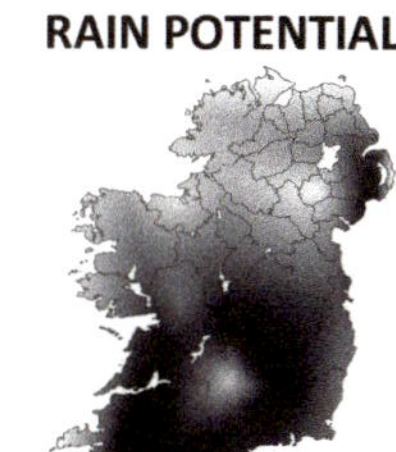

FROST/SNOW

OCTOBER 11 FRIDAY

The depression continues to linger over Ireland with winds easing to mostly moderate and in an east to southeast flow, but remaining blustery in the north of the country. Widespread showers continue everywhere, with some heavy falls in the northeast and southeast. Some brief sunny spells between the showers, mostly in the southwest.

Belfast: Strong blustery winds, isolated showers, some sunny spells, colder
Dublin: Cloudy, occasional squally showers, windy at times
Cork: Cloudy, rain, gusty winds at times
Galway: Partly cloudy, light squally showers, fluctuating winds

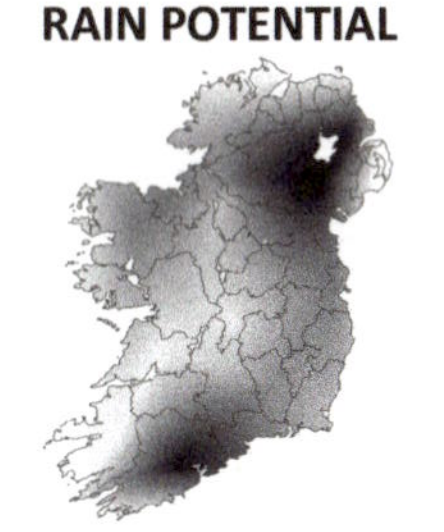

OCTOBER 12 SATURDAY

The depression that has bought the recent wet spell drifts off the country to be situated to the northeast of Ireland, directing a cooler and light to moderate north to northwest flow, but stronger in the north. Mostly sunny everywhere, with brief light showers in the west and northern coasts at first, spreading the band of showers into most other districts for a time, then clearing almost everywhere, except for another band of rain briefly brushing the northern coasts. Chance of fog in the north and east.

Belfast: Mostly sunny, brief morning showers, warmer, fluctuating winds
Dublin: Mostly sunny and dry, mild day cool night, moderate shifting breezes
Cork: Sunny, isolated showers, mild day cool night, moderate shifting breezes
Galway: Mostly sunny and dry, fluctuating breezes

OCTOBER 13 SUNDAY

Low pressure continues to direct a light northwesterly flow throughout, with chance of overnight fog patches in many districts, particularly in the north and west. Chance of some isolated overnight showers, mostly in the southwest, clearing to a mostly dry and sunny day throughout.

Belfast: Sunny, isolated showers, winds easing to light, chance of overnight fog patches
Dublin: Sunny, mostly dry, lighter breezes, chance of overnight misty patches
Cork: Mostly sunny and dry, moderate breezes
Galway: Partly cloudy, isolated showers, light breezes, chance of fog

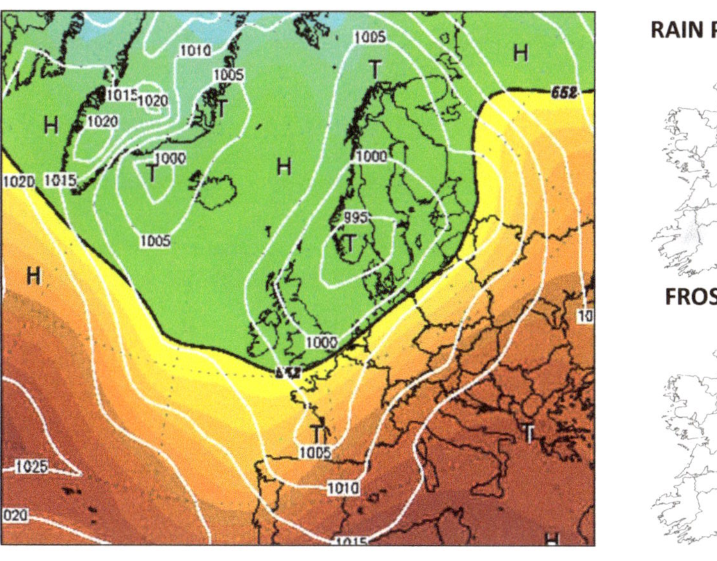

OCTOBER 14 MONDAY

A shower embedded low pressure trough maintains a light to moderate northwesterly flow over the country. Cold overnight with chance of pockets of frost in inland parts. A changeable day with a mix of sunny spells and passing showers in most districts, with a wet front in the evening bringing a band of rain, with some heavy falls likely western to northwestern parts.

Belfast: Partly cloudy, cooler, rain, moderate winds freshening later in the day
Dublin: Partly cloudy, dry, moderate breezes
Cork: Partly cloudy, light showers, mild day cool night, moderate breezes
Galway: Partly cloudy, heavy rain, cold overnight, light breezes, chance of hail showers

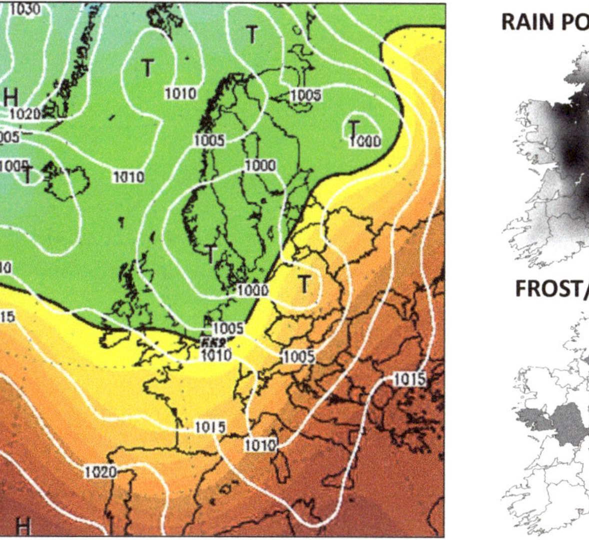

OCTOBER **15** TUESDAY

A depression to the northwest of Ireland directs a series of rain embedded fronts and moderate to fresh west to southwest winds over the country. Dull and gloomy almost everywhere, with widespread rain, occasionally heavy, throughout, easing later in the day. Some brief sunny breakthroughs mostly in the north and west.

Belfast: Cloudy, light showers, windy
Dublin: Warmer, cloudy, rain, freshening winds
Cork: Overcast, rain, windy with occasional gusts
Galway: Overcast, rain, cold overnight, strong onshore winds

RAIN POTENTIAL

FROST/SNOW

OCTOBER **16** WEDNESDAY

The depression that brought widespread rain yesterday drifts in a northeastward tract with its centre situated to the north of Ireland. Moderate to fresh southwest winds prevail, veering later to northwest and blustery at times. A changeable day with a mix of sunny spells and showers, some heavier for a time in northern parts.

Belfast: Partly cloudy, light showers, strong gusty winds, possible gales
Dublin: Mix of showers and passing sunny spells, cooler, strong blustery winds
Cork: Strong and blustery winds, mix of showers and sunny spells, cooler
Galway: Mostly cloudy, dry, strong blustery winds with chance of gales

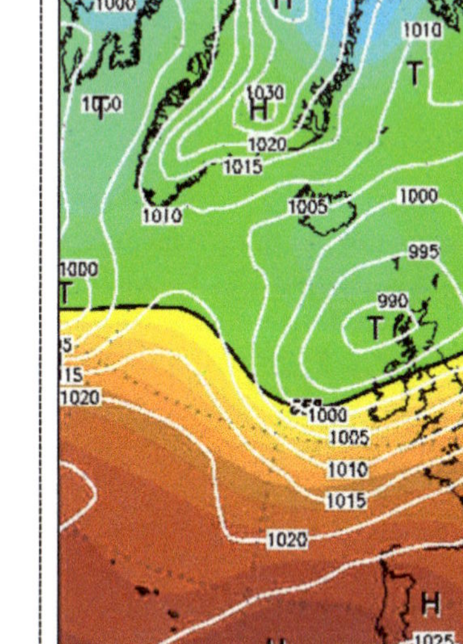

RAIN POTENTIAL

FROST/SNOW

OCTOBER **17** THURSDAY

A complex of depressions sit to the northwest and northeast of Ireland, while a high pressure ridge flows south of the country. Fresh northwest winds prevail, stronger with occasional gusts at times along the coasts, and up to gale strength for a time in the north. Dry with brief sunny spells in most districts, while outbreaks of showers likely mostly around the northern and western coasts.

Belfast: Cloudy and dry, strong winds with gale gusts likely
Dublin: Cloudy, dry, increasing threats of rain, strong winds, chance of gale gusts
Cork: Cloudy, dry, strong winds with occasional gusts at times
Galway: Unsettled and stormy winds, cloudy, dry with increasing threats of rain

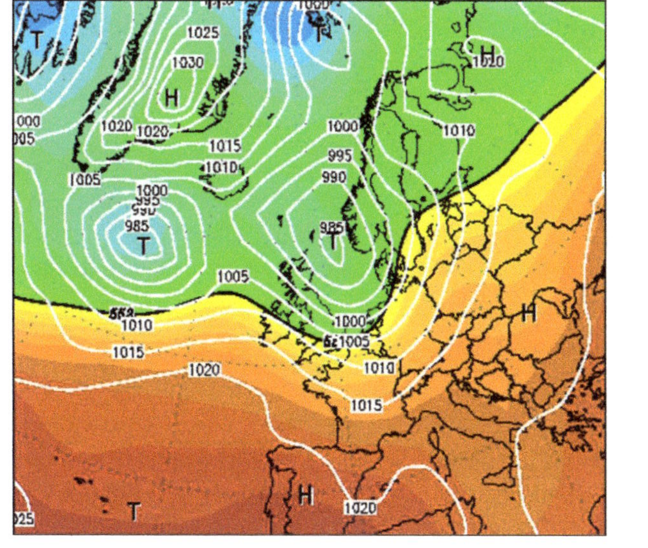

RAIN POTENTIAL

FROST/SNOW

OCTOBER **18** FRIDAY

Depressions to the northeast and northwest of Ireland continue to bring unsettled conditions throughout the country, with moderate to fresh southwest to westerly winds prevailing, unsettled and stronger around the northern and western coasts. Widespread cloud, dull and gloomy, with a stormy air throughout. Rain bands spread from the south to the north during the day, with heavy falls mostly in the southwest. Mostly dry and chance of some brief sunny spells in the far northwest.

Belfast: Overcast, rain, cooler, strong gusty winds, chance of gales for a time
Dublin: Stormy air, overcast, rain, strong and blustery winds to near gales
Cork: Dull and gloomy air, heavy rain, windy
Galway: Overcast, light showers, strong winds

RAIN POTENTIAL

FROST/SNOW

OCTOBER **19** SATURDAY

A depression to the northwest of Ireland directs a moderate westerly flow over the country, stronger for a time along the northwest coasts. A weak frontal trough crosses the country overnight bringing some isolated showers here and there, mostly in the northwest, and squally for a time, particularly in the south and east. Mostly dry and sunny day throughout.

Belfast: Dry, mild, sunny, strong blustery winds
Dublin: Sunny, isolated showers, cold overnight, strong blustery winds at times
Cork: Stormy air, sunny, light squally showers
Galway: Sunny, dry, warmer day cooler night, strong gusty winds, possible gales

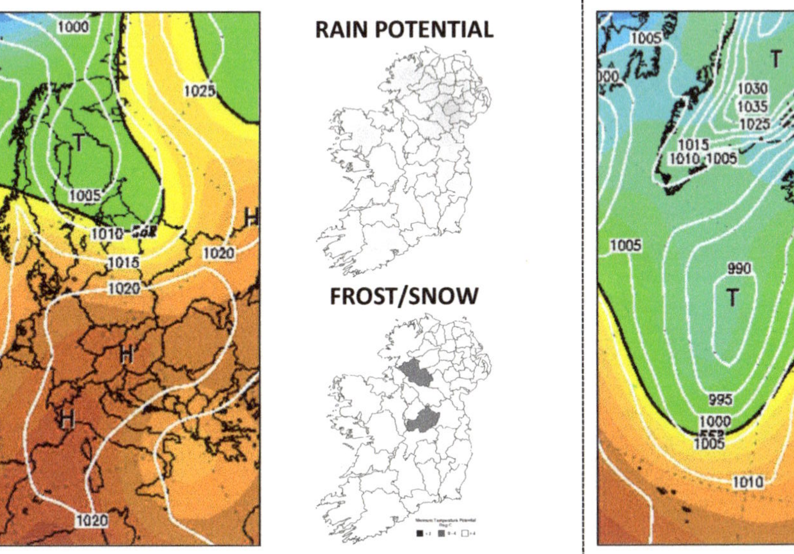

RAIN POTENTIAL

FROST/SNOW

OCTOBER **20** SUNDAY

A fresh to strong southerly flow prevails directed from a large depression off the western coast of Ireland. Mostly dry with passing sunny spells throughout the country at first, before a wet front spreads cloud and rain into the south of the country at first, moving up into northern parts by evening. Some outbreaks of rain are likely to be heavy particularly in southern parts.

Belfast: Partly cloudy, outbreaks of rain, windy at times, cold overnight
Dublin: Mix of showers and sunny spells, cold overnight, chance of overnight fog
Cork: Cloudy, heavy rain, cooler, lighter breezes strengthening later in the day
Galway: Partly cloudy, rain, mild day cold night, moderate winds strengthening later

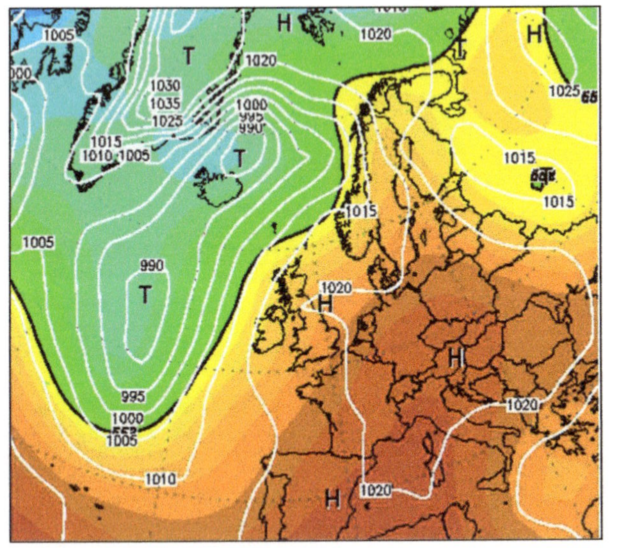

RAIN POTENTIAL

FROST/SNOW

OCTOBER 21 MONDAY

The depression centred to the northwest of Ireland directs a moderate south to southeast flow over the country, freshening to strong and blustery for a time, particularly along the northern coasts, before easing with a southwest change. A stormy air prevails everywhere, with a mix of sunny spells and outbreaks of rain in most districts, heaviest in the northern half of the country, lighter in the southern parts.

Belfast: Stormy air, partly cloudy, light showers, strong winds, gales possible at times
Dublin: Partly cloudy, warmer, rain, strong winds with occasional gusts
Cork: Strong and blustery winds, sunny day, some squally showers
Galway: Strong blustery winds, mix of sunny spells and showers, cold overnight

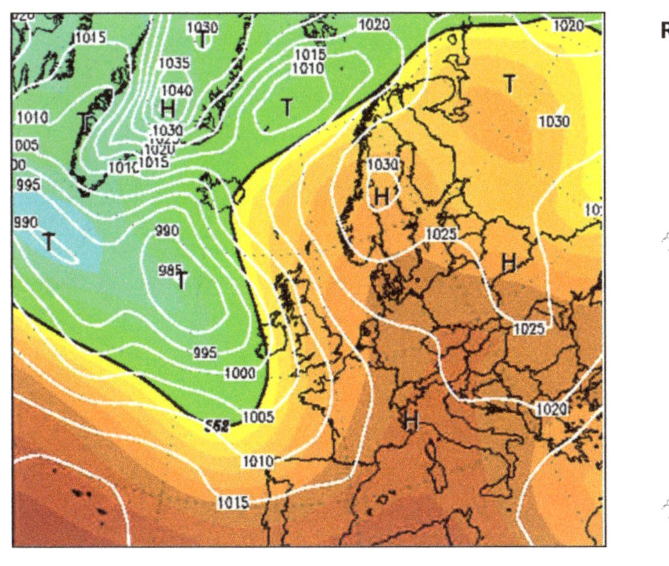

RAIN POTENTIAL

FROST/SNOW

OCTOBER 22 TUESDAY

The depression to the northwest of Ireland intensifies strengthening winds throughout the country, with a fresh southwesterly flow strengthening to gales at times. A changeable day with overnight rain clearing to a mostly dry day with passing sunny spells everywhere. Chance of some isolated lingering showers mostly in the eastern half of the country.

Belfast: Partly cloudy, dry, strong winds with gale gusts likely
Dublin: Partly cloudy, dry, blustery at times
Cork: Partly cloudy, light showers, warmer, fluctuating windy spells
Galway: Partly cloudy, dry, milder, strong winds with gale gusts possible

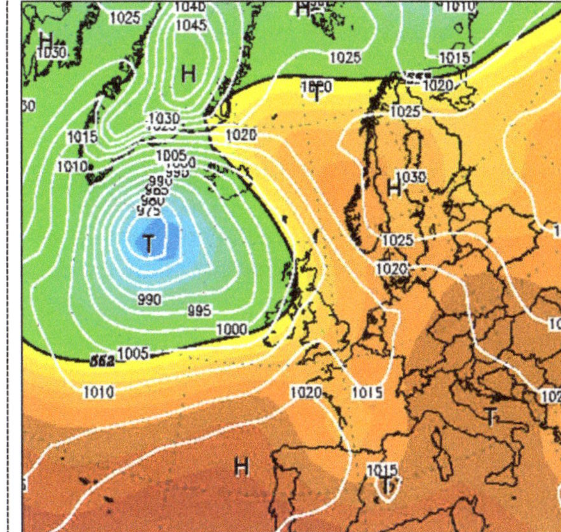

RAIN POTENTIAL

FROST/SNOW

OCTOBER **23** WEDNESDAY

A high pressure ridges over the east of Ireland while a large depression lingers to the west of the country, with a moderate southerly airflow prevailing throughout, but stronger in the west. Mostly cloudy and dry everywhere with the occasional threat of rain.

Belfast: Dry, cloudy, strong blustery winds at times
Dublin: Cloudy, dry with occasional threats of rain, moderate breezes
Cork: Overcast, mostly dry with passing threats of rain, moderate breezes
Galway: Mostly cloudy, dry, warmer, strong winds at times

RAIN POTENTIAL

FROST/SNOW

OCTOBER **24** THURSDAY

The high pressure ridge maintains a moderate south to southeasterly flow over the country, while low pressure to the west brings fresher southerly breezes along the western and northern coasts. Mostly dry with occasional sunny spells everywhere, although mostly dull and gloomy in the east.

Belfast: Partly cloudy, dry, blustery with winds easing later in the day
Dublin: Partly cloudy, mostly dry with odd threat of rain, breezy spells
Cork: Overcast, dry with odd threat of rain, moderate breezes
Galway: Partly cloudy, dry, fresh winds easing later, chance of overnight mist

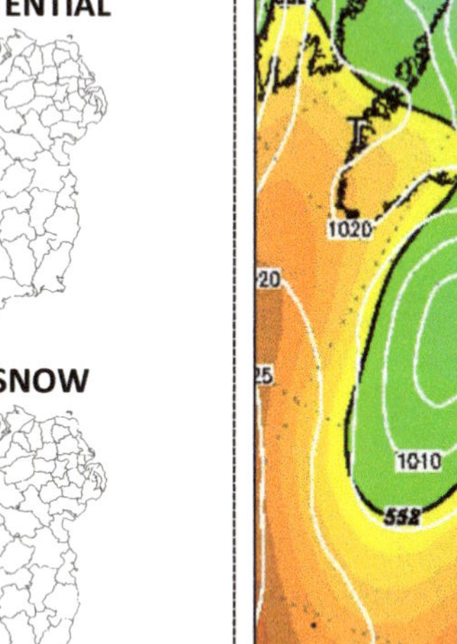

RAIN POTENTIAL

FROST/SNOW

OCTOBER 25 FRIDAY

Anticyclonic conditions continue to prevail throughout the country with lighter southeasterly breezes prevailing. Milder everywhere with scattered misty patches in sheltered districts, possibly briefly turning to passing drizzle spells overnight. Mostly dry and warm throughout the day, with brief sunny breakthroughs, mostly in the north, west, and south. Dull and gloomy with passing threats of rain in the east.

Belfast: Cloudy and dry, moderate winds
Dublin: Overcast, dry with passing threats of rain, moderate breezes
Cork: Widespread cloud, dull and gloomy, dry, mild, moderate breezes
Galway: Cloudy, dry, warmer, moderate breezes, chance of misty patches

RAIN POTENTIAL

FROST/SNOW

OCTOBER 26 SATURDAY

The high pressure ridge drifts to the northeast of Ireland, while low pressure lingers to the south of the country, with Ireland laying in a trough between the two with a mild southeasterly air prevailing. Widespread patches of mist or fog overnight which will be slow to clear in many districts. Widespread cloud everywhere and mostly dry with passing threats of rain, except for the chance of some spotty showers briefly in the central north.

Belfast: Dull and gloomy air, dry with increasing threats of rain, light breezes
Dublin: Overcast, mostly dry with increasing threats of rain, cooler, light breezes
Cork: Overcast, dry, occasional breezy spells
Galway: Overcast, cooler, dry with increasing threats of rain, windy spells

RAIN POTENTIAL

FROST/SNOW

OCTOBER **27** SUNDAY

A large depression to the southwest of Ireland directs a rain embedded trough over the country starting in the south at first and spreading northwards throughout the country. Southeasterly breezes prevail, light at first, allowing for patches of fog and mist overnight, then freshening with the passage of rain bands that bring widespread outbreaks of rain and drizzle. Some rain likely to be heavy at times, particularly in a line from the southeast to the northwest.

Belfast: Overcast, rain, windy spells
Dublin: Overcast, rain, breezy spells
Cork: Cloudy, rain, warmer, freshening breezy spells
Galway: Overcast, rain, windy spells

RAIN POTENTIAL

FROST/SNOW

OCTOBER **28** MONDAY

The depression moves to the west of Ireland with the showery troughs continuing to bring a mix of misty drizzles and scattered showers throughout, with brief sunny breakthroughs mostly in the north and east.
Winds generally light to moderate in a southeasterly flow, but fresher with occasional gusts mostly around the southern coastal fringes.

Belfast: Unsettled air, cloudy, spotty showers, some breezy spells
Dublin: Cloudy, dry, moderate breezes, chance of brief misty drizzle patches
Cork: Overcast, scattered showers, unsettled windy spells
Galway: Overcast, light showers, fresh to strong winds

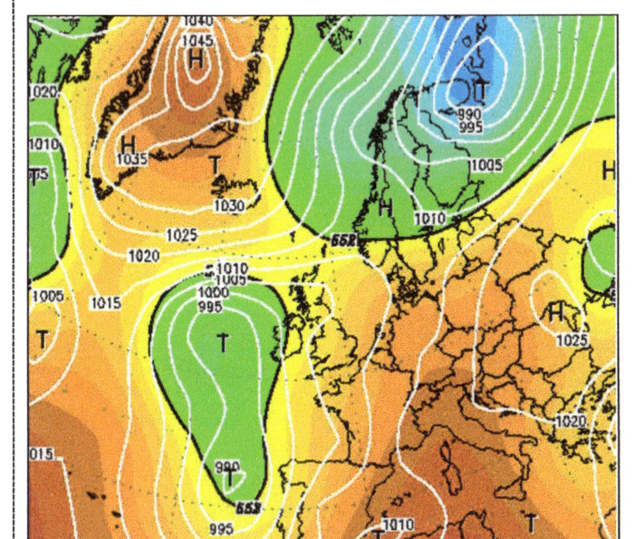

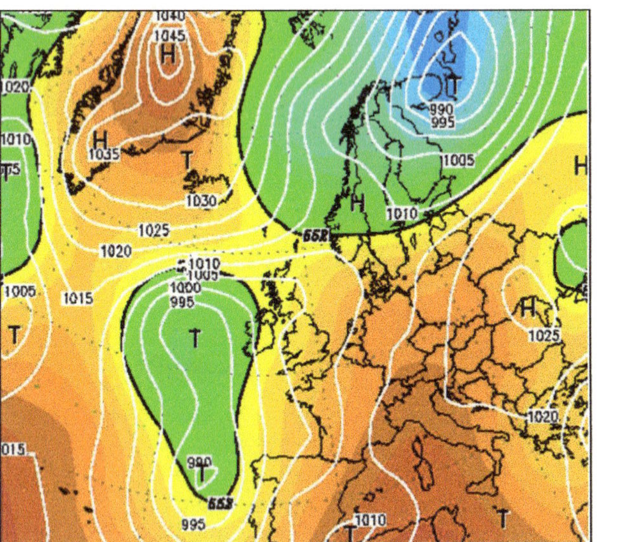

RAIN POTENTIAL

FROST/SNOW

OCTOBER 29 TUESDAY

The depression slowly drifts eastward crossing to the immediate south of the country, directing light to moderate southeasterly winds becoming more variable as the day progresses. After a mostly dry night with scattered fogs and misty patches in many districts, the frontal troughs on the edge of the depression moves onto the country bringing occasional showers in most districts during the day, more persistent in the south and eastern coastal districts. Rain activity likely to clear to fog or mist in many parts by evening.

Belfast: Dull and gloomy air, rain, fluctuating windy spells
Dublin: Shifting windy spells, overcast, scattered showers, cooler
Cork: Overcast, rain, strong winds with occasional gusts
Galway: Unsettled winds, dull and gloomy, scattered showers

RAIN POTENTIAL

FROST/SNOW

OCTOBER 30 WEDNESDAY

The depression expands its low pressure centre to cover almost all of Ireland, bringing light and variable breezes throughout, with further patches of rain and drizzle in most districts overnight, easing later to widespread patches of mist or fog. Mostly dry and overcast day in the north of the country, while mostly dry with passing sunny breakthroughs elsewhere. Chance of some lingering showers in the east during the day.

Belfast: Cloudy, cooler, windy, cooler
Dublin: Breezy, cloudy, dry with odd threat of rain, mild day cool night
Cork: Partly cloudy, light showers, fluctuating beezes
Galway: Overcast, dry with passing threats of rain, fluctuating breezes

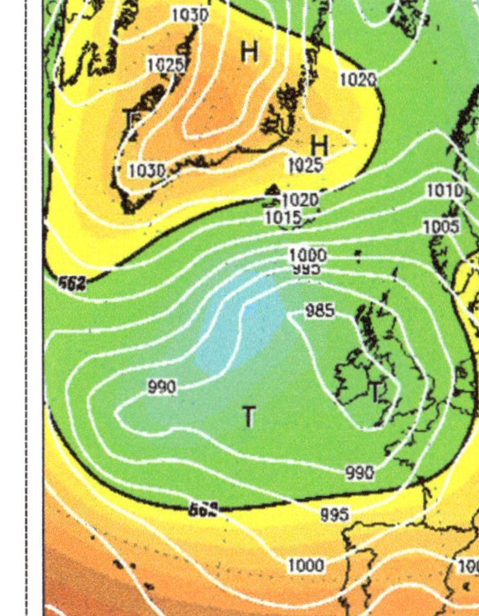

RAIN POTENTIAL

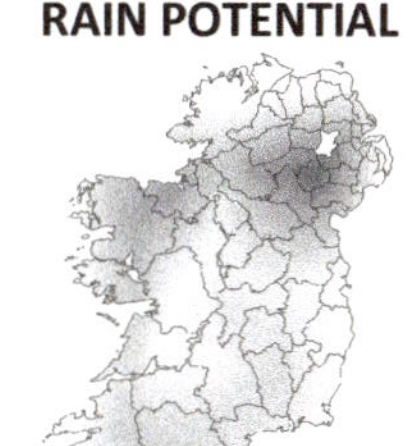

FROST/SNOW

OCTOBER **31** THURSDAY

Low pressure continues to prevail over Ireland with light southeasterly breezes at first, strengthening later with a southerly change, then backing to southwesterly and gusty. A stormy air prevails throughout with thunderstorms and rain in many districts, including heavy rain in the south of the country, and lighter rain with some sunny breakthroughs in the north and east.

Belfast: Mix of brief sunny spells and scattered showers, windy with occasional gusts
Dublin: Partly cloudy, light showers, mild, unsettled winds with occasional gusts
Cork: Stormy air, cloudy, heavy rain, strong and blustery onshore winds
Galway: Cloudy, light showers, unsettled breezes with occasional gusts

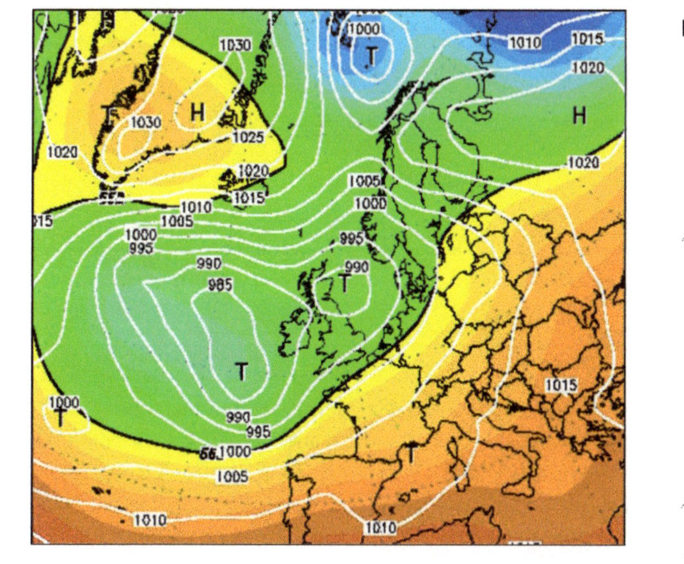

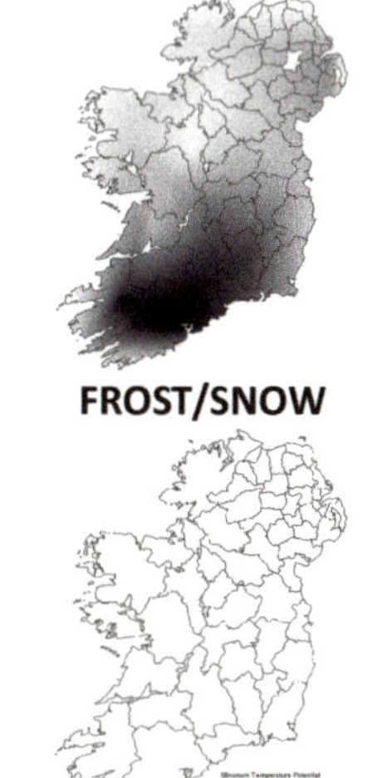

NOVEMBER

PHASES OF THE MOON

2nd	New Moon
5th	Southern Declination
9th	First Quarter
13th	Crossing Equator
14th	Perigee #6
15th	Full Moon
18th	Northern Declination
23rd	Third Quarter
26th	Apogee
26th	Crossing Equator

1	2	3	4	5	6	7	8	9	10	11	12	13	14	15	16	17	18	19	20	21	22	23	24	25	26	27	28	29	30
Fri	Sat	Sun	Mon	Tue	Wed	Thur	Fri	Sat	Sun	Mon	Tue	Wed	Thur	Fri	Sat	Sun	Mon	Tue	Wed	Thur	Fri	Sat	Sun	Mon	Tue	Wed	Thur	Fri	Sat
	N			V				1Q				XhN	P6	F			^					3Q			A XhS				

MONTHLY SUMMARY

1st-10th Low pressure over Ireland during this period likely to bring some rain activity on a daily basis in many districts. Unusually mild maximum temperatures in the north of the country.
2nd-4th Cooler spell, with chance of some ground frosts in inland parts and isolated in the southeast. Some good sunny spells.
4th A large depression to the southwest of the country extends a series of fronts over Ireland, bringing widespread heavy rain, and including the chance of localised flooding in some districts in the southeast.
6th-9th Mostly overcast spell.
10th A high pressure ridges to the west of Ireland while a depression sits to the north, resulting in unsettled winds in many districts with the chance of reaching gale force for a time in the northwest.

Over the next four weeks high pressure dominates the first couple of weeks, bringing mostly dry and sunny days with light winds, followed by depressions bringing unsettled conditions over the second half of this outlook period. Rainfall amounts will be above normal almost everywhere during this look-ahead period, with some rainfall expected almost daily over the second to fourth weeks ahead, with the first week likely to be mostly dry. Some districts, particularly in the southern half of Ireland, may see up to twice as much rainfall than is normal for this time of year. Sunshine hours will also be above normal during this outlook period almost everywhere, except in some western and northern districts, with especially bright and sunny conditions over the first week, particularly in the east. Temperatures will generally be around half to one degree above the average for this time of year, despite a colder start to this outlook period due to the anticyclonic dominance bringing clearer skies, which will allow for colder overnight temperatures and include the development of ground frosts in many districts. Another cool spell can also be expected into the third week ahead, including the development of wintry showers of hail and possible snowfalls. After light winds during the first week ahead, stronger winds are likely to prevail for the remainder of this outlook period, especially during the third and fourth weeks, with the chance of gales in the fourth week. During the next four weeks there is likely to be around 20-25 days when some rain may be measured in the west and southwest, and around 13-18 elsewhere. Thunder may be heard on 6 days, hail showers on 5 days, gales on 1 day, and snow on 3 days. Widespread fogs can be expected on 2 days and frosts in more than 10 days in inland areas, while up to 6 days around the coasts, and possibly up to 20-22 days in inland parts in the southeast.

11th-12th Very bright and sunny spell throughout, especially in the east. Scattered sub-zero overnight minimas likely in many parts of the west and east.
11th-16th Anticyclonic. Dry and settled conditions throughout with light winds to mostly calm air prevailing. Some cloudy spells in the west and north with only a few frost pockets overnight, while mostly sunny in the east and south brings cold nights and widespread ground frosts. The development of fog becomes more widespread towards the end of this period.
12th-13th Overnight frosts expected mostly in inland counties and in northern parts.
16th-20th Unusually warm spell.
17th-24th High pressure moves away in a southerly tract, allowing low pressure to advance a series of rain bands across the country in a southwest flow. Rainfall amounts will be generally light except for some heavier rain in the north and west at times. Winds will strengthen to strong for a time towards the end of this spell.
20th-22nd Very dull and gloomy spell with overcast conditions prevailing.
25th-30th Unsettled weather. Cool and blustery westerly winds prevail and heavy rain at times.
25th Widespread rain, with heavy falls in the southeast that may cause some localised flooding.
26th-28th Cooler maximum temperatures settle into single digits during this spell. Chance of snow showers in some districts.
27th-28th Chance of scattered thunderstorms.
27th-30th Wintry showers likely, some of hail.
29th Strong winds with gusts, possibly to gale force at times. Heavy rain in a number of districts in the west and south of the country.

IRELAND RAINFALL ESTIMATES

November 2024

d=mainly dry, n= nonrecordable, l=light shrs, s=significant shrs, r=rain, h=hvy falls

		ULSTER PROVINCE					CONNAUGHT PROVINCE						LEINSTER PROVINCE				MUNSTER PROVINCE						
		Alderg	Hillsbo	Armag	Loughl	Malin H	Sligo	Ardtar	Belmul	Drums	Galway	Roscoi	Markre	Edende	Kilkenr	Dublin	Shannc	Newpo	Tralee	Killarne	Cork	All	Moon
1st	Fri	l	d	s	l	l	l	l	l	l	l	l	l	d	d	n	n	r	l	l	d	39	N
2nd	Sat	d	d	d	d	r	l	d	l	d	s	d	s	d	d	d	d	d	d	l	d	22	
3rd	Sun	n	l	d	d	d	n	d	d	l	d	s	d	l	s	d	d	d	s	r	d	24	
4th	Mon	h	h	l	h	h	h	h	h	h	h	h	h	h	h	h	h	h	h	h	h	644	
5th	Tue	h	r	h	s	s	s	s	l	r	l	r	s	l	l	l	l	s	l	l	l	136	V
6th	Wed	l	l	s	l	l	l	l	s	s	l	l	l	l	l	l	l	h	r	d	l	70	
7th	Thu	l	d	l	d	d	d	d	s	n	l	d	l	d	d	n	d	l	l	d	l	16	
8th	Fri	n	l	d	l	s	s	l	l	l	l	l	n	n	d	d	d	r	l	n	s	38	
9th	Sat	l	l	l	s	r	l	s	s	l	s	l	s	l	n	l	s	s	s	l	l	57	1Q
10th	Sun	s	d	l	d	d	d	d	l	d	d	n	d	d	d	n	d	d	d	d	d	6	
11th	Mon	d	d	d	d	d	d	d	d	d	d	d	d	d	d	d	d	d	d	d	d	0	
12th	Tue	d	d	d	d	d	d	d	d	d	d	d	d	d	d	d	d	d	d	d	d	0	XhN
13th	Wed	d	d	d	d	d	d	d	d	d	d	d	d	d	d	d	d	d	d	d	d	0	
14th	Thur	d	d	d	d	l	d	d	d	d	d	d	d	d	d	d	d	l	d	d	d	2	P6
15th	Fri	d	d	d	d	d	d	d	d	d	d	d	d	d	d	d	d	d	d	d	d	0	F
16th	Sat	d	d	d	s	d	l	s	d	l	d	d	s	l	d	d	d	r	d	d	d	28	
17th	Sun	l	l	n	l	s	s	l	r	l	s	l	s	l	l	s	l	s	l	d	l	67	
18th	Mon	l	d	l	l	s	l	l	l	d	d	n	l	d	l	d	l	d	l	l	s	26	^
19th	Tue	d	d	d	s	n	l	s	n	l	l	l	l	d	d	d	d	l	d	l	d	16	
20th	Wed	l	s	d	s	s	s	s	s	s	l	s	s	l	s	l	l	l	r	s	s	88	
21st	Thur	l	l	l	s	s	l	s	s	l	d	l	s	d	l	d	l	l	d	l	d	44	
22nd	Fri	l	l	l	l	l	l	l	l	l	s	l	l	d	n	n	l	r	l	l	d	46	
23rd	Sat	l	n	l	l	l	l	l	l	l	l	l	d	l	d	l	l	l	l	s	n	29	3Q
24th	Sun	l	s	d	l	s	s	l	s	l	l	l	l	l	r	d	s	l	s	r	l	73	
25th	Mon	h	h	l	h	s	h	h	s	r	r	r	r	h	h	h	r	h	r	r	h	371	XhS
26th	Tue	r	s	h	s	r	s	s	r	r	h	r	r	r	s	d	s	h	r	r	l	209	A
27th	Wed	s	l	r	s	l	s	s	l	l	h	s	s	s	n	l	r	h	r	r	r	152	
28th	Thu	l	l	l	s	h	s	s	s	s	s	s	s	d	d	l	s	s	l	l	d	81	
29th	Fri	s	s	l	h	r	h	h	h	h	r	r	r	s	r	s	r	r	h	h	r	280	
30th	Sat	s	l	s	s	s	s	s	s	l	s	s	s	l	l	d	l	r	l	r	d	86	
	Estimate:	112	123	100	129	156	132	129	138	129	164	132	128	103	115	74	112	200	163	187	124	2648	
	Average:	80	88	75	119	109	128	130	134	102	120	129	128	88	86	73	94	170	168	121	120	2263	
	Trend:	wtr	wtr	wtr	av	wtr	av	av	av	wtr	wtr	av	av	wtr	wtr	av	wtr	wtr	av	wtr	av	wtr	

IRELAND SUNSHINE ESTIMATES

November 2024

F=fine (8-12 hours of sunshine), pc= partly cloudy (4-7 hours), c=cloudy (1-3 hours), o=overcast (0 hours)

		ULSTER PROVINCE					CONNAUGHT PROVINCE						LEINSTER PROVINCE				MUNSTER PROVINCE						
		Alderg	Hillsbo	Armag	Loughl	Malin H	Sligo	Ardtarı	Belmul	Drums	Galway	Roscoı	Markre	Edende	Kilkenı	Dublin	Shannc	Newpo	Tralee	Killarne	Cork	All	Moon
1st	Fri	c	c	c	c	c	o	c	o	c	c	c	c	c	pc	pc	c	c	pc	pc	pc	56	N
2nd	Sat	pc	pc	pc	pc	pc	pc	pc	pc	pc	pc	pc	pc	pc	pc	pc	pc	pc	c	c	c	119	
3rd	Sun	pc	pc	pc	pc	pc	c	pc	c	pc	c	c	pc	pc	pc	pc	pc	pc	pc	pc	pc	96	
4th	Mon	o	o	o	o	o	o	o	o	o	o	o	o	o	o	o	o	o	o	o	o	0	
5th	Tue	pc	pc	pc	pc	pc	pc	pc	pc	pc	pc	pc	pc	pc	pc	pc	pc	pc	pc	pc	pc	130	V
6th	Wed	o	o	o	o	o	o	o	o	o	o	o	o	o	o	c	o	o	o	o	o	9	
7th	Thu	o	o	o	o	o	o	o	o	o	o	o	o	o	o	o	o	o	o	o	o	1	
8th	Fri	c	c	c	c	c	o	c	o	c	o	o	c	c	c	c	o	o	o	o	o	16	
9th	Sat	c	c	c	c	c	o	c	o	c	o	o	c	c	c	c	o	o	o	o	o	16	1Q
10th	Sun	c	c	c	c	c	c	c	c	c	c	c	c	c	pc	c	pc	pc	pc	pc	pc	75	
11th	Mon	F	F	F	F	F	F	F	F	F	F	F	F	F	F	F	pc	pc	pc	pc	pc	163	
12th	Tue	F	F	F	F	F	F	F	F	pc	F	F	F	pc	pc	F	F	pc	pc	pc	pc	157	XhN
13th	Wed	pc	pc	pc	pc	pc	pc	pc	pc	pc	pc	pc	pc	pc	pc	pc	pc	pc	pc	pc	pc	131	
14th	Thur	pc	pc	pc	pc	pc	c	pc	c	pc	pc	pc	pc	pc	F	pc	F	F	F	F	F	126	P6
15th	Fri	o	o	o	o	o	o	o	o	c	c	c	o	c	pc	o	pc	pc	F	F	F	51	F
16th	Sat	pc	pc	pc	pc	pc	c	pc	c	pc	c	c	pc	pc	F	F	pc	pc	pc	pc	pc	106	
17th	Sun	o	o	o	o	o	o	o	o	o	o	o	o	o	o	o	o	o	o	o	o	1	
18th	Mon	c	c	c	c	c	c	c	c	c	c	c	c	c	o	c	o	o	o	o	o	27	^
19th	Tue	pc	pc	pc	pc	pc	pc	pc	pc	pc	pc	pc	pc	pc	pc	pc	pc	pc	pc	pc	pc	132	
20th	Wed	o	o	o	o	o	o	o	o	o	o	o	o	o	o	o	o	o	o	o	o	0	
21st	Thur	o	o	o	o	o	o	o	o	o	o	o	o	o	o	o	o	o	o	o	o	12	
22nd	Fri	o	o	o	o	o	o	o	o	o	o	o	o	o	o	o	o	o	o	o	o	7	
23rd	Sat	c	c	c	c	c	c	c	c	c	c	c	c	c	c	c	c	c	c	c	c	51	3Q
24th	Sun	o	o	o	o	o	c	o	c	o	o	o	o	o	o	o	o	o	o	o	o	10	
25th	Mon	o	o	o	o	o	o	o	o	o	o	o	o	o	o	o	o	o	o	o	o	0	XhS
26th	Tue	c	c	c	c	c	c	c	c	c	c	c	c	c	pc	c	c	c	pc	pc	pc	71	A
27th	Wed	c	c	c	c	c	c	c	c	c	c	c	c	c	pc	c	c	pc	pc	pc	pc	69	
28th	Thu	pc	pc	pc	pc	pc	c	pc	c	pc	c	c	pc	pc	pc	pc	c	pc	pc	pc	pc	94	
29th	Fri	o	o	o	o	o	o	o	o	o	o	o	o	o	o	o	o	o	o	o	o	3	
30th	Sat	pc	pc	pc	pc	pc	c	pc	c	pc	c	c	pc	pc	pc	pc	pc	pc	pc	pc	pc	89	
Estimate hours:		89	89	89	89	89	72	89	72	97	78	75	89	97	105	106	84	95	105	105	105	1817	
Average hours:		59	59	57	52	45	51	49	52	55	60	72	65	70	67	72	58	50	53	52	65	1161	
Trend:		more	more	more	more	more	more	more	more	more	more	av	more	more	more	more	more	more	more	more	more	more	

NOVEMBER **1** FRIDAY

A large depression moves across Ireland in an unsettled southwest flow, backing westerly later in the day. A changeable day with a mix of thunderstorms, showers, and occasional sunny spells in most districts, although mostly dry around the eastern coastal fringes.

Belfast: Cloudy, mostly dry, mild day cool night, fluctuating winds
Dublin: Unsettled blustery winds at times, partly cloudy, isolated showers
Cork: Partly cloudy, mostly dry with odd threat of rain, unsettled windy spells
Galway: Cloudy, light showers, mild day cold night, windy spells

RAIN POTENTIAL

FROST/SNOW

NOVEMBER **2** SATURDAY

A large depression over Ireland weakens with winds easing to gentle throughout. Some brief overnight showers mostly in the west and north with overnight persistent fog or mists elsewhere, with cloud clearing to mostly sunny by noon throughout. Chance of some late spotty showers along the northern coasts, and very few sunny spells in the far south. Evening fog patches likely to return, as well as some pockets of light overnight frosts mostly in the east and inland parts.

Belfast: Mostly sunny, cooler, dry, light shifting breezes
Dublin: Partly cloudy, dry, mild day cold night, fluctuating breezes
Cork: Cloudy, dry, cooler, light breezes to mostly calm air, chance of fog
Galway: Mix of showers and sunny spells, mild day cold night, light breezes

RAIN POTENTIAL

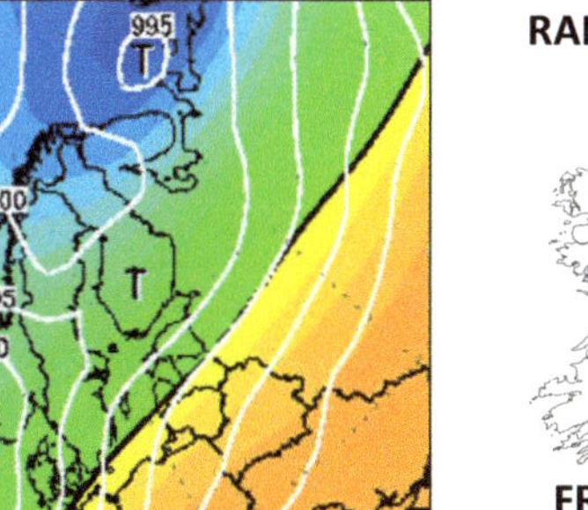

FROST/SNOW

NOVEMBER 3 SUNDAY

A complex of depressions move over Ireland with light to moderate southwest winds prevailing at first, backing southerly later in the day. A changeable day with a mix of sunny spells and threats of rain building everywhere, with some outbreaks of spotty showers mostly in the eastern half of the country. Chance of slight ground frosts overnight mostly in central parts.

Belfast: Partly cloudy, brief showers, squally winds at times
Dublin: Partly cloudy, dry, mild day cold night, shifting breezy spells
Cork: Partly cloudy, slightly warmer, mostly dry, increasing threats of rain, windy spells
Galway: Light breezes to mostly calm air, mostly cloudy, dry

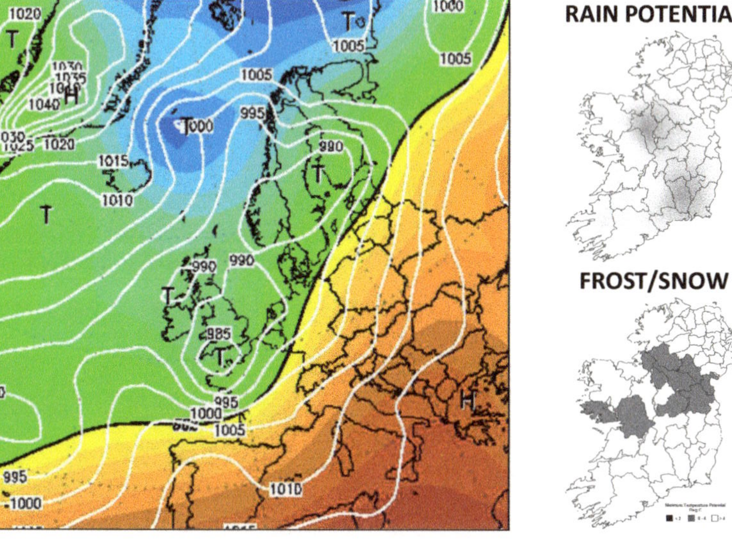

RAIN POTENTIAL

FROST/SNOW

NOVEMBER 4 MONDAY

A depression to the southwest of Ireland moves northwards in a light to moderate southeasterly flow at first, becoming strong and blustery with a northeasterly change later in the day. A stormy air prevails throughout the country with dull and gloomy skies and widespread heavy rain, starting in the south at first and moving up to the north of the country as the day progresses.

Belfast: Overcast, heavy rain, unsettled windy spells
Dublin: Overcast, heavy rain, milder, breezy spells
Cork: Stormy air, overcast, heavy rain, unsettled winds
Galway: Overcast, heavy rain, colder, unsettled and changeable winds

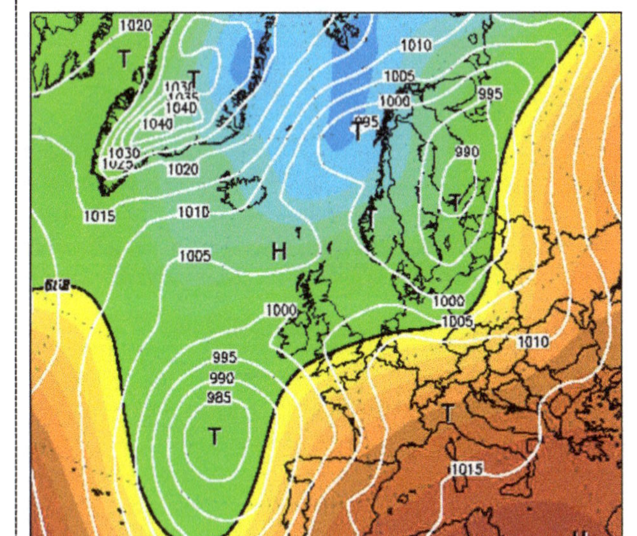

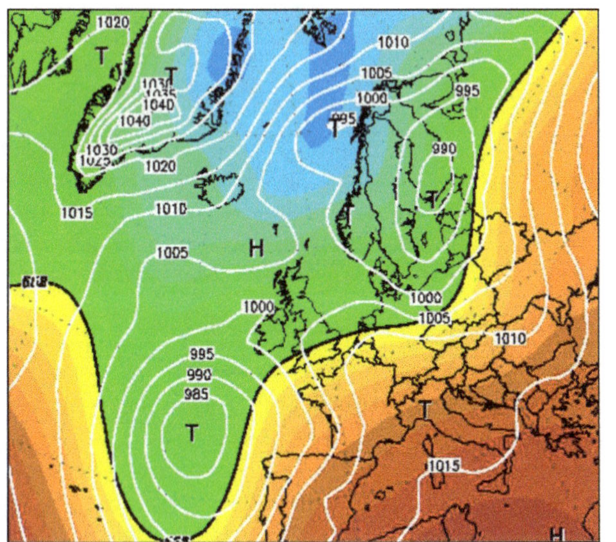

RAIN POTENTIAL

FROST/SNOW

NOVEMBER 5 TUESDAY

An intense depression covers the country bringing very strong and gusty winds everywhere, with gale gusts likely, easing later in the day. The stormy air continues to bring widespread rain activity at first, before contracting later to persisting mostly in the north, with some heavy falls still likely in the northeast. Increasing sunny spells develop behind the rain.

Belfast: Overnight rain, some sunny spells, mild day cool night, stormy air
Dublin: Mostly sunny, brief showers, fluctuating winds
Cork: Mostly sunny, light overnight showers, mild day cold night, fluctuating breezes
Galway: Unsettled winds, mix of showers and sunny spells, slightly warmer

RAIN POTENTIAL

FROST/SNOW

NOVEMBER 6 WEDNESDAY

The intense depression that brought stormy conditions over the past couple of days continues drifts to the northeast of Ireland while another depression to the west of Ireland approaches, with a low pressure trough crosses the country in a light southwest flow at first, becoming unsettled later in the day. Another band of scattered showers and drizzle crosses the country at first, before clearing to mostly dry and cloudy throughout.

Belfast: Overcast, spotty showers, cooler, fluctuating breezes
Dublin: Cloudy, light showers, moderate breezes
Cork: Overcast, light showers, moderate breezes
Galway: Overcast, light showers, moderate breezes, mild day cool night

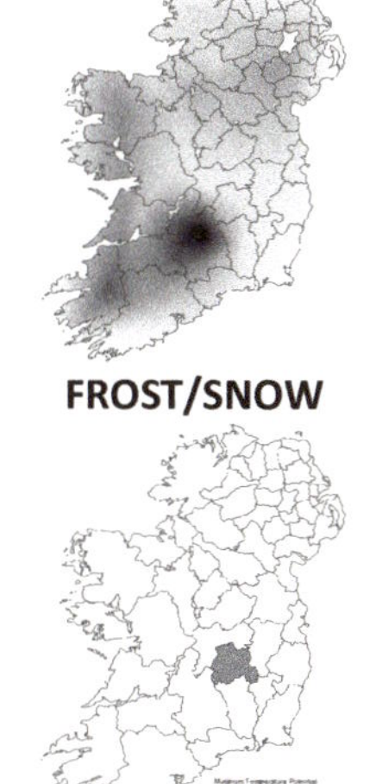

RAIN POTENTIAL

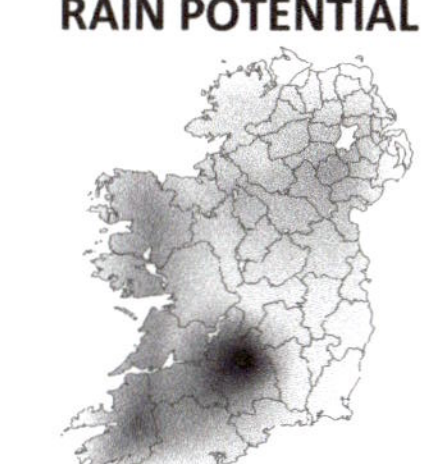

FROST/SNOW

NOVEMBER 7 THURSDAY

Low pressure over Ireland extends a light to moderate southwesterly flow at first, backing west to northwesterly later in the day ahead of a high pressure ridge to the southeast of the country. Dull and gloomy throughout and mostly dry in the east, with a brief spell of showers or passing drizzle patches elsewhere. Chance of fog patches overnight in the northern, inland, and southern parts.

Belfast: Overcast, blustery air easing later, mostly dry with passing threats of rain
Dublin: Overcast, isolated showers, strong breezes at times
Cork: Overcast, isolated showers, milder, fresh breezes
Galway: Overcast, light showers, fresh to strong winds

RAIN POTENTIAL

FROST/SNOW

NOVEMBER 8 FRIDAY

A high pressure system ridges over the country bringing a light west to southwest flow, to a mostly calm air throughout. Mostly dry with patches of fog overnight, and some frost patches in inland parts. Mostly dry and cloudy throughout the day, except for some brief outbreaks of light rain or drizzly spells in over the western half of the country.

Belfast: Cloudy, mild day cool night, some drizzly showers, moderate breezes
Dublin: Cloudy, colder, dry, light breezes to mostly calm air, chance of fog patches
Cork: Overcast, light showers, mostly calm air, chance of misty patches
Galway: Overcast, occasional showers, mild day cold night, chance of fog patches

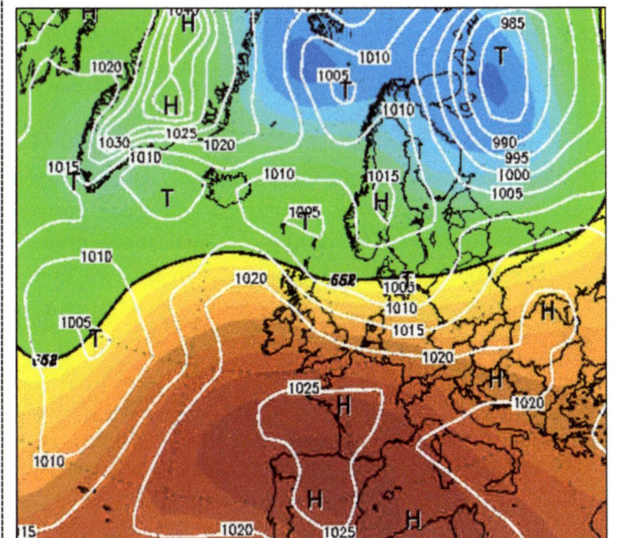

RAIN POTENTIAL

FROST/SNOW

NOVEMBER 9 SATURDAY

A high pressure ridge sits to the east of Ireland while a depression nears the northwest of the country. Mostly light southwest winds prevail at first, but building to gale force in the north and west later in the day. Widespread cloud throughout, with a band of rain crossing the country from the northwest to the southeast, bringing scattered showers into most districts, heavier in the northwest.

Belfast: Cloudy, some showers, unsettled winds strengthening later in the day
Dublin: Cloudy warmer, isolated showers, light breezes freshening later in the day
Cork: Moderate breezes, dull and gloomy air, isolated showers
Galway: Overcast, light showers, moderate breezes

RAIN POTENTIAL

FROST/SNOW

NOVEMBER 10 SUNDAY

An intense depression to the north of Ireland brings low pressure throughout the country with fresh to strong westerly winds throughout, and up to gale force over the northern half of the country. After some isolated showers likely in some northern and western districts overnight, a mostly dry day can be expected everywhere with increasing sunny spells in the west and south.

Belfast: Stormy winds, mostly cloudy and dry
Dublin: Mostly cloudy, isolated showers, colder, windy spells
Cork: Partly cloudy, colder, dry, moderate breezes
Galway: Partly cloudy, dry, mild day cool night, moderate breezes

RAIN POTENTIAL

FROST/SNOW

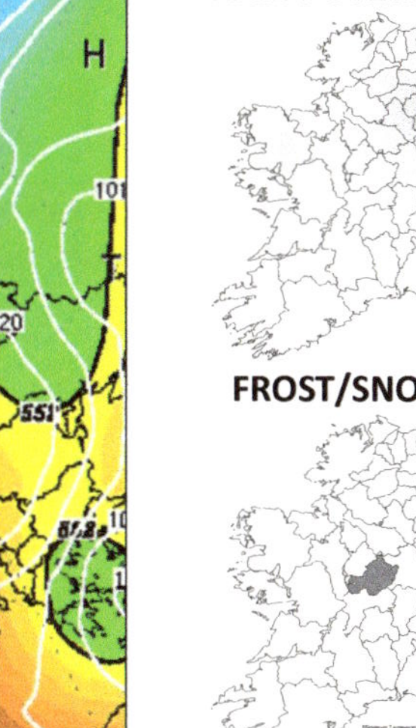

NOVEMBER **11** MONDAY

A large high sits to the northwest of Ireland directing a cold and light northerly flow throughout, with widespread frosts arriving in most districts. Dry and sunny everywhere. An intense depression to the northeast of Ireland may see some residual isolated showers briefly lingering around the northern coasts overnight.

Belfast: Sunny, dry, cooler, light northerly flow
Dublin: Sunny, dry, colder with chance of overnight frosts, moderate breezes
Cork: Mostly sunny and dry, cold northerly breezes
Galway: Sunny, colder, dry, light breezes to mostly calm air, chance of frosts overnight

RAIN POTENTIAL

FROST/SNOW

NOVEMBER **12** TUESDAY

A large high sits over Ireland bringing a cold and light northerly air throughout. Dry and mostly sunny everywhere with frosts in most districts, and scattered patches of fog in the north and west. Frosts may be severe in numerous inland and central districts. Chance of some isolated coastal showers overnight in the far north.

Belfast: Sunny, dry, cold overnight, light shifting breezes, chance of overnight fog
Dublin: Sunny, dry, cold with frosts, light shifting breezes
Cork: Partly cloudy, dry, cool, moderate to light shifting breezes
Galway: Sunny, dry, frosts, gentle breezes, chance of fog

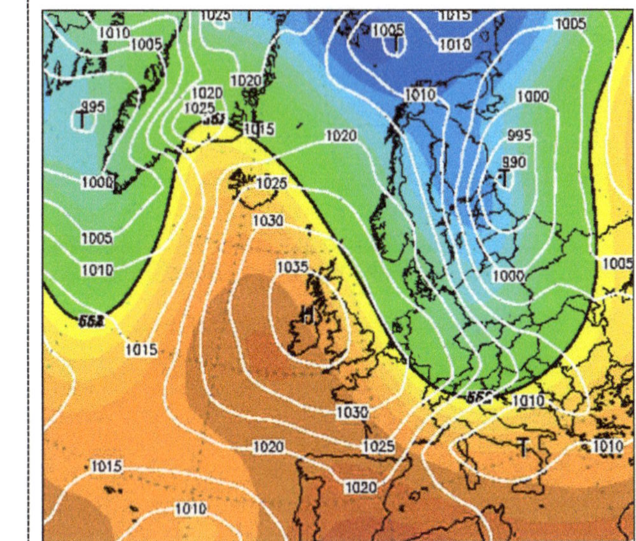

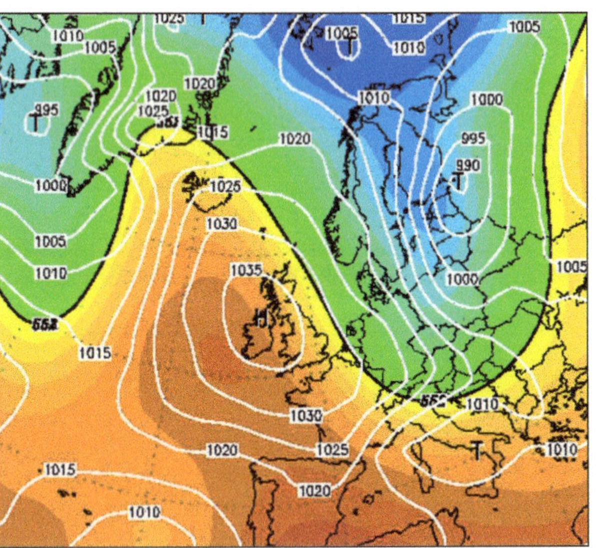

RAIN POTENTIAL

FROST/SNOW

NOVEMBER 13 WEDNESDAY

An anticyclonic system remains stalled over Ireland with light winds everywhere of variable directions. Dry and mostly sunny throughout with scattered fogs mostly in the west and north, and widespread frosts, with severe frosts in many parts, except in the far south.

Belfast: Mostly sunny, dry, cool, light breezes, chance of frost and fog overnight
Dublin: Mostly sunny, dry, frosts, shifting breezy spells
Cork: Partly cloudy, dry and cool, light breezes
Galway: Mostly calm and serene air, partly cloudy, dry, frost and fog patches likely

RAIN POTENTIAL

FROST/SNOW

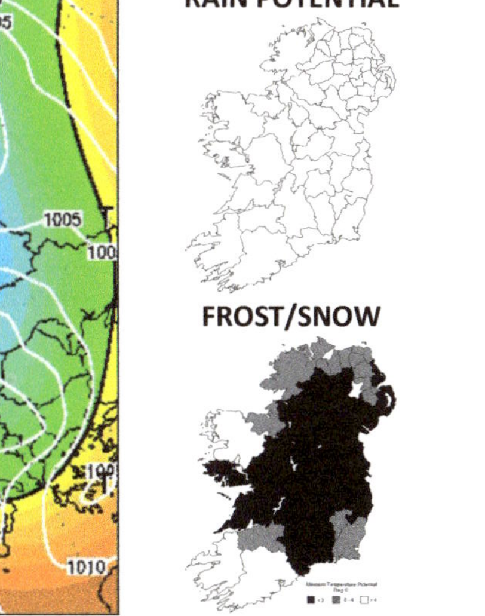

NOVEMBER 14 THURSDAY

A large anticyclonic system remains stalled over Ireland bringing dry and mostly sunny conditions throughout. Winds generally light to mostly calm everywhere, and fog patches likely in western districts. Frosts in most districts, except in the far northern and northwestern coastal regions, and severe frosts likely in the midlands.

Belfast: Partly cloudy, colder with frosts, light shifting breezes
Dublin: Mostly sunny, dry, slightly warmer, light breezes
Cork: Sunny, dry, cool, gentle shifting breezes, chance of shallow fog patches
Galway: Partly cloudy, cold overnight, gentle breezes, chance of frost and fog patches

RAIN POTENTIAL

FROST/SNOW

NOVEMBER 15 FRIDAY

The large anticyclonic system lingering over Ireland begins to weaken, with a gentle westerly flow prevailing. Mostly dry throughout, except for the chance of some isolated drizzle patches overnight. Sunny in the south, with increasing cloud in the west and east and some patches of fog here and there. Mostly overcast in the north, where freshening winds are likely to develop by the end of the day. Milder throughout, with light frosts likely mostly in inland parts to the southeast.

Belfast: Milder, dull and gloomy, dry, unsettled winds, blustery later in the day
Dublin: Mostly calm, cloudy and dry, milder, chance of fog patches
Cork: Sunny, dry, mostly calm air, chance of shallow fog patches
Galway: Cloudy, dry, pleasant serene air, chance of fog patches

RAIN POTENTIAL

FROST/SNOW

NOVEMBER 16 SATURDAY

Anticyclonic conditions continue, with light southerly breezes prevailing, backing westerly and strengthening in the north. Cold overnight with patches of fog and frosts in many districts, with fog patches slow to clear in inland parts. Mostly dry with occasional sunny spells throughout, but cloudier in the west and north. Some developing afternoon showers in western and northern districts.

Belfast: Unsettled and changeable, partly cloudy, dry, gusty winds at times
Dublin: Light breezes to mostly calm, sunny, mild day cool night, chance of fog patches
Cork: Mostly sunny and dry, gentle shifting breezes, possible fog
Galway: Cloudy, dry, cooler, gentle shifting breezes, chance of fog

RAIN POTENTIAL

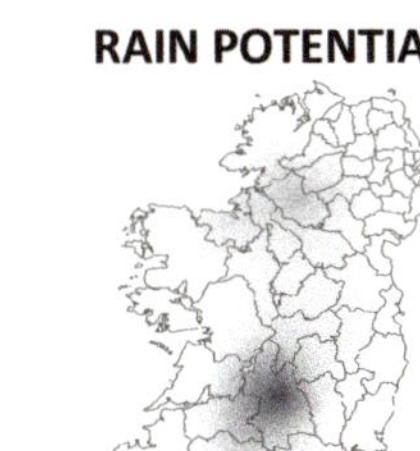

FROST/SNOW

NOVEMBER 17 SUNDAY

The high pressure ridge slips to the southeast of Ireland while low pressure moves into the west and north, with light to moderate southerly winds prevailing, turning southwesterly later in the day. Widespread cloud everywhere brings a milder day throughout. A wet front crosses the country in the northwest at first, moving southeasterly through the country, bringing spotty showers and drizzle outbreaks in most districts, heavier and more frequent in the west.

Belfast: Overcast, light showers, moderate winds
Dublin: Overcast, occasional showers, moderate to light breezes
Cork: Overcast, light showers, gentle breezes, mists and drizzle patches likely
Galway: Overcast, milder, occasional showers and drizzle patches, light breezes

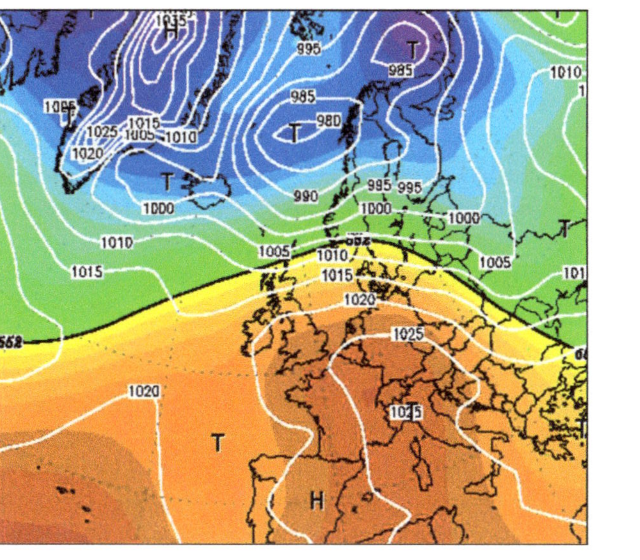

RAIN POTENTIAL

FROST/SNOW

NOVEMBER 18 MONDAY

Low pressure lies over Ireland with light to moderate westerly breezes prevailing, freshening later particularly in the north, where winds may reach gale force by evening. Overnight patches of mist and fog mostly around the coasts, and widespread cloud and only short sunny spells throughout the day. Mostly dry in the east with light showers in elsewhere.

Belfast: Cloudy, dry, milder, moderate breezes
Dublin: Cloudy, mostly dry with odd threat of rain, moderate breezes
Cork: Mostly calm air, dull and gloomy, light showers, fog patches possible
Galway: Cloudy, dry, gentle breezes

RAIN POTENTIAL

FROST/SNOW

NOVEMBER 19 TUESDAY

Low pressure contracts northwards away from Ireland, allowing a high pressure system to ridge up over the country, with a cool and light northwesterly breeze to a mostly calm air prevailing. Mostly dry with good sunny spells throughout, as well as frosts in many districts, and scattered fog patches likely in the west. Overnight showers mostly in northern parts possible.

Belfast: Dry, cooler northwest breezes, some sunny spells
Dublin: Mostly sunny and dry, colder, frosts likely overnight, light shifting breezes
Cork: Partly cloudy, dry, colder, light shifting breezes
Galway: Partly cloudy, brief showers, light breezes, chance of frost and fog patches

RAIN POTENTIAL

FROST/SNOW

NOVEMBER 20 WEDNESDAY

High pressure drifts over the eastern half of Ireland while a depression to the northwest of the country pushes down along the northwest. Moderate southerly breezes prevail at first, later strengthening with a southwest to westerly change. Dull and gloomy throughout with rain spreading everywhere as the day progresses, clearing later in the day but remaining overcast with lingering threats of rain.

Belfast: Overcast, scattered showers, freshening winds, gale gusts possibly developing
Dublin: Overcast, warmer, isolated showers, light breezes freshening later in the day
Cork: Overcast, scattered showers, fresh onshore breezes
Galway: Overcast, light showers, mild day cold night, moderate breezes

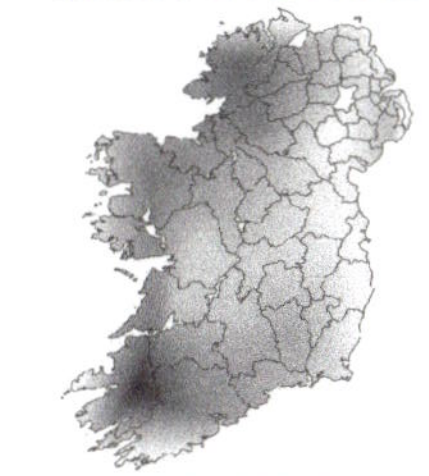

RAIN POTENTIAL

FROST/SNOW

NOVEMBER **21** THURSDAY

An intense depression to the north of Ireland directs strengthening west to northwest winds over the northern half of the country, up to gale force at times, while the high pressure ridge to the south brings lighter northwesterly breezes. Overcast everywhere with a stormy air at first. Showers will be heavy in the north, lighter in central parts, and more isolated in the south.

Belfast: Stormy air, overcast, light squally showers, chance of gales
Dublin: Overcast, dry with odd threat of rain, cool, blustery at times
Cork: Overcast, colder, mostly dry, odd threat of rain, fresh breezy spells at times
Galway: Overcast, mostly dry with increasing threats of rain, windy spells

RAIN POTENTIAL

FROST/SNOW

NOVEMBER **22** FRIDAY

The high pressure ridge over the country maintains a steady west to northwesterly flow, while a large depression to the northwest and another to the northeast bring a strong northwesterly flow along the far northern coasts. Dull and gloomy everywhere, with some light showers in most districts overnight, and a further belt of rain in the afternoon crosses the country from the west to the east, becoming more isolated as it moves eastwards, remaining mostly dry along the east and southeast coastal fringes.

Belfast: Unsettled and stormy air, overcast, cooler, light squally showers
Dublin: Overcast, isolated showers, fresh breezes
Cork: Overcast, dry with odd threat of rain, breezy spells
Galway: Overcast, warmer, windy spells, scattered showers

RAIN POTENTIAL

FROST/SNOW

NOVEMBER 23 SATURDAY

An unsettled and stormy air prevails as an intense depression to the northwest of Ireland directs a very strong westerly flow throughout the country, with winds up to gale force at times, particularly in the northwest. Cloudy and with outbreaks of light showers and drizzle patches almost everywhere, although remaining mostly dry along the east and southeast coastal fringes.

Belfast: Storm, cloudy, isolated showers
Dublin: Cloudy, isolated showers, strong gusty winds
Cork: Cloudy, isolated showers, strong gusty winds
Galway: Cloudy, light showers, strong and blustery winds at times

RAIN POTENTIAL

FROST/SNOW

NOVEMBER 24 SUNDAY

A large depression north of Ireland continues to bring strong and unsettled westerly winds throughout the country. Dull and gloomy, with widespread cloud, and squally showers everywhere, more persistent at times in the north of the country. Winds may reach gale force along the northern and western coasts.

Belfast: Overcast, scattered showers, cool, strong gusty winds with gale gusts at times
Dublin: Overcast, dry, cooler, strong gusty winds at times, rain threatening
Cork: Unsettled and blustery air, overcast, scattered showers
Galway: Overcast, scattered showers, strong gusty winds, chance of gales

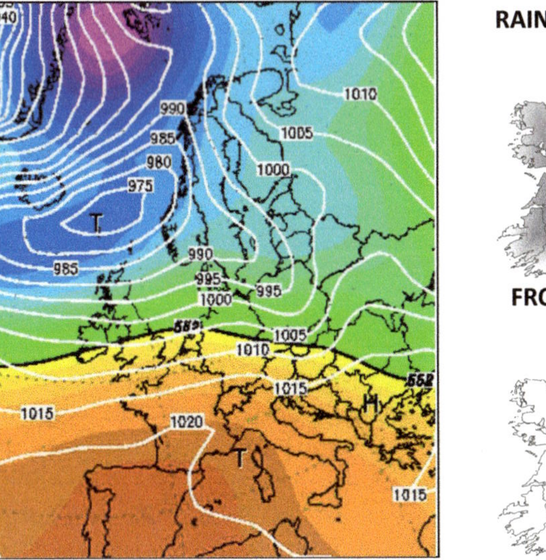

NOVEMBER 25 MONDAY

The depression to the north of Ireland maintains a storm influence over the country with strong winds in a south to southeast direction, up to gale force for a time. Widespread and heavy rain everywhere, particularly persistent and heavy in the south and east. Thunderstorm activity is likely in the north.

Belfast: Stormy air, overcast, heavy rain, cooler, chance of thunderstorms and gales
Dublin: Overcast, heavy rain, windy spells, warmer day cooler night
Cork: Overcast, heavy rain, colder, fresh winds at times
Galway: Unsettled and blustery air, overcast, rain, colder

RAIN POTENTIAL

FROST/SNOW

NOVEMBER 26 TUESDAY

The northern depression moves onto the country bringing unsettled and changeable winds, predominately in a cold, northwesterly flow. Bands of wintry rain move across the country from the west and northwest, with some heavy showers likely in the north and west. Some showers may be of hail and there is a chance of snow in the west for a brief time. Some good sunny spells likely in the south of the country.

Belfast: Cloudy, colder, rain, gusty winds
Dublin: Cloudy, colder, mostly dry with odd threat of rain, unsettled windy spells
Cork: Partly cloudy, colder, light showers, fluctuating windy spells
Galway: Cloudy, cold, heavy rain, fluctuating winds, chance of snow and gales

RAIN POTENTIAL

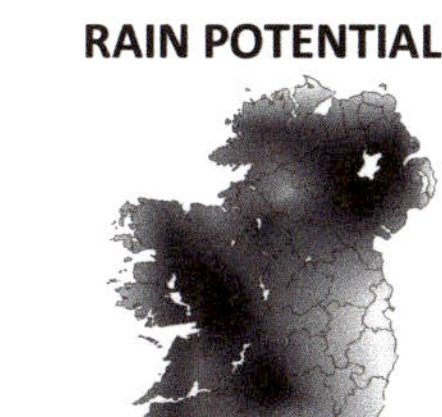

FROST/SNOW

NOVEMBER 27 WEDNESDAY

The depression to the north of Ireland continues to bring a stormy air across the country, with a cold westerly flow prevailing. Widespread wintry showers in all districts, with heavy precipitation in the west and southwest. Chance of light snow and gale force gusts for a time in the north and west. Longer sunny spells in the south.

Belfast: Cloudy, brief wintry showers, strong winds, gales possible, chance of light snow
Dublin: Cold, mostly cloudy, light passing showers, strong winds at times
Cork: Partly cloudy, rain, cold, chance of wintry showers, strengthening winds
Galway: Cloudy, heavy wintry rain, cold, blustery, snow and gales possible

RAIN POTENTIAL

FROST/SNOW

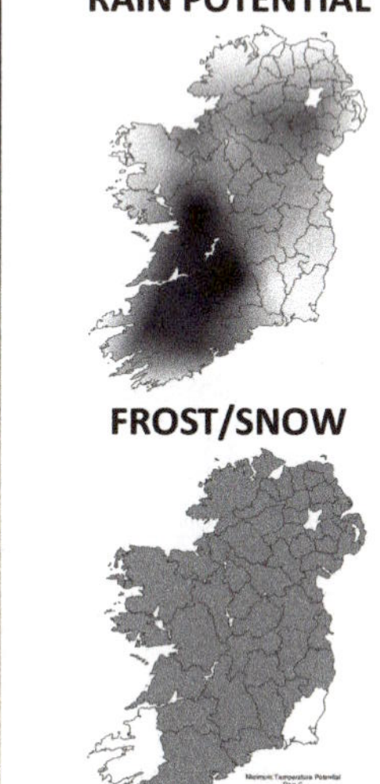

NOVEMBER 28 THURSDAY

The depression over Ireland maintains a stormy air throughout with strong and blustery northwest to southwest winds prevailing everywhere, up to gale force for a time, particularly in the western half of the country. Some sunny spells in most parts, interspersed with the passage of wet fronts bringing wintry showers almost everywhere, some of hail or snow, except in the far south and eastern coasts. Frosts will be widespread overnight.

Belfast: Storm winds, some sunny spells, cold, wintry showers, chance of snow flurries
Dublin: Partly cloudy, cold, brief spotty showers, strong winds with gale gusts possible
Cork: Mostly sunny, dry, cold, chance of overnight frosts, blustery with possible gales
Galway: Stormy air, cloudy, wintry showers, chance of snow, hail showers and gales

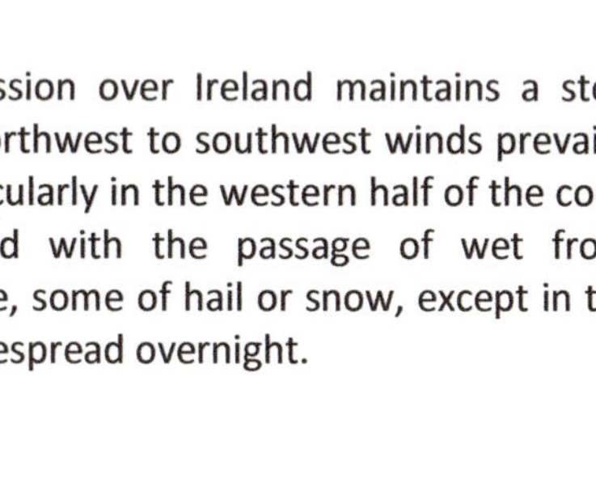

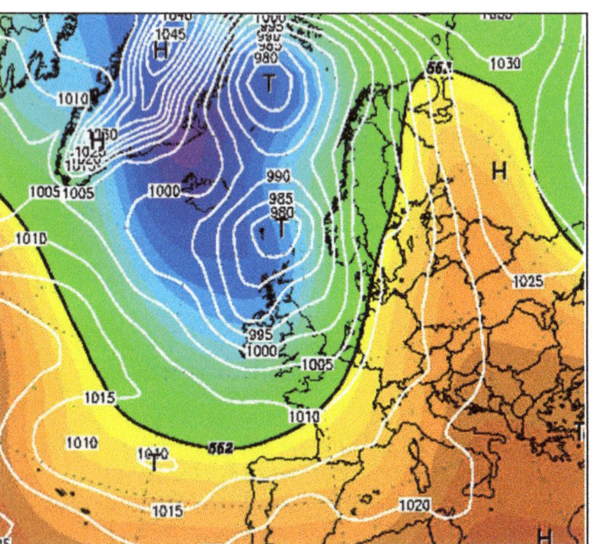

RAIN POTENTIAL

FROST/SNOW

NOVEMBER **29** FRIDAY

Low pressure prevails to the north of Ireland while the high pressure ridge sits to the south of the country. A cold southwesterly flow prevails throughout, strong and blustery and at times up to gale force around the northern and western coasts. Widespread cloud, dull and gloomy and with wintry rain everywhere. Very cold overnight temperatures will see frosts almost everywhere, and the chance of wintry showers may fall as snow, mostly in the west.

Belfast: Overcast, colder, blustery, chance of gales, outbreaks of wintry rain likely
Dublin: Overcast, warmer day cold night, scattered showers, moderate winds
Cork: Overcast, wintry rain, mild day cold night, possible frosts, windy spells
Galway: Overcast, wintry rain, blustery, chance of frost and snow showers

RAIN POTENTIAL

FROST/SNOW

NOVEMBER **30** SATURDAY

An intense depression to the north of Ireland returns a stormy air across the country with strong gusty northwest winds everywhere, with winds up to storm force at times. Mostly cloudy in the north while some sunny spells elsewhere, and scattered showers in most districts, particularly in the north and west, but remaining mostly dry along the coastal fringes to the east and southeast.

Belfast: Storm, milder, brief squally wintry showers
Dublin: Stormy air, partly cloudy, mostly dry, odd threat of rain, gale winds likely
Cork: Partly cloudy, cold, mostly dry, threats of rain, gales to storm force winds likely
Galway: Stormy air, mostly cloudy, passing wintry showers, chance of snow

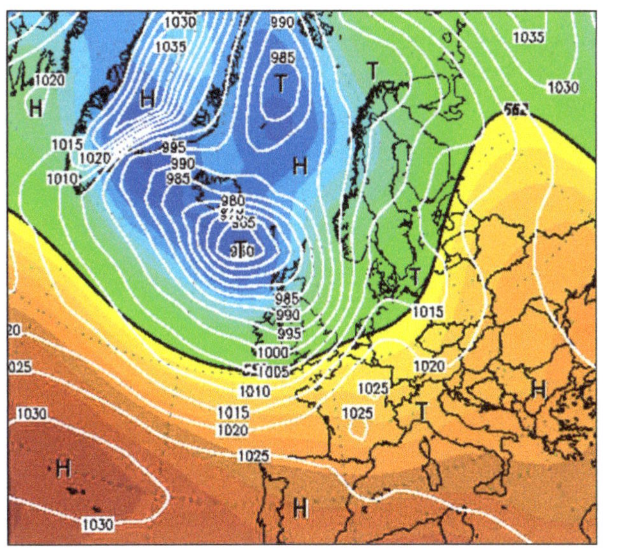

RAIN POTENTIAL

FROST/SNOW

DECEMBER

PHASES OF THE MOON

1st	New Moon
2nd	Southern Declination
8th	First Quarter
10th	Crossing Equator
12th	Perigee #11
15th	Full Moon
15th	Northern Declination
22nd	Third Quarter
23rd	Crossing Equator
24th	Apogee
30th	New Moon
30th	Southern Declination

1	2	3	4	5	6	7	8	9	10	11	12	13	14	15	16	17	18	19	20	21	22	23	24	25	26	27	28	29	30	31
Sun	Mon	Tue	Wed	Thur	Fri	Sat	Sun	Mon	Tue	Wed	Thur	Fri	Sat	Sun	Mon	Tue	Wed	Thur	Fri	Sat	Sun	Mon	Tue	Wed	Thur	Fri	Sat	Sun	Mon	Tue
N	V						1Q		XhN		P11			F ^							3Q	XhS	A						N V	

MONTHLY SUMMARY

1st Wintry showers likely, some of hail.
1st-10th Unsettled weather from last month continues with heavy rain and frequent strong westerly winds at first, easing later with south to southwest change. Scattered showers almost everywhere on each day.
2nd-3rd Widespread rain across the country, heavy at times in the west and south.
3rd-5th Chance of thunderstorms, mostly in the southwest.
5th-6th Widespread fog patches possible.
6th-7th Strong and blustery winds. Chance of thunderstorms.
8th-10th Mild maximum temperatures during this spell.
10th An intense depression passes near the northwest coast bringing stormy conditions over the country. Widespread rain throughout, with heavy falls in the south and southwest that may cause some localised flooding. Winds will be unsettled and blustery, possibly to gale force for a time.

The next four weeks ahead will bring variable conditions, with wet and windy conditions to start, and followed later by calmer and drier weather, before unsettled conditions return. Low pressure situated close to the northwest of the country in the first few days starts this outlook period, followed by more stable conditions for a couple of weeks, and ending this outlook period with another round of low pressure and widespread unsettled weather. Rainfall will be above normal everywhere during this outlook period, and particularly wetter than normal in the midlands, western and northern parts where some districts may see up to twice as much rainfall than normal for this time of year. Sunshine hours will vary considerable with many parts in the north and east of the country possibly seeing up to twice as much sunshine as the west and south. Some parts in the far northwest may see up to twice as many sunshine hours as is normal for this outlook period. Temperatures will be milder than normal during most of the coming four weeks, with the most seasonal temperatures for this time of year during a calmer and colder spell between Christmas Day and the New Year's period. Frosts will be well below average with only around half the normal number expected, and some coastal areas may not see any. The first half of this outlook period will see both maximum and minimum temperatures up to 3°C above the norm for this time of year and any air frost temperatures are likely to be confined to the few days either side of the New Year. The next four weeks are likely to be unusually windier than normal for this time of year, particularly around the north and northwest. During this outlook period, there is likely to be around 15-23 days with rain, thunderstorms on 11 days, hail on 8 days and gales on 4-5 days. There may be 1 tornado seen during this outlook period, but no snow. Fogs may be observed on 9 days, and frosts only on 3-5 days, mostly in inland areas.

11th-27th A series of deep depressions to the northwest of the country directs a strong southwest flow and frequent outbreaks of rain or showers each day, sometimes heavy falls. Milder than normal temperatures on most days.
11th-14th An intense Atlantic depression situated near the northwest of Ireland pushes a series of wet fronts, scattered thunderstorms with hail showers and occasional gale gusts across the country. Rain likely to be very heavy at times and possibly bring some pockets of localised flooding. Maximum temperatures will be generally milder than normal for this time of year.
12th-14th Widespread heavy rain likely in most districts.
17th-23nd Stormy air. Thunderstorms, possible hail showers, and strong blustery winds with chance of gale gusts likely. Chance of heavy rain possibly causing some localised flooding in the Midlands and in the western river catchments.
23rd-24th Widespread heavy rain. Milder temperatures.
25th-4th Jan 2025 Spell of calmer and colder weather.
26th-31st Mostly sunny spell after widespread fogs lift.
28th-30th Cooler spell with widespread dense fogs expected.
28th-5th January 2025 High pressure prevails. An anticyclonic system drifts slowly towards the country during this period bringing a calmer spell with very light winds, frequent cloud, regular scattered fogs, and a mostly dry spell with only light showers now and then. Sunnier once the fogs lift. Much cooler than normal.
29th-2nd January 2025 Widespread frosts likely, severe in the north Midlands to the southeast.

IRELAND RAINFALL ESTIMATES
December 2024

d=mainly dry, n= nonrecordable, l=light shrs, s=significant shrs, r=rain, h=hvy falls

		ULSTER PROVINCE					CONNAUGHT PROVINCE						LEINSTER PROVINCE				MUNSTER PROVINCE						
		Alderg	Hillsbo	Armag	Loughl	Malin I	Sligo	Ardtarı	Belmul	Drums	Galway	Roscoı	Markre	Edende	Kilkenı	Dublin	Shannc	Newpo	Tralee	Killarne	Cork	All	Moon
1st	Sun	l	l	l	s	l	s	s	l	l	l	l	s	l	l	d	l	l	l	n	l	46	N
2nd	Mon	s	r	n	r	s	r	r	h	r	r	r	r	s	r	r	s	h	h	r	r	220	V
3rd	Tue	l	l	l	r	3s	r	r	r	s	r	s	l	s	l	n	s	r	s	r	s	144	
4th	Wed	l	s	l	l	s	l	l	s	s	s	l	s	s	s	s	l	s	h	l	r	89	
5th	Thur	l	d	l	s	r	r	s	r	l	s	d	s	d	l	d	l	s	l	l	n	69	
6th	Fri	s	s	l	s	l	r	s	s	l	l	s	s	s	r	l	s	s	r	h	r	153	
7th	Sat	s	l	s	l	l	s	l	s	l	l	n	l	l	s	l	s	l	s	r	l	72	
8th	Sun	l	d	l	l	l	s	l	s	l	s	l	l	n	l	l	s	l	r	h	l	61	1Q
9th	Mon	d	d	d	s	d	r	s	l	l	s	l	l	d	d	d	l	s	l	r	l	60	XhN
10th	Tue	s	r	d	h	r	h	h	h	h	r	r	r	r	h	s	s	s	h	h	h	340	
11th	Wed	s	s	r	s	s	r	s	s	r	r	r	s	s	l	d	l	r	r	h	s	149	
12th	Thu	s	r	l	h	s	h	h	h	h	h	h	r	r	h	l	r	h	h	h	h	444	P11
13th	Fri	r	s	r	s	r	s	s	r	s	s	s	l	s	l	r	s	r	l	s	s	142	
14th	Sat	s	s	s	s	l	s	s	s	r	h	r	s	s	s	l	r	h	r	h	r	214	
15th	Sun	s	s	s	r	s	r	r	s	s	r	l	s	l	s	r	h	r	s	r	r	162	F ^
16th	Mon	s	s	l	r	s	r	r	r	h	r	h	h	r	s	l	l	h	r	h	s	193	
17th	Tue	s	s	s	s	r	s	s	r	r	s	r	l	r	s	r	r	s	r	r	s	182	
18th	Wed	l	l	s	s	s	s	s	r	r	s	s	s	l	d	d	l	r	s	r	l	113	
19th	Thur	l	s	d	s	s	r	s	s	s	h	s	s	s	s	d	s	r	l	r	l	114	
20th	Fri	s	s	s	s	r	s	s	h	s	h	r	n	s	s	s	r	r	h	h	r	211	
21st	Sat	s	l	s	s	r	s	s	s	r	s	s	s	l	n	s	s	r	s	r	l	126	
22nd	Sun	s	l	s	s	s	r	s	s	s	l	l	s	l	l	d	h	s	l	s	l	116	3Q XhS
23rd	Mon	l	s	l	s	h	r	s	h	s	s	r	r	l	s	d	n	h	s	h	s	173	
24th	Tue	s	s	s	s	r	s	s	s	s	s	s	r	r	r	r	r	r	h	h	s	199	A
25th	Wed	l	d	s	l	s	l	l	l	d	l	l	l	l	r	s	s	l	r	l	h	87	
26th	Thu	d	d	d	s	s	l	s	l	n	l	d	l	d	d	d	s	s	l	l	n	34	
27th	Fri	l	l	d	l	s	l	l	s	l	s	l	l	l	l	l	l	s	r	s	s	66	
28th	Sat	l	d	l	d	d	d	d	l	n	d	n	d	n	d	d	l	n	s	d	d	10	
29th	Sun	d	d	d	d	d	d	d	d	d	d	n	d	n	n	d	d	n	d	d	d	2	
30th	Mon	d	d	d	d	d	d	d	d	d	d	d	d	d	d	d	d	d	d	d	d	1	N V
31st	Tue	d	d	d	n	d	d	n	d	d	d	d	n	d	d	d	d	s	d	d	d	3	
Estimate:		109	118	99	197	217	255	197	257	204	236	185	167	127	132	94	181	278	311	426	206	3995	
Average:		79	90	77	124	116	126	128	137	110	123	138	126	95	84	73	104	180	155	112	133	2309	
Trend:		wtr	wtr	wtr	wtr	wtr	wtr	wtr	wtr	wtr	wtr	wtr	wtr	wtr	wtr	wtr	wtr	wtr	wtr	wtr	wtr	wtr	

IRELAND SUNSHINE ESTIMATES
December 2024

F=fine (8-12 hours of sunshine), pc= partly cloudy (4-7 hours), c=cloudy (1-3 hours), o=overcast (0 hours)

		ULSTER PROVINCE					CONNAUGHT PROVINCE						LEINSTER PROVINCE				MUNSTER PROVINCE						
		Alderg	Hillsbo	Armag	Loughl	Malin H	Sligo	Ardtar	Belmul	Drums	Galway	Roscoı	Markre	Edende	Kilkenı	Dublin	Shannc	Newpo	Tralee	Killarne	Cork	All	Moon
1st	Sun	pc	pc	pc	pc	pc	c	pc	c	pc	pc	pc	pc	pc	pc	pc	pc	pc	pc	pc	pc	95	N
2nd	Mon	o	o	o	o	o	o	o	o	o	o	o	o	o	o	o	o	o	o	o	o	0	V
3rd	Tue	c	c	c	c	c	c	c	c	pc	c	c	c	pc	pc	pc	c	pc	pc	pc	pc	72	
4th	Wed	o	o	o	o	o	o	o	o	o	o	o	o	o	o	o	o	o	o	o	o	3	
5th	Thur	c	c	c	c	c	c	c	c	c	c	c	c	c	pc	pc	pc	pc	c	c	c	73	
6th	Fri	c	c	c	c	c	c	c	c	c	c	c	c	c	c	pc	c	c	o	o	o	43	
7th	Sat	c	c	c	c	c	o	c	o	c	c	c	c	c	pc	pc	pc	pc	pc	pc	pc	66	
8th	Sun	c	c	c	c	c	o	c	o	c	o	o	c	c	c	pc	o	c	c	c	c	39	1Q
9th	Mon	c	c	c	c	c	o	c	o	o	o	o	c	o	o	c	o	o	o	o	o	14	XhN
10th	Tue	o	o	o	o	o	o	o	o	o	o	o	o	o	o	o	o	o	o	o	o	0	
11th	Wed	c	c	c	c	c	c	c	c	c	c	c	c	c	c	pc	c	c	c	c	c	52	
12th	Thu	c	c	c	c	c	c	c	c	c	c	c	c	c	c	pc	c	c	c	c	c	52	P11
13th	Fri	o	o	o	o	o	o	o	o	o	o	o	o	o	c	o	o	o	c	c	c	14	
14th	Sat	o	o	o	o	o	o	o	o	o	o	o	o	o	o	o	o	o	o	o	o	6	
15th	Sun	c	c	c	c	c	o	c	o	c	o	o	c	c	c	c	o	c	c	c	c	26	F ^
16th	Mon	o	o	o	o	o	o	o	o	c	c	c	o	c	c	o	pc	c	c	c	c	37	
17th	Tue	o	o	o	o	o	o	o	o	o	o	o	o	o	c	o	o	o	c	c	c	13	
18th	Wed	c	c	c	c	c	c	c	c	c	c	c	c	c	c	pc	c	c	c	c	c	64	
19th	Thur	pc	pc	pc	pc	pc	c	pc	c	pc	pc	c	pc	pc	pc	pc	pc	pc	pc	pc	pc	92	
20th	Fri	o	o	o	o	o	o	o	o	o	o	o	o	o	o	o	o	o	o	o	o	0	
21st	Sat	pc	pc	pc	pc	pc	c	pc	c	pc	c	c	pc	pc	pc	pc	c	c	pc	pc	pc	81	
22nd	Sun	c	c	c	c	c	c	c	c	c	c	c	c	c	c	c	o	c	o	o	o	30	3Q XhS
23rd	Mon	o	o	o	o	o	o	o	o	o	o	o	o	o	o	o	o	o	o	o	o	0	
24th	Tue	o	o	o	o	o	o	o	o	o	o	o	o	o	o	o	o	o	o	o	o	0	A
25th	Wed	o	o	o	o	o	o	o	o	o	o	o	o	o	o	o	o	o	o	o	o	2	
26th	Thu	pc	pc	pc	pc	pc	c	pc	c	pc	c	c	pc	pc	pc	pc	c	pc	pc	pc	pc	101	
27th	Fri	o	o	o	o	o	o	o	o	o	o	o	o	o	o	o	o	o	o	o	o	2	
28th	Sat	c	c	c	c	c	pc	c	pc	c	pc	pc	c	c	c	o	pc	pc	pc	pc	pc	78	
29th	Sun	o	o	o	o	o	o	o	o	o	o	o	o	o	o	o	o	o	c	c	c	5	
30th	Mon	c	c	c	c	c	o	c	o	c	o	o	c	c	c	pc	o	c	o	o	o	35	N V
31st	Tue	pc	pc	pc	pc	pc	c	pc	c	pc	c	c	pc	pc	c	pc	pc	pc	c	c	c	70	
Estimate hours:		60	60	60	60	60	35	60	35	67	45	40	60	67	73	85	54	64	61	61	61	1166	
Average hours:		39	39	34	34	28	34	34	33	33	36	42	47	45	46	53	44	30	36	36	48	774	
Trend:		more	more	more	more	more	av	more	av	more	av	av	more	more	more	more	av	more	more	more	more	more	

DECEMBER **1** SUNDAY

An intense depression to the north of Ireland maintains a very strong and blustery northwest flow throughout the country. Unsettled everywhere with a mix of sunny spells and occasional wintry showers. Winds may reach gale force at times, particularly over the northern half of the country.

Belfast: Storm, isolated wintry showers, some brief sunny spells, gales likely
Dublin: Partly cloudy, strong gusty winds with gales likely, cold and dry
Cork: Partly cloudy, cold, light wintry showers, strong gusty winds
Galway: Partly cloudy, light showers, windy with gale gusts likely

RAIN POTENTIAL

FROST/SNOW

DECEMBER **2** MONDAY

A trough between two depressions to the north of Ireland direct a light to moderate westerly flow throughout, with the occasional gusty spells. A band of rain will move across the country from the west at first and spread swiftly eastward to bring widespread rain throughout, followed by a brief dry spell before another band of rain is likely to cross the country from the southwest to the north. Rain will heavy at times many places, particularly in the west and southwest.

Belfast: Overcast, rain, cold overnight, winds easing but still unsettled
Dublin: Overcast, cold, rain, windy
Cork: Overcast, warmer, rain, fluctuating breezes
Galway: Overcast, rain, mild day cold night, unsettled breezes

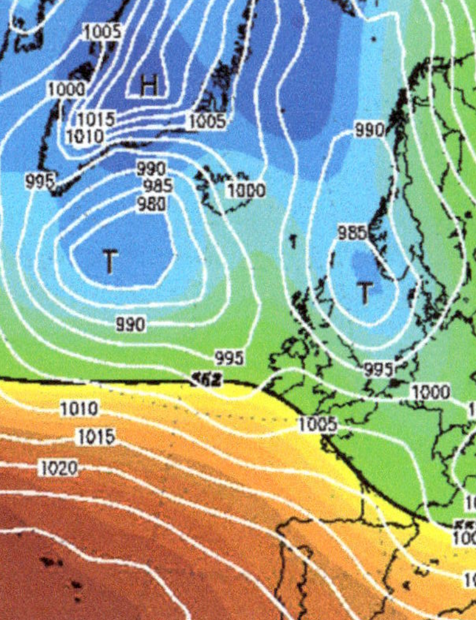

RAIN POTENTIAL

FROST/SNOW

DECEMBER **3** TUESDAY

A depression to the northwest of Ireland extends a cold and wet trough over the country in a fresh to strong and occasionally gust northwest to westerly flow. Mostly dry with some sunny spells in the east, while a mix of sunny spells and scattered showers can be expected elsewhere. Rain may be heavy for a time in the northwest of the country, and gales likely around the coasts.

Belfast: Cloudy, strong gusty winds, possible gales, some brief driving showers
Dublin: Partly cloudy, isolated showers, winds possibly to storm force for a time
Cork: Strong and blustery winds, some sunny spells, light overnight wintry showers
Galway: Cloudy, cool, rain, strong northwest winds

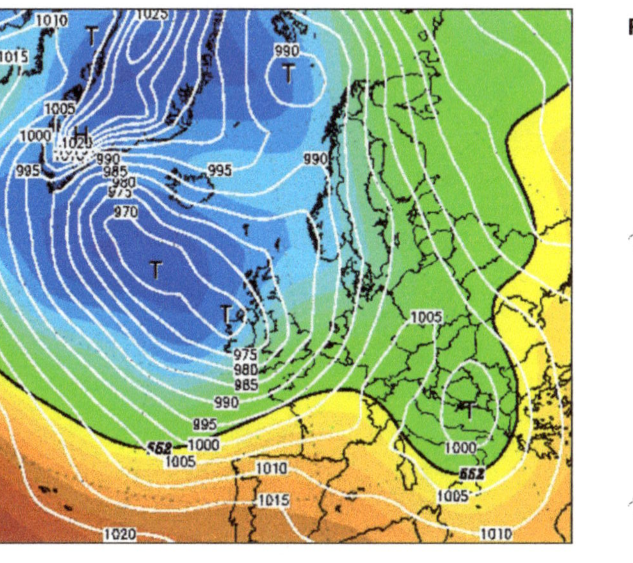

RAIN POTENTIAL

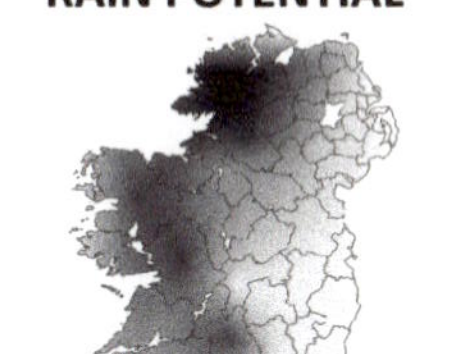

FROST/SNOW

DECEMBER **4** WEDNESDAY

A large depression brings a breezy northwest flow across the country, lighter in the east and strong in the northwest. Overcast, cold, and unsettled everywhere, with scattered wintry showers throughout, more isolated in the east, frequent in the west, and heavy at times in the northwest. Colder, with widespread overnight frosts expected to develop.

Belfast: Overcast, scattered showers, cold overnight, strong unsettled winds
Dublin: Overcast, scattered wintry showers, blustery, gale winds likely at times
Cork: Overcast, overnight rain, colder, windy with gusts
Galway: Dull and gloomy air, scattered showers, cold overnight, windy

RAIN POTENTIAL

FROST/SNOW

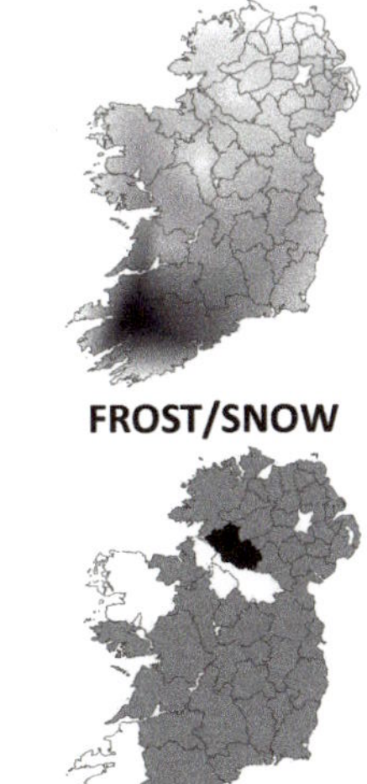

DECEMBER 5 THURSDAY

A large depression lingers over Ireland with moderate southwest winds prevailing. A mixed day of some sunny spells, longer in the east and midlands, and occasional showers mostly in the western half of the country. Rain likely to be heavy for a time in the northwest. Chance of thunderstorms and some snow in the west of the country. Overnight frosts likely to be fairly extensive except around the coastal fringes in the north and west.

Belfast: Stormy air, mostly cloudy and dry, passing threats of rain
Dublin: Partly cloudy and dry, cool, winds easing to moderate
Cork: Cloudy, isolated showers, moderate breezes
Galway: Mostly cloudy, light wintry showers, chance of thunderstorms and snow

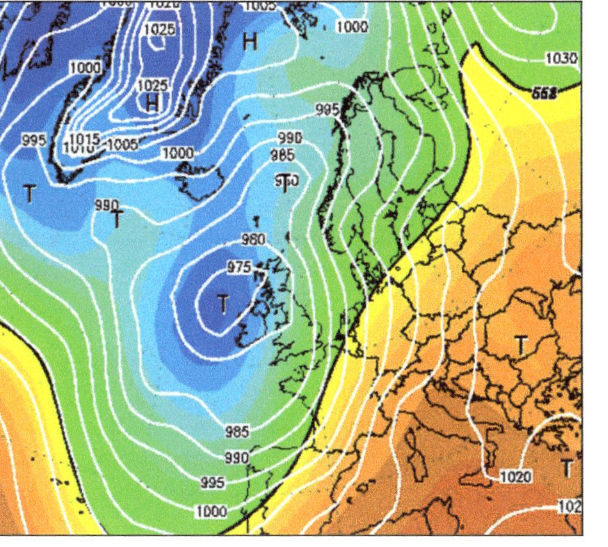

RAIN POTENTIAL

FROST/SNOW

DECEMBER 6 FRIDAY

Low pressure and fresh to strong southwesterly winds prevail throughout the country. A changeable day with a mix of some sunny spells and passing bands of scattered showers. Rain likely to be heavy for a time in some parts over the southern half of the country. Overnight frosts mostly confined to the east and inland districts.

Belfast: Cloudy, rain, blustery at times, cold overnight
Dublin: Partly cloudy, brief showers, windy
Cork: Overcast, warmer, rain, windy spells
Galway: Cloudy, light showers, cold overnight, fresh winds

RAIN POTENTIAL

FROST/SNOW

DECEMBER **7** SATURDAY

An intensifying depression to the west of Ireland brings a stormy air throughout, with fresh to strong southerly winds with gale gusts at times. Bands of heavy showers cross the country moving in from the west and spreading to the east. Chance of thunderstorms, particularly in the west. Milder temperatures with few frosts.

Belfast: Cloudy, milder, passing squalls, storm winds likely
Dublin: Stormy air, partly cloudy, brief showers, gale gusts to storm force possible
Cork: Partly cloudy, light showers, strong gusty winds, possible gales
Galway: Stormy air, cloudy, squally showers, thunderstorms and storm winds likely

RAIN POTENTIAL

FROST/SNOW

DECEMBER **8** SUNDAY

The large and intense depression lingering to the west of Ireland drifts slowly northwards, while directing strong and blustery south to southwest winds throughout, up to gale force at times. Dull and gloomy in the west, with some sunny breakthroughs elsewhere. A mostly dry day in the north, with occasional showers in the west, and isolated showers in the east and south.

Belfast: Cloudy, dry, strong winds to gale force, possible storm gusts
Dublin: Partly cloudy, colder, isolated showers, strong gusty winds, gales likely
Cork: Cloudy, strong blustery winds with gusts, light showers, colder
Galway: Overcast, light showers, strong winds with gale gusts likely

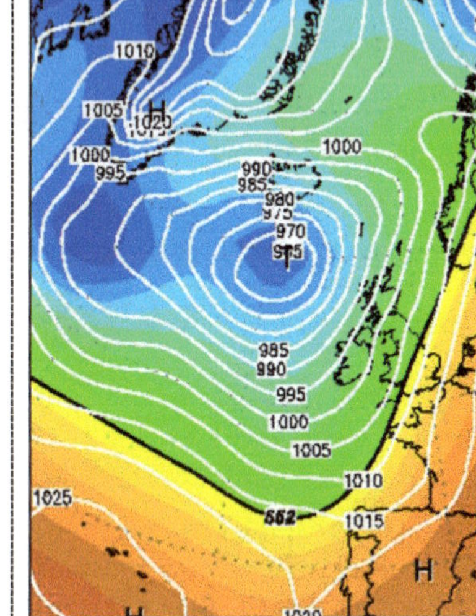

RAIN POTENTIAL

FROST/SNOW

DECEMBER **9** MONDAY

The large depression to the northwest of Ireland continues to bring fresh to strong and unsettled southwest winds across the country, up to gale force and with severe gusts at times. Very dull and gloomy with occasional showers in the south and western half of the country, while mostly dry and with some brief sunny spells elsewhere.

Belfast: Cloudy, dry, strong gusty winds, possible gale force at times
Dublin: Cloudy, dry, gusty winds, gales likely
Cork: Overcast, windy, squally showers
Galway: Overcast, scattered showers, strong gusty winds, chance of gales

RAIN POTENTIAL

FROST/SNOW

DECEMBER **10** TUESDAY

The depression drifts southwards to be centred just off the western coasts of Ireland intensifying winds and storm activity throughout the country. Very strong and up to gale force south to southwest winds prevail with severe gusts at times. Overcast and stormy with thunderstorms and widespread rain throughout. Rain likely to be very heavy in the west and south and possibly cause disruption to local areas.

Belfast: Stormy air, overcast, rain, gales likely, chance of thunderstorms
Dublin: Overcast, scattered showers, warmer, strong and blustery winds, gales possible
Cork: Stormy air, overcast, heavy rain, chance of gales and thunderstorms
Galway: Overcast, rain, strong blustery winds to gales likely

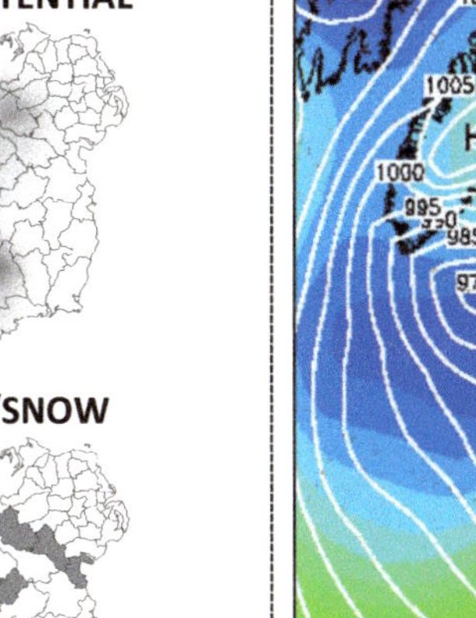

RAIN POTENTIAL

FROST/SNOW

DECEMBER **11** WEDNESDAY

The depression to the northwest of Ireland directs a stormy air across the country, with fresh to strong southwest winds prevailing, up to gale force at times particularly in the north and west. A changeable day throughout as bands of thunderstorms and showers move up through the country from the southwest and west to the north and east as the day progresses, with just the occasional sunny spell between the showers. Chance of some snow overnight mostly in western parts.

Belfast: Cloudy, blustery with chance of gale gusts, light showers
Dublin: Partly cloudy, mostly dry, threats of rain, colder overnight, strong gusty winds
Cork: Cloudy, colder, light showers, strong gusty winds, gales likely
Galway: Cloudy, rain, strong blustery winds, chance of snow and storm wind gusts

RAIN POTENTIAL

FROST/SNOW

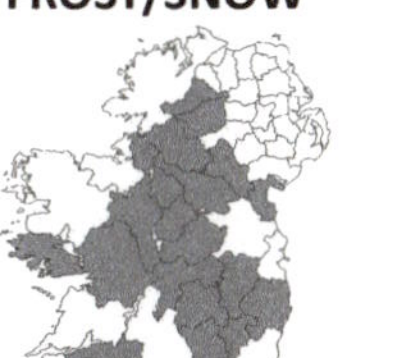

DECEMBER **12** THURSDAY

The depression to the north of Ireland directs a strong south to southwest flow over the country, with a stormy air bringing winds up to gale force at times. Widespread rain throughout, with particularly heavy rain in the south and west. Some brief sunny breakthroughs likely between the bands of rain. Chance of thunderstorms. Milder temperatures.

Belfast: Cloudy, cooler, rain, strong gusty winds, gales likely
Dublin: Partly cloudy, strong gusty winds, squally showers, gales likely
Cork: Unsettled and blustery winds, cloudy, heavy rain
Galway: Cloudy, heavy rain, strong blustery winds

RAIN POTENTIAL

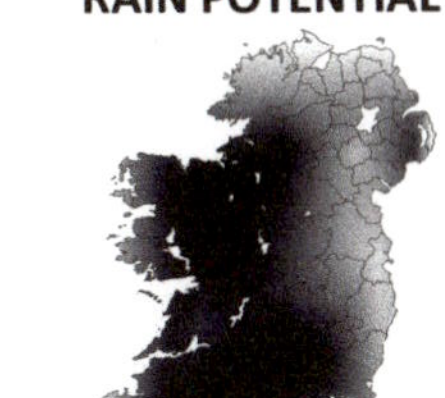

FROST/SNOW

DECEMBER **13** FRIDAY

The depression to the north of Ireland intensifies bringing strong blustery south to southwest winds throughout and up to storm force at times in many districts. Widespread cloud and rain everywhere, with the chance of hail showers here and there, as well as possible thunderstorms. Some snowfalls may occur in central parts of the country.

Belfast: Stormy, overcast, scattered showers, cold overnight
Dublin: Storm, overcast, rain, gales likely with winds possibly to storm force at times
Cork: Stormy air, cloudy, squally showers, gale gusts likely, chance of storm force winds
Galway: Overcast, rain, winds to storm force at times

DECEMBER **14** SATURDAY

The storm conditions continue throughout as the depression drifts to the immediate north of Ireland changing the gusty winds to westerlies and up to storm force around the northern coasts for a time. Dull and gloomy everywhere with widespread and persistent rain, particularly in the south of the country, where rainfall may be heavy at times.

Belfast: Overcast, stormy air, scattered showers, persistent drizzles
Dublin: Overcast, mild day cool night, light squally showers, strong stormy winds
Cork: Overcast, warmer, rain, strong gusty winds, gales likely
Galway: Stormy air, overcast, heavy rain, winds to storm force at times likely

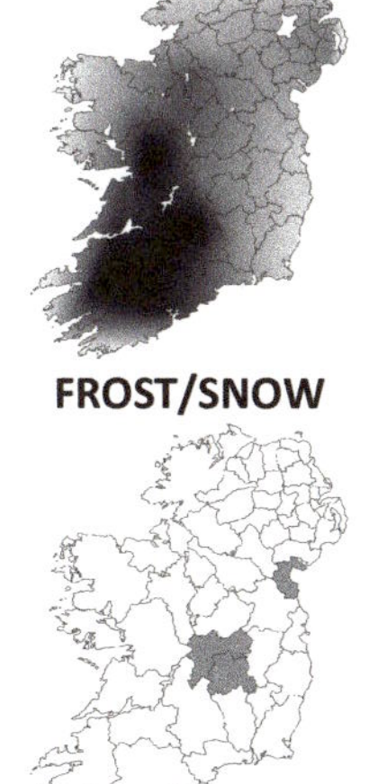

DECEMBER 15 SUNDAY

A large depression remains almost stalled to the north of Ireland and maintains a very strong and gusty southwest flow throughout the country, with winds to gale force in many districts from time to time. Widespread scattered showers prevail with some brief sunny outbreaks between the bands of rain. Rain heavy at times in some western districts.

Belfast: Cloudy, milder, rain, strong winds to gale force
Dublin: Cloudy, rain, strong winds, gales likely
Cork: Cloudy, cooler, blustery winds, gales likely, rain
Galway: Colder, rain, dull and gloomy air, strong gusty winds

RAIN POTENTIAL

FROST/SNOW

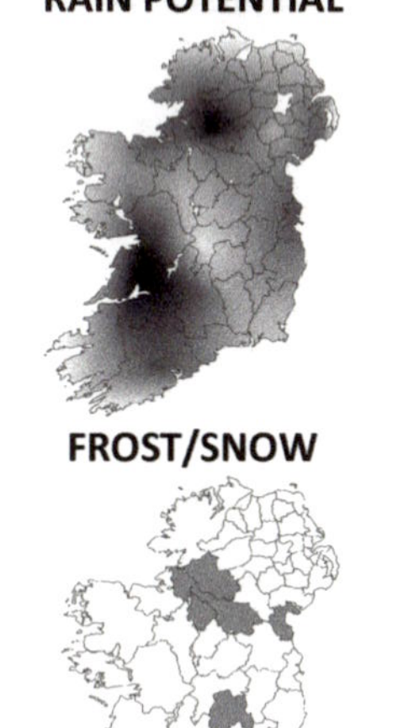

DECEMBER 16 MONDAY

A couple of depressions to the north and northwest of Ireland continue to bring unsettled conditions throughout, with fresh to strong northwest winds prevailing, possibly up to gale force at times. Dull and gloomy over the northern half of the country while some sunny spells break through in the southern half. Bands of scattered showers cross the country from the west and spread to the east throughout the day. Rain likely to be heavy in some western districts and possibly accompanying thunderstorm activity.

Belfast: Overcast, cooler, rain, strong gusty winds, possible gales
Dublin: Overcast, strong winds, possible gales, brief squalls likely
Cork: Partly cloudy, occasional showers, strong and gusty winds, gales possible
Galway: Cloudy, rain, blustery, chance of thunderstorms and gale winds at times

RAIN POTENTIAL

FROST/SNOW

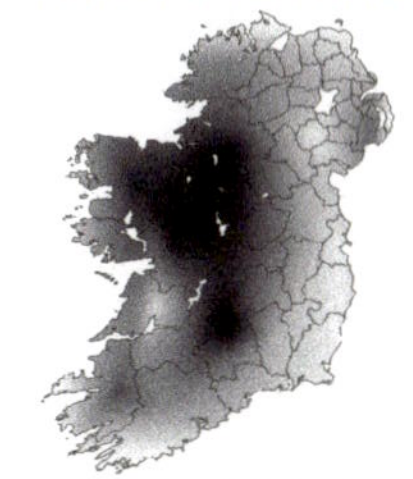

DECEMBER 17 TUESDAY

Stormy conditions continue to linger over the country as the depression to the northwest of Ireland intensifies. Strong west to southwest winds prevail, up to gale force at times in the southwest, and possibly storm force for a time, with winds easing with a northwest change later in the day. Dull and gloomy almost everywhere and with thundery bands of rain bringing widespread showers throughout, heavier in the east.

Belfast: Overcast, scattered showers, blustery, gale force likely, chance of storm force
Dublin: Overcast, cooler, rain, strong gusty winds to gale force at times
Cork: Cloudy, squally showers, strong winds with gale gusts, possibly to storm force
Galway: Stormy air, overcast, scattered showers, thunderstorms and gales possible

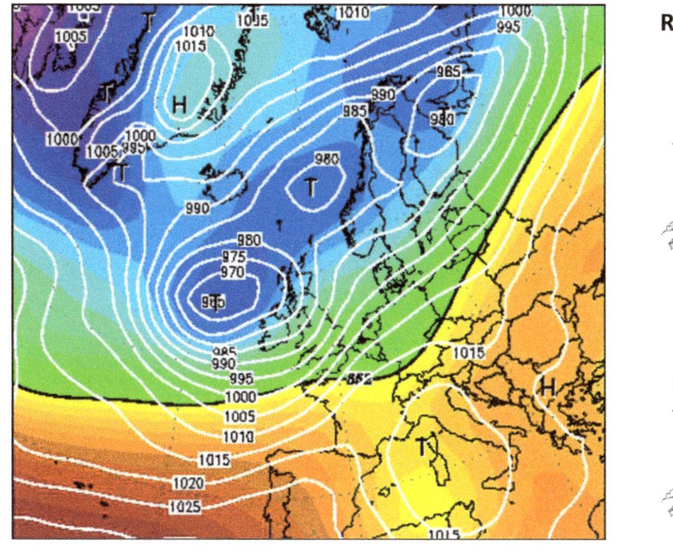

RAIN POTENTIAL

FROST/SNOW

DECEMBER 18 WEDNESDAY

The depression drifts slowly northwards away from Ireland bringing a colder northwest flow across the country, with blustery winds throughout and up to gale force at times in the west and east. Cloudy everywhere with passing wintry showers in most districts, although mostly dry with threats of rain in the east. Chance of hail showers here and there, and possible snow for a time in the west.

Belfast: Cloudy, colder, strong gusty winds, passing showers
Dublin: Colder, dry, some cloudy spells, strong blustery winds, possible gale gusts
Cork: Cloudy, blustery, cold, light showers
Galway: Cloudy, cold, scattered showers, blustery, chance of gales and snow likely

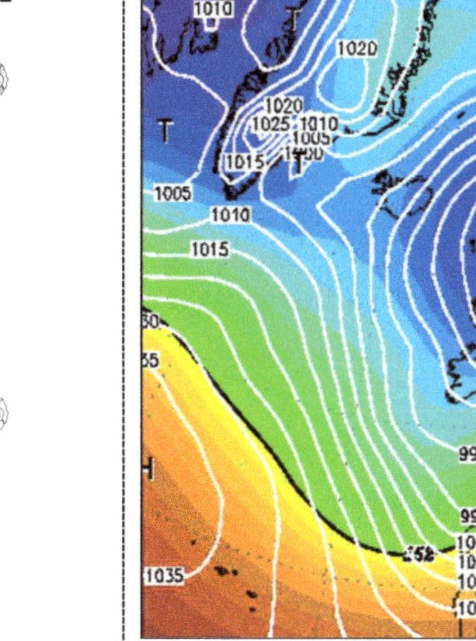

RAIN POTENTIAL

FROST/SNOW

DECEMBER **19** THURSDAY

The depression to the north of Ireland continues to drift away with the northwest winds easing to moderate to fresh throughout, with occasional gusts along the coastal fringes. Cooler everywhere with a mixed day of wintry showers and some sunny spells everywhere. Overnight frosts in most districts except along the western and far northern coasts, as well as the chance of snow at times in the west.

Belfast: Changeable, mix of wintry showers and sunny outbreaks, gale gusts likely
Dublin: Partly cloudy, cold and dry, strong winds
Cork: Unsettled and blustery air, cold, partly cloudy, some brief showers
Galway: Partly cloudy, cold, heavy rain, windy with gale gusts likely, chance of snow

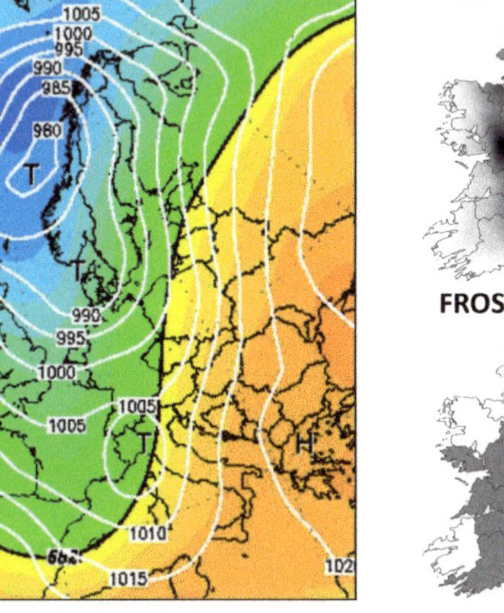

RAIN POTENTIAL

FROST/SNOW

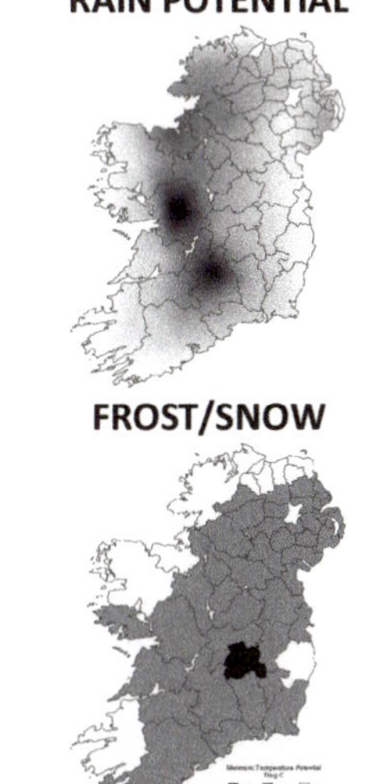

DECEMBER **20** FRIDAY

Another intense storm system far to the northwest of Ireland begins to move closer while a high pressure ridge moves up onto the southern half of the country. Strong and blustery southwest winds prevail, up to gale force at times in the west. Overcast and gloomy with widespread wintry showers throughout, with very heavy rain at times in the west and southwest. Chance of snow in the west again.

Belfast: Storm, overcast, squally rain and wintry showers
Dublin: Overcast, drizzly showers, warmer, strong winds, gales likely
Cork: Overcast, warmer, rain, blustery winds to gale force at times
Galway: Overcast, heavy rain, stormy winds, chance of snow

RAIN POTENTIAL

FROST/SNOW

DECEMBER 21 SATURDAY

A very large depression northwest of Ireland dominates the country once more with fresh to strong west to southwest winds prevailing, up to gale force at times and with storm gusts likely. Some sunny spells in most districts between bands of frequent and often wintry showers, heaviest in the west and north. Chance of thunderstorms, hail, and snow, mostly in the west of the country.

Belfast: Partly cloudy, milder, squalls, gusty winds, possibly to storm force at times
Dublin: Partly cloudy, occasional showers, mild day cold night, stormy winds
Cork: Partly cloudy, light showers, mild day cold night, strong forceful winds
Galway: Cloudy, cold, showers, stormy air, chance of thunderstorms, snow, and hail

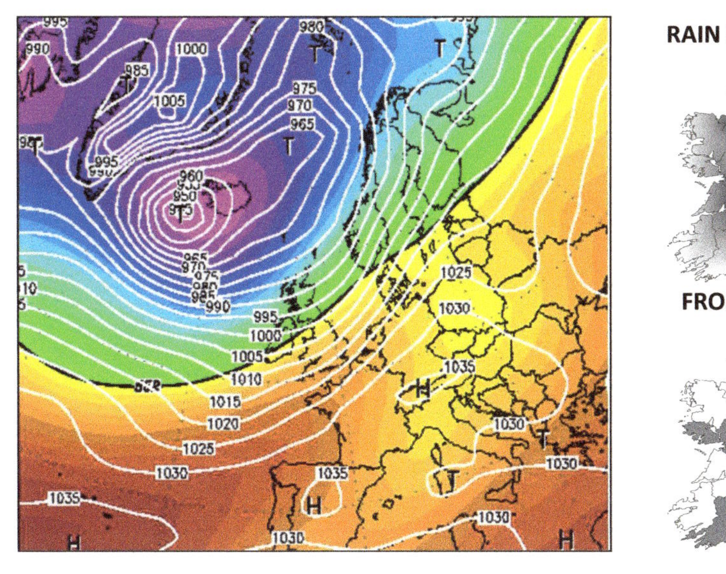

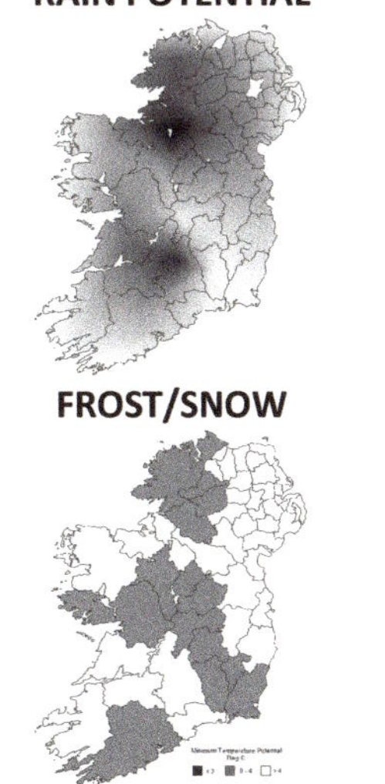

DECEMBER 22 SUNDAY

The large depression north of Ireland maintains a strong and blustery southwest flow throughout. A mixed day of some brief sunny spells and passing bands of wintry showers in most districts, although dull and gloomy in the south. Rain bands likely to cross the country from the northwest and spread in a southeasterly tract. Chance of thunderstorms and hail here and there, and possible snow once more in the west.

Belfast: Cloudy, cooler, light showers, strong gusty winds with gale gusts likely
Dublin: Cloudy, cooler, dry, winds likely to storm force for a time
Cork: Overcast, spotty showers, strong winds
Galway: Cloudy, wintry showers, gusty winds, thunderstorms, gales and snow likely

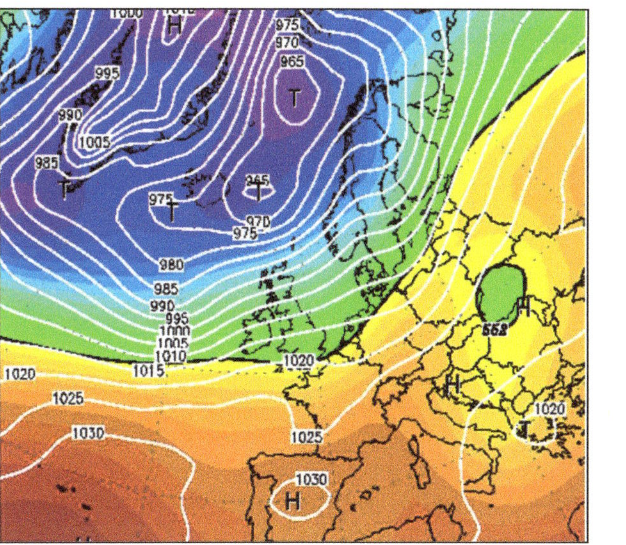

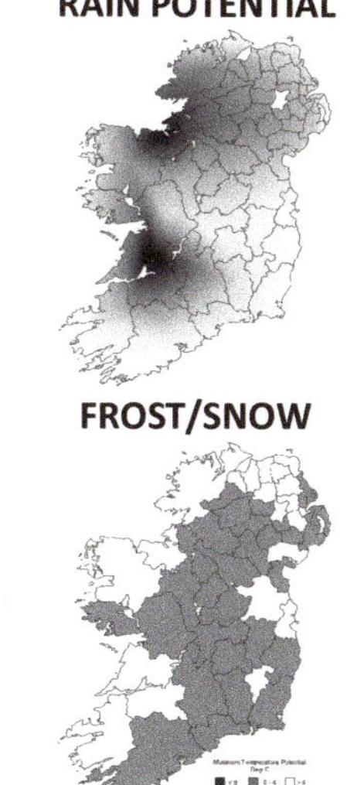

DECEMBER **23** MONDAY

Low pressure and a stormy air continues to dominate the country in a strong southwesterly flow. Dull and gloomy everywhere with outbreaks of rain in most districts, particularly heavy in the west and northwest, and spreading from the northwest to the southeast. Mostly dry with to isolated showers in the eastern coastal fringes areas.

Belfast: Blustery and stormy air, overcast, milder, scattered showers
Dublin: Overcast, warmer, dry, increasing threats of rain, gusty winds, gales likely
Cork: Overcast, scattered showers, blustery winds, gales likely
Galway: Overcast, stormy winds, driving showers

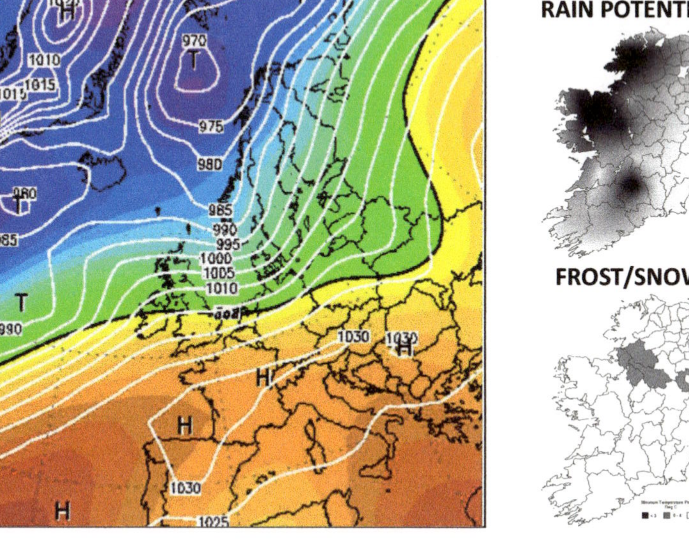

RAIN POTENTIAL

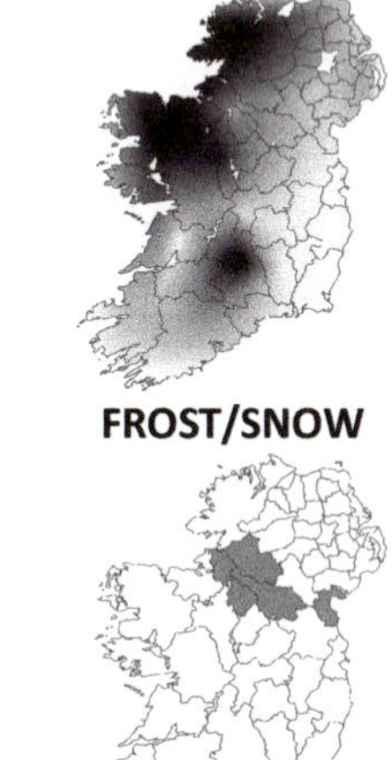

FROST/SNOW

DECEMBER **24** TUESDAY

The depression to the north of Ireland continues to bring further dull and gloomy conditions throughout the country, with outbreaks of showers moving in from the Atlantic and spreading everywhere as the day progresses, with some heavy falls through central parts. Fresh to blustery southwest winds prevail ahead of the front throughout, possibly up to gale force at times, and then easing as the front passes.

Belfast: Milder, overcast, scattered showers, strong gusty winds with gale gusts likely
Dublin: Overcast, rain, blustery winds, possible gales to storm force for a time
Cork: Overcast and stormy air, gales likely, squally showers
Galway: Overcast, scattered showers, strong gusty winds with gales likely

RAIN POTENTIAL

FROST/SNOW

DECEMBER **25** WEDNESDAY

The depression to the north of Ireland continues to direct a fresh west to southwest flow and a dull and gloomy stormy air throughout, with winds lighter in the south. A wet front crosses the country from the northwest at first and spreading slowly down through the country into the southeast, bringing heavy rain at times in to the west and central parts, and lighter rain elsewhere. Chance of further light snow in the west once more.

Belfast: Overcast, mostly dry, threats of rain, cooler, winds easing to moderate
Dublin: Overcast, colder, rain, blustery at times
Cork: Colder, overcast and gloomy air, heavy rain, winds easing
Galway: Overcast, cold, moderate winds, brief showers, light snow possible

RAIN POTENTIAL

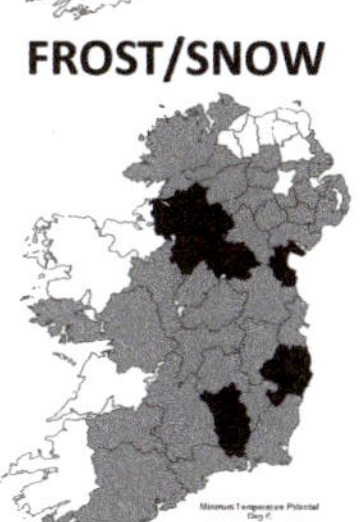

FROST/SNOW

DECEMBER **26** THURSDAY

A high pressure system ridges to the south of the country with winds easing to a light to moderate southwest to westerly flow, with only some blustery spells likely to linger around the northwestern coasts. A mixed day throughout. Mostly dry, cold, and with occasional sunny spells in the north and east, while a mix of sunny spells and isolated showers persist in the west and south. Frosts are likely to be widespread overnight. Chance of some light snow once more in some western parts.

Belfast: Partly cloudy, dry, cold, fluctuating winds
Dublin: Partly cloudy, cold, dry, fluctuating breezes
Cork: Partly cloudy, cold, isolated showers, light shifting breezes
Galway: Partly cloudy, light showers, cold, moderate breezes, chance of snow

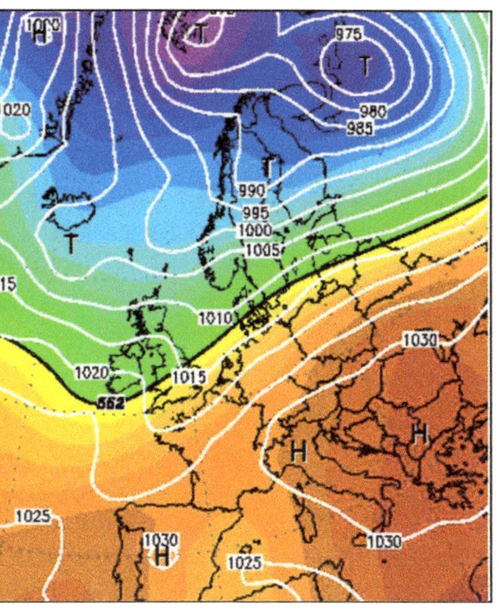

RAIN POTENTIAL

FROST/SNOW

DECEMBER **27** FRIDAY

A large anticyclonic system centred to the immediately south of Ireland directs a gentle southwest to westerly flow over the country and widespread cloud throughout. Some showers in all districts, at first in the west spreading to the east as the day progresses, and light and drizzly on and off most of the day. Mild along the western and far northern coasts, with overnight frosts elsewhere. Some wintry showers may fall as light snow in the west of the country.

Belfast: Overcast, light wintry showers, cold, moderate breezes
Dublin: Overcast, cold, light rain, moderate breezes
Cork: Overcast, cool, light showers, moderate breezes
Galway: Overcast, cold, scattered wintry showers, moderate breezes, chance of snow

RAIN POTENTIAL

FROST/SNOW

DECEMBER **28** SATURDAY

Anticyclonic. Light winds prevail, mostly in a northerly flow and widespread frosts throughout with the chance of overnight patches of fog in many districts. Dull and gloomy in the east with some sunny spells in most other districts. Some light showers in the south, west, and northeast of the country, with the chance of some of them being wintry and falling as snow flurries in some northern parts.

Belfast: Cloudy, cold and dry, shifting breezy spells, chance of overnight snow flurries
Dublin: Overcast, cold, dry with odd threat of rain, light breezes
Cork: Partly cloudy, mostly dry, cold, moderate to light breezes, chance of fog
Galway: Partly cloudy, cold, dry, gentle breezes, fog patches likely

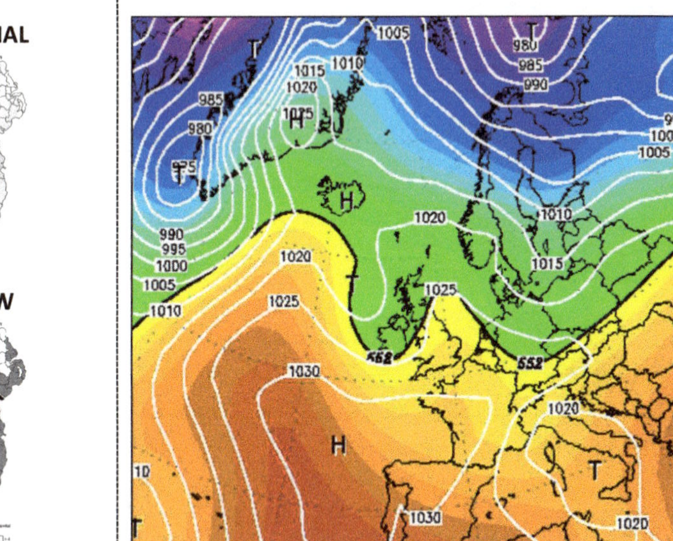

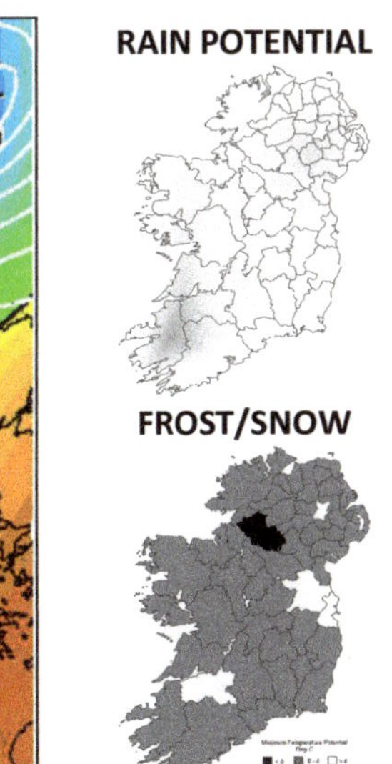

DECEMBER 29 SUNDAY

A large anticyclonic system sits over Ireland bringing a calm and serene day with widespread cloud and passing threats of rain throughout. Scattered fogs in many places and frosts will be widespread and severe in sheltered parts. Chance of snow flurries once more in northern parts.

Belfast: Overcast, cold, dry with threats of rain, chance of snow, fog possible
Dublin: Mostly calm, dull and gloomy air, cold, frosts and fog patches likely
Cork: Cloudy, cold, mostly dry, calm air, fog patches likely
Galway: Cold, frosts, dull and gloomy air, mostly dry, fog likely

RAIN POTENTIAL

FROST/SNOW

DECEMBER 30 MONDAY

A large anticyclonic system continues to linger over Ireland bringing a very cold day with a dry day and a mostly calm air throughout. Widespread cloud prevails everywhere, although some sunny spells likely to break through in the east and north for a time. Frosts and fogs will be widespread with freezing fogs and severe frosts in places, particularly severe in inland parts and in the northeast. Chance of light snow being observed in some northern parts.

Belfast: Mostly calm air, cloudy, cold, mostly dry, chance of fog and light snow
Dublin: Cold, frosts, some sunny spells, dry, calm atmosphere, fog likey
Cork: Overcast, cold, mostly calm air, fog patches likely
Galway: Overcast, dry, threats of rain, gentle breezes, frosts and fog patches

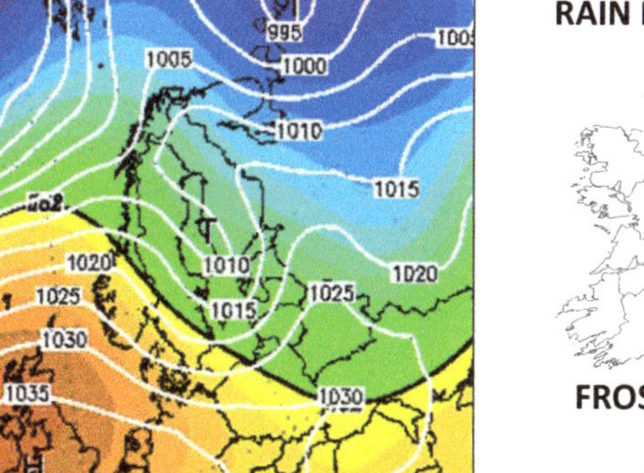

RAIN POTENTIAL

FROST/SNOW

DECEMBER 31 TUESDAY

The large anticyclonic system over Ireland strengthens while a depression to the northwest of the country pushes against the high pressure ridge near the northwest. Mostly dry with occasional sunny spells throughout, except for the chance of some drizzle patches developing in the northwest and possible snow flurries in the north. Cold, with overnight fog patches and with widespread frosts in most districts, and some freezing fog patches in the north.

Belfast: Partly cloudy, cold, severe frosts, dry, light breezes, fog and snow flurries likely
Dublin: Dry, occasional sunny spells, cold, chance of fog and frost patches overnight
Cork: Cloudy, dry, light shifting breezes
Galway: Mostly cloudy, dry, cold with frost and fog patches likely overnight

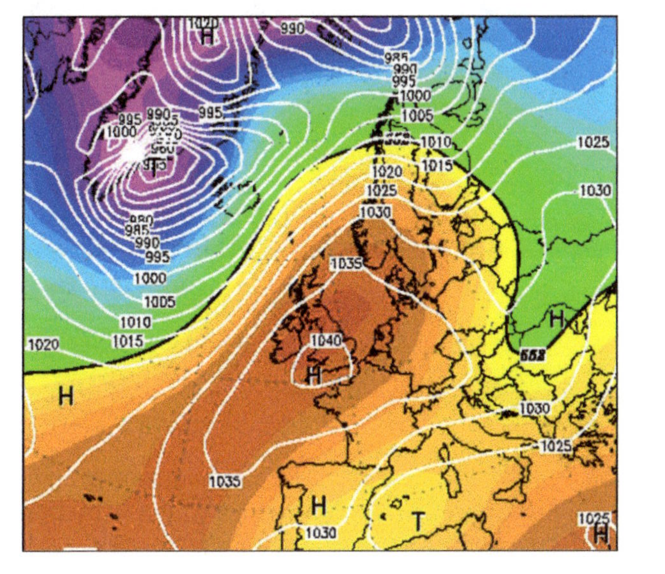

RAIN POTENTIAL

FROST/SNOW

APPENDICES

RAINFALL EXPECTATIONS

WETTER OR DRIER BY 10MM THAN LONG-TERM AVERAGES

	Ulster Province					Connaught Province						Leinster Province				MUNSTER PROVINCE					
Date	Aldergrove	Hillsborough	Armagh	Lough Navar	Malin Head	Sligo	Ardtarmon	Belmullet	Drumsna	Galway	Roscommon	Markree Castle	Edenderry	Kilkenny	Dublin	Shannon	Newport	Tralee	Killarney	Cork	All
JAN	drr	drr	drr	drr	drr	drr	drr	w tr	av	av	drr	drr	drr	drr	drr	drr	av	w tr	w tr	drr	drr
FEB	drr	drr	drr	drr	drr	drr	drr	drr	drr	drr	drr	drr	drr	drr	drr	drr	drr	drr	drr	drr	drr
MAR	av	w tr	drr	w tr	drr	av	drr	drr	drr	drr	drr	av	drr	drr	av	drr	drr	drr	drr	drr	drr
APR	w tr	w tr	w tr	w tr	w tr	w tr	av	w tr	w tr	w tr	w tr	w tr	av	av	av	w tr	w tr	w tr	w tr	w tr	w tr
MAY	w tr	w tr	w tr	w tr	w tr	w tr	w tr	w tr	w tr	w tr	w tr	w tr	w tr	w tr	w tr	w tr	w tr	w tr	w tr	w tr	w tr
JUNE	drr	drr	drr	drr	drr	drr	drr	drr	drr	drr	drr	drr	drr	drr	drr	drr	drr	drr	drr	drr	drr
JULY	drr	drr	drr	drr	drr	drr	drr	drr	drr	drr	drr	drr	drr	drr	drr	drr	drr	drr	drr	drr	drr
AUG	drr	drr	av	drr	drr	drr	drr	drr	drr	drr	drr	drr	drr	drr	drr	drr	drr	drr	drr	drr	drr
SEP	w tr	w tr	w tr	av	w tr	w tr	av	av	w tr	w tr	w tr	w tr	w tr	av	av	w tr	w tr	w tr	w tr	w tr	w tr
OCT	w tr	w tr	w tr	av	drr	av	av	w tr	w tr	w tr	drr	av	w tr	w tr	w tr	w tr	av	w tr	w tr	w tr	w tr
NOV	w tr	w tr	w tr	av	w tr	av	av	av	w tr	w tr	av	av	w tr	w tr	av	w tr	w tr	av	w tr	av	w tr
DEC	w tr	w tr	w tr	w tr	w tr	w tr	w tr	w tr	w tr	w tr	w tr	w tr	w tr	w tr	w tr	w tr	w tr	w tr	w tr	w tr	w tr
Overall	av	av	av	av	av	av	drr	w tr	w tr	w tr	av	av	av	av	av	av	av	w tr	w tr	av	w tr
year	861	944	812	1179	1107	1260	1284	1245	1047	1193	1289	1260	971	840	758	978	1607	1583	1131	1228	22578
estimates	917	995	825	1167	1155	1271	1130	1362	1222	1385	1204	1176	938	890	723	969	1568	1750	1879	1176	23700

	Ulster Province					Connaught Province						Leinster Province				MUNSTER PROVINCE					
Date	Aldergrove	Hillsborough	Armagh	Lough Navar	Malin Head	Sligo	Ardtarmon	Belmullet	Drumsna	Galway	Roscommon	Markree Castle	Edenderry	Kilkenny	Dublin	Shannon	Newport	Tralee	Killarney	Cork	All
WINTER	drr	drr	drr	drr	drr	drr	drr	av	drr	drr	drr	drr	drr	drr	drr	drr	drr	drr	w tr	drr	drr
SPRING	w tr	w tr	av	w tr	av	w tr	av	w tr	w tr	w tr	w tr	w tr	w tr	av	w tr	av	av	w tr	w tr	av	w tr
SUMMER	drr	drr	drr	drr	drr	drr	drr	drr	drr	drr	drr	drr	drr	drr	drr	drr	drr	drr	drr	drr	drr
AUTUMN	w tr	w tr	w tr	av	w tr	w tr	av	w tr	w tr	w tr	av	av	w tr	w tr	w tr	w tr	w tr	w tr	w tr	w tr	w tr
Overall	av	av	av	av	av	av	drr	w tr	w tr	w tr	av	av	av	av	av	av	av	w tr	w tr	av	w tr
year	861	944	812	1179	1107	1260	1284	1245	1047	1193	1289	1260	971	840	758	978	1607	1583	1131	1228	22578
estimates	917	995	825	1167	1155	1271	1130	1362	1222	1385	1204	1176	938	890	723	969	1568	1750	1879	1176	23700

RAINFALL EXPECTATIONS

WETTER OR DRIER BY 25% THAN LONG-TERM AVERAGES

	Ulster Province					Connaught Province						Leinster Province				MUNSTER PROVINCE					
Date	Aldergrove	Hillsborough	Armagh	Lough Navar	Malin Head	Sligo	Ardtarnon	Belmullet	Drumsna	Galway	Roscommon	Markree Castle	Edenderry	Kilkenny	Dublin	Shannon	Newport	Tralee	Killarney	Cork	All
JAN	drr	drr	drr	av	av	av	av	av	av	av	av	av	drr	drr	drr	av	av	av	w tr	drr	av
FEB	drr	drr	drr	drr	drr	drr	drr	av	drr	drr	drr	drr	drr	drr	drr	drr	drr	drr	drr	drr	drr
MAR	av	av	av	av	av	av	av	drr	av	drr	drr	av	av	drr	av	drr	drr	drr	av	drr	av
APR	w tr	w tr	w tr	w tr	w tr	av	av	w tr	w tr	w tr	w tr	av	av	av	av	w tr	w tr	w tr	w tr	w tr	w tr
MAY	w tr	w tr	w tr	w tr	w tr	w tr	w tr	w tr	w tr	w tr	w tr	w tr	w tr	w tr	w tr	w tr	w tr	w tr	w tr	w tr	w tr
JUNE	drr	drr	drr	av	av	drr	drr	drr	drr	drr	drr	drr	drr	drr	drr	drr	drr	drr	drr	drr	drr
JULY	drr	drr	drr	drr	drr	drr	drr	drr	drr	drr	drr	drr	drr	drr	drr	drr	drr	drr	drr	drr	drr
AUG	av	av	av	drr	av	drr	drr	drr	drr	drr	drr	drr	av	drr	drr	drr	drr	av	av	av	av
SEP	w tr	w tr	av	av	av	av	av	av	w tr	w tr	w tr	av	w tr	av	av	av	av	w tr	w tr	av	w tr
OCT	w tr	av	w tr	av	av	av	av	w tr	av	av	av	av	av	w tr	w tr	w tr	av	av	w tr	w tr	av
NOV	w tr	w tr	w tr	av	w tr	av	av	av	w tr	w tr	av	av	av	w tr	av	av	av	av	w tr	av	av
DEC	w tr	w tr	w tr	w tr	w tr	w tr	w tr	w tr	w tr	w tr	w tr	w tr	w tr	w tr	w tr	w tr	w tr	w tr	w tr	w tr	w tr
Overall	av	av	av	av	av	av	av	av	av	av	av	av	av	av	av	av	av	av	w tr	av	av
year	861	944	812	1179	1107	1260	1284	1245	1047	1193	1289	1260	971	840	758	978	1607	1583	1131	1228	22578
estimates	917	995	825	1167	1155	1271	1130	1362	1222	1385	1204	1176	938	890	723	969	1568	1750	1879	1176	23700

	Ulster Province					Connaught Province						Leinster Province				MUNSTER PROVINCE					
Date	Aldergrove	Hillsborough	Armagh	Lough Navar	Malin Head	Sligo	Ardtarnon	Belmullet	Drumsna	Galway	Roscommon	Markree Castle	Edenderry	Kilkenny	Dublin	Shannon	Newport	Tralee	Killarney	Cork	All
WINTER	drr	drr	drr	drr	drr	drr	drr	av	av	av	drr	av	drr	drr	drr	drr	av	av	av	drr	drr
SPRING	w tr	w tr	av	w tr	av	av	av	av	w tr	av	av	av	w tr	av	w tr	av	av	w tr	w tr	av	w tr
SUMMER	drr	drr	drr	drr	av	drr	drr	drr	drr	drr	drr	drr	drr	drr	drr	drr	drr	drr	drr	drr	drr
AUTUMN	w tr	w tr	w tr	av	av	av	av	av	w tr	w tr	av	av	av	w tr	av	av	av	av	w tr	w tr	av
Overall	av	av	av	av	av	av	av	av	av	av	av	av	av	av	av	av	av	av	w tr	av	av
year	861	944	812	1179	1107	1260	1284	1245	1047	1193	1289	1260	971	840	758	978	1607	1583	1131	1228	22578
estimates	776	804	752	1169	1125	1170	1025	1230	1027	1257	1056	1179	790	746	695	915	1667	1490	1688	1176	21738

RAINFALL ESTIMATION SUMMARIES

% WETTER OR DRIER OF LONG-TERM AVERAGES

	Ulster Province					Connaught Province						Leinster Province				Munster Province					
Date	Aldergrove	Hillsborough	Armagh	Lough Navar	Malin Head	Sligo	Ardtamon	Belmullet	Drumsna	Galway	Roscommon	Markree Castle	Edenderry	Kilkenny	Dublin	Shannon	Newport	Tralee	Killarney	Cork	All
JANUARY	-36%	-42%	-35%	-20%	-11%	-11%	-18%	16%	2%	8%	-22%	-14%	-51%	-41%	-59%	-15%	1%	7%	70%	-37%	-11%
	drr	drr	drr	drr	drr	drr	drr	w tr	w tr	w tr	drr	drr	drr	drr	drr	drr	w tr	w tr	w tr	drr	drr
FEBRUARY	-55%	-60%	-54%	-48%	-59%	-51%	-49%	-17%	-44%	-43%	-59%	-39%	-58%	-49%	-46%	-51%	-29%	-45%	-32%	-75%	-47%
	drr	drr	drr	drr	drr	drr	drr	drr	drr	drr	drr	drr	drr	drr	drr	drr	drr	drr	drr	drr	drr
MARCH	9%	19%	-19%	13%	-12%	-9%	-22%	-28%	-12%	-35%	-31%	3%	-17%	-26%	-5%	-36%	-32%	-34%	-22%	-52%	-19%
	w tr	w tr	drr	w tr	drr	drr	drr	drr	drr	drr	drr	w tr	drr	drr	drr	drr	drr	drr	drr	drr	drr
APRIL	51%	57%	36%	59%	26%	23%	5%	25%	60%	63%	26%	15%	-3%	9%	18%	32%	31%	28%	80%	26%	33%
	w tr	w tr	w tr	w tr	w tr	w tr	w tr	w tr	w tr	w tr	w tr	w tr	drr	w tr	w tr	w tr	w tr	w tr	w tr	w tr	w tr
MAY	62%	80%	34%	48%	33%	55%	27%	83%	108%	63%	97%	43%	110%	53%	76%	56%	37%	117%	256%	51%	73%
	w tr	w tr	w tr	w tr	w tr	w tr	w tr	w tr	w tr	w tr	w tr	w tr	w tr	w tr	w tr	w tr	w tr	w tr	w tr	w tr	w tr
JUNE	-44%	-55%	-35%	-21%	-23%	-45%	-34%	-36%	-40%	-25%	-53%	-42%	-57%	-57%	-54%	-39%	-33%	-43%	-43%	-73%	-42%
	drr	drr	drr	drr	drr	drr	drr	drr	drr	drr	drr	drr	drr	drr	drr	drr	drr	drr	drr	drr	drr
JULY	-66%	-66%	-64%	-72%	-36%	-68%	-75%	-46%	-57%	-66%	-58%	-64%	-47%	-25%	-51%	-64%	-69%	-45%	-61%	-54%	-58%
	drr	drr	drr	drr	drr	drr	drr	drr	drr	drr	drr	drr	drr	drr	drr	drr	drr	drr	drr	drr	drr
AUGUST	-13%	-19%	-3%	-38%	-14%	-47%	-39%	-50%	-34%	-33%	-33%	-33%	-16%	-31%	-31%	-58%	-57%	-11%	-18%	-19%	-31%
	drr	drr	drr	drr	drr	drr	drr	drr	drr	drr	drr	drr	drr	drr	drr	drr	drr	drr	drr	drr	drr
SEPTEMBER	49%	73%	22%	5%	16%	22%	-2%	0%	88%	94%	34%	10%	27%	5%	15%	15%	12%	50%	95%	22%	32%
	w tr	w tr	w tr	w tr	w tr	w tr	drr	av	w tr	w tr	w tr	w tr	w tr	w tr	w tr	w tr	w tr	w tr	w tr	w tr	w tr
OCTOBER	35%	15%	47%	-2%	-15%	7%	-8%	37%	10%	18%	-15%	-2%	14%	88%	37%	28%	4%	6%	93%	58%	20%
	w tr	w tr	w tr	drr	drr	w tr	drr	w tr	w tr	w tr	drr	drr	w tr	w tr	w tr	w tr	w tr	w tr	w tr	w tr	w tr
NOVEMBER	40%	40%	34%	9%	44%	3%	-1%	3%	26%	36%	2%	-1%	17%	34%	1%	19%	18%	-3%	55%	4%	17%
	w tr	w tr	w tr	w tr	w tr	w tr	drr	w tr	w tr	w tr	w tr	drr	w tr	w tr	w tr	w tr	w tr	drr	w tr	w tr	w tr
DECEMBER	38%	31%	28%	59%	86%	103%	54%	87%	86%	92%	34%	33%	34%	58%	29%	74%	54%	100%	280%	55%	73%
	w tr	w tr	w tr	w tr	w tr	w tr	w tr	w tr	w tr	w tr	w tr	w tr	w tr	w tr	w tr	w tr	w tr	w tr	w tr	w tr	w tr
WINTER	-29%	-33%	-33%	-31%	-28%	-26%	-32%	-3%	-16%	-24%	-37%	-17%	-52%	-47%	-36%	-39%	-14%	-14%	24%	-56%	-25%
	drr	drr	drr	drr	drr	drr	drr	drr	drr	drr	drr	drr	drr	drr	drr	drr	drr	drr	w tr	drr	drr
SPRING	39%	49%	16%	37%	12%	20%	1%	20%	47%	24%	24%	18%	29%	11%	32%	13%	6%	30%	88%	4%	25%
	w tr	w tr	w tr	w tr	w tr	w tr	w tr	w tr	w tr	w tr	w tr	w tr	w tr	w tr	w tr	w tr	w tr	w tr	w tr	w tr	w tr
SUMMER	-39%	-45%	-32%	-44%	-24%	-53%	-50%	-45%	-43%	-41%	-47%	-46%	-39%	-38%	-45%	-54%	-54%	-33%	-40%	-47%	-43%
	drr	drr	drr	drr	drr	drr	drr	drr	drr	drr	drr	drr	drr	drr	drr	drr	drr	drr	drr	drr	drr
AUTUMN	41%	41%	35%	3%	14%	10%	-4%	15%	37%	46%	5%	2%	19%	46%	18%	21%	11%	15%	80%	30%	23%
	w tr	w tr	w tr	w tr	w tr	w tr	drr	w tr	w tr	w tr	w tr	w tr	w tr	w tr	w tr	w tr	w tr	w tr	w tr	w tr	w tr
OVERALL	6%	5%	2%	-1%	4%	1%	-12%	9%	17%	16%	-7%	-7%	-3%	6%	-5%	-1%	-2%	11%	66%	-4%	5%
	w tr	w tr	w tr	drr	w tr	w tr	drr	w tr	w tr	w tr	drr	drr	drr	w tr	drr	drr	drr	w tr	w tr	drr	w tr

ESTIMATED SUNSHINE HOURS TREND

TRENDS BY MONTHS FOR SELECTED TOWNS

	Ulster Province						Connaught Province							Leinster Province					MUNSTER PROVINCE						
Date	Aldergrove	Hillsborough	Armagh	Lough Navar	Malin Head	Ulster Overall	Sligo	Ardtarmon	Belmullet	Drumsna	Galway	Roscommon	Connaught Overall	Markree Castle	Edenderry	Kilkenny	Dublin	Leinster Overall	Shannon	Newport	Tralee	Killarney	Cork	Munster Overall	Ireland Overall
Dec 2023	av	av	snnr	av	av	av	av	snnr	av	av	snnr	cldr	av	av	av	snnr	av	av	snnr	snnr	av	av	snnr	snnr	snnr
Jan	av	av	av	cldr	cldr	cldr	av	av	av	cldr	av	cldr	cldr	av	av	av	av	av	av	av	av	av	av	av	cldr
Feb	cldr	av	av	cldr	cldr	cldr	av	av	av	cldr	av	cldr	cldr	av	av	av	av	av	av	av	av	av	av	av	cldr
Winter	av	av	av	cldr	cldr	cldr	av	av	av	cldr	av	cldr	cldr	cldr	cldr	av	av	cldr	av	snnr	av	av	av	av	cldr
Mar	av	av	av	av	av	av	av	av	av	av	av	av	av	av	av	av	av	av	snnr	snnr	snnr	snnr	av	snnr	snnr
Apr	av	av	av	av	av	av	av	av	cldr	av	av	av	av	av	av	av	av	av	av	av	av	av	av	av	average
May	cldr	cldr	av	cldr	cldr	cldr	cldr	av	cldr	av	cldr	cldr	cldr	av	av	av	av	av	av	av	cldr	cldr	cldr	cldr	cldr
Spring	av	av	snnr	cldr	cldr	av	av	av	cldr	av	av	cldr	cldr	av	av	av	av	av	av	snnr	av	av	cldr	av	cldr
June	snnr	snnr	snnr	snnr	snnr	snnr	snnr	snnr	snnr	snnr	snnr	snnr	snnr	snnr	snnr	snnr	snnr	snnr	snnr	snnr	snnr	snnr	snnr	snnr	snnr
July	snnr	snnr	snnr	snnr	snnr	snnr	snnr	snnr	snnr	snnr	snnr	snnr	snnr	snnr	snnr	snnr	snnr	snnr	snnr	snnr	snnr	snnr	snnr	snnr	snnr
Aug	av	av	av	av	av	cldr	cldr	av	cldr	snnr	cldr	cldr	cldr	av	av	snnr	av	av	av	snnr	snnr	snnr	av	snnr	average
Summer	snnr	snnr	snnr	snnr	snnr	snnr	snnr	snnr	snnr	snnr	snnr	snnr	snnr	snnr	snnr	snnr	snnr	snnr	snnr	snnr	snnr	snnr	snnr	snnr	snnr
Sep	snnr	snnr	snnr	snnr	snnr	snnr	snnr	snnr	av	snnr	snnr	av	snnr	snnr	snnr	snnr	snnr	snnr	snnr	snnr	snnr	snnr	av	snnr	snnr
Oct	av	av	av	snnr	snnr	snnr	snnr	snnr	av	snnr	snnr	av	snnr	av	av	av	av	snnr	snnr	snnr	av	snnr	av	snnr	snnr
Nov	snnr	snnr	snnr	snnr	snnr	snnr	snnr	snnr	av	snnr	av	av	snnr	snnr	snnr	snnr	snnr	snnr	snnr	snnr	snnr	snnr	snnr	snnr	snnr
Autumn	snnr	snnr	snnr	snnr	snnr	snnr	snnr	snnr	snnr	snnr	snnr	av	snnr	snnr	snnr	snnr	snnr	snnr	snnr	snnr	snnr	snnr	snnr	snnr	snnr
Dec	snnr	snnr	snnr	snnr	snnr	snnr	av	snnr	av	snnr	av	av	snnr	av	snnr	snnr	snnr	snnr	av	snnr	snnr	snnr	av	snnr	snnr
Year	snnr	snnr	snnr	snnr	snnr	snnr	snnr	snnr	av	snnr	snnr	av	snnr	snnr	snnr	snnr	snnr	snnr	snnr	snnr	snnr	snnr	snnr	snnr	snnr

ESTIMATED SUNSHINE HOURS

	Ulster Province					Connaught Province						Leinster Province				Munster Province							
Date	Aldergrove	Hillsborough	Armagh	Lough Navar	Malin Head	Sligo	Ardtarmon	Belmullet	Drumsna	Galway	Roscommon	Markree Castle	Edenderry	Kilkenny	Dublin	Shannon	Newport	Tralee	Killarney	Cork	Estimated Totals	Av Totals	Month Trend
Jan	25	38	38	19	13	27	38	31	15	36	19	33	41	44	48	40	42	32	32	41	654	946	less
Feb	43	53	53	38	33	46	53	69	29	62	31	45	52	56	60	54	55	56	56	52	997	1268	less
Mar	95	105	105	90	85	115	105	118	90	122	89	90	98	112	96	127	119	133	133	129	2158	1987	more
Apr	145	145	145	145	145	131	145	131	147	141	136	145	147	149	159	152	150	138	138	138	2871	2957	average
May	169	169	169	169	169	151	169	151	167	156	153	169	167	165	187	161	163	144	144	144	3235	3611	less
Jun	236	236	236	236	236	214	236	214	249	224	219	236	249	263	258	234	248	267	267	267	4826	3173	more
Jul	211	211	211	211	211	201	211	201	215	207	204	211	215	219	220	213	216	218	218	218	4244	2841	more
Aug	126	126	126	126	126	88	126	88	148	105	97	126	148	171	163	122	147	180	180	180	2696	2793	average
Sep	142	142	142	142	142	127	142	127	144	136	131	142	144	147	157	144	146	137	137	137	2804	2219	more
Oct	104	104	104	104	104	100	104	100	104	105	103	104	104	104	108	110	107	100	100	100	2068	1653	more
Nov	89	89	89	89	89	72	89	72	97	78	75	89	97	105	106	84	95	105	105	105	1817	1161	more
Dec	60	60	60	60	60	35	60	35	67	45	40	60	67	73	85	54	64	61	61	61	1166	774	more
Estimate	1444	1476	1476	1427	1411	1307	1476	1337	1472	1416	1296	1449	1530	1609	1646	1495	1552	1572	1572	1572	29535	25381	more
Average	1314	1314	1192	1272	1230	1200	1204	1278	1134	1276	1355	1280	1327	1284	1438	1273	1140	1245	1237	1391	25381		
Trend	more	more	more	more	more	more	more	av	more	more	av	more	more	more	more	more	more	more	more	more	more		

ESTIMATED AVERAGE MONTHLY MAXIMUM TEMPERATURES

	Ulster Province					Connaught Province						Leinster Province				MUNSTER PROVINCE					
	Aldergrove	Hillsborough	Armagh	Lough Navar	Malin Head	Sligo	Ardtarmon	Belmullet	Drumsna	Galway	Roscommon	Markree Castle	Edenderry	Kilkenny	Dublin	Shannon	Newport	Tralee	Killarney	Cork	All
Jan	8.7	8.7	8.7	8.4	8.6	8.7	8.6	9.7	8.7	8.9	8.7	8.7	8.3	9.0	8.9	9.1	10.1	8.3	10.3	8.7	**8.9**
Feb	7.9	7.9	7.9	8.1	8.4	8.6	8.4	9.6	8.7	8.4	8.7	8.6	7.8	8.1	8.2	8.5	9.2	7.8	9.6	7.8	**8.4**
Mar	8.4	8.4	8.4	8.4	7.4	7.6	8.4	9.3	8.1	8.1	8.1	7.6	7.4	8.2	7.5	8.5	9.3	7.4	9.0	7.5	**8.1**
Apr	11.5	11.5	11.5	11.0	10.3	10.8	12.1	11.2	11.3	10.9	11.3	10.8	11.5	12.2	11.5	11.6	11.4	11.5	12.1	10.9	**11.3**
May	15.4	15.4	15.4	15.2	13.1	13.4	15.9	14.2	14.8	14.2	14.8	13.4	14.9	15.6	14.8	15.2	15.0	14.9	15.4	13.7	**14.7**
Jun	19.0	19.0	19.0	18.3	15.5	17.0	20.1	16.6	18.6	17.4	18.6	17.0	19.1	19.6	18.4	18.5	18.2	19.1	18.8	17.4	**18.3**
Jul	21.7	21.7	21.7	20.7	17.9	18.7	22.7	18.8	21.2	19.5	21.2	18.7	21.4	21.9	20.1	21.3	20.3	21.4	21.4	19.7	**20.6**
Aug	20.8	20.8	20.8	19.8	17.7	17.9	20.9	18.5	19.3	18.9	19.3	17.9	21.0	21.7	20.6	20.0	19.6	21.0	20.1	19.9	**19.8**
Sep	19.1	19.1	19.1	18.8	17.8	18.4	19.7	18.2	18.5	17.9	18.5	18.4	19.4	19.8	19.0	19.3	18.6	19.4	19.7	18.3	**18.8**
Oct	17.0	17.0	17.0	16.6	16.0	16.0	17.1	16.6	15.9	16.1	15.9	16.0	16.6	17.1	16.5	16.6	16.6	16.6	16.6	15.4	**16.5**
Nov	12.3	12.3	12.3	12.2	11.9	11.9	11.9	12.7	11.7	11.9	11.7	11.9	11.5	12.6	12.1	12.3	12.4	11.5	13.0	11.8	**12.1**
Dec	9.7	9.7	9.7	9.1	10.1	9.4	9.4	10.4	9.3	9.9	9.3	9.4	9.1	10.3	10.2	10.2	10.3	9.1	11.2	9.6	**9.8**
Year	14.3	14.3	14.3	13.9	12.9	13.2	14.6	13.8	13.8	13.5	13.8	13.2	14.0	14.7	14.0	14.3	14.3	14.0	14.8	13.4	13.9

ESTIMATED AVERAGE MONTHLY MINIMUM TEMPERATURES

	Ulster Province					Connaught Province						Leinster Province				MUNSTER PROVINCE					
	Aldergrove	Hillsborough	Armagh	Lough Navar	Malin Head	Sligo	Ardtarmon	Belmullet	Drumsna	Galway	Roscommon	Markree Castle	Edenderry	Kilkenny	Dublin	Shannon	Newport	Tralee	Killarney	Cork	All
Jan	3.4	2.7	2.7	1.2	4.1	4.1	4.1	5.0	2.4	2.5	2.8	4.1	2.3	2.8	3.3	4.3	4.5	6.6	6.6	4.4	**3.7**
Feb	2.5	1.3	1.3	0.7	3.7	3.7	3.7	4.3	1.4	1.8	2.1	3.7	1.8	1.5	1.8	2.9	3.8	5.1	5.1	3.6	**2.8**
Mar	2.2	0.7	0.7	-0.2	3.8	2.4	2.4	2.9	0.6	0.6	1.2	2.4	1.1	1.2	1.3	1.4	2.8	3.2	3.2	1.5	**1.8**
Apr	4.2	3.6	3.6	1.6	4.8	4.8	4.8	5.2	3.3	2.9	3.4	4.8	3.4	4.1	3.4	4.9	5.0	6.3	6.3	4.5	**4.2**
May	7.2	6.7	6.7	4.6	7.6	7.7	7.7	8.0	6.3	6.1	6.9	7.7	6.8	6.9	6.4	7.5	8.6	7.9	7.9	7.1	**7.1**
Jun	9.5	9.1	9.1	7.1	10.0	9.7	9.7	9.7	8.5	7.4	8.8	9.7	8.5	8.1	8.3	9.4	10.4	9.4	9.4	9.2	**9.1**
Jul	11.5	11.2	11.2	9.0	11.8	11.9	11.9	11.4	11.0	10.1	11.2	11.9	10.6	11.0	10.8	12.0	11.9	12.3	12.3	11.5	**11.3**
Aug	13.1	12.9	12.9	11.6	13.3	13.5	13.5	13.3	12.6	11.3	12.7	13.5	12.3	12.0	11.9	12.9	13.7	13.5	13.5	12.0	**12.8**
Sep	11.5	11.1	11.1	8.8	11.9	12.3	12.3	12.4	10.4	9.7	10.7	12.3	10.3	10.7	10.7	12.1	11.5	13.2	13.2	11.5	**11.4**
Oct	10.3	9.6	9.6	7.8	11.1	10.7	10.7	10.9	9.2	8.8	9.4	10.7	8.8	9.6	9.2	10.8	10.9	12.0	12.0	10.0	**10.1**
Nov	6.2	5.0	5.0	3.2	7.3	6.8	6.8	7.3	3.9	3.9	4.5	6.8	3.6	4.0	4.7	5.9	6.7	7.7	7.7	5.6	**5.6**
Dec	4.6	3.8	3.8	1.5	5.5	4.8	4.8	6.2	3.3	3.1	3.3	4.8	3.2	3.4	4.3	5.2	5.2	7.0	7.0	4.3	**4.4**
Year	7.2	6.5	6.5	4.7	7.9	7.7	7.7	8.1	6.1	5.7	6.4	7.7	6.1	6.3	6.3	7.4	7.9	8.7	8.7	7.1	7.0

ESTIMATED MAXIMUM AVERAGE TEMPERATURES – QUICK REFERENCE

MONTHLY

	Ulster Province						Connaught Province							Leinster Province					MUNSTER PROVINCE						
	Aldergrove	Hillsborough	Armagh	Lough Navar	Malin Head	Ulster Overall	Sligo	Ardtarmon	Belmullet	Drumsna	Galway	Roscommon	Connaught Overall	Markree Castle	Edenderry	Kilkenny	Dublin	Leinster Overall	Shannon	Newport	Tralee	Killarney	Cork	Munster Overall	Overall
Jan	wmr	wmr	wmr	wmr	a-w	wmr	a-w	a-w	wmr	wmr	a-w	wmr	wmr	wmr	wmr	wmr	wmr	wmr	a-w	wmr	a-c	a-w	wmr	wmr	wmr
Feb	a-w	a-w	av	a-w	a-w	a-w	wmr	wmr	wmr	wmr	av	wmr	wmr	wmr	av	av	a-w	a-w	av	wmr	clr	av	av	av	a-w
Mar	a-c	a-c	clr	av	clr	a-c	av	av	clr	clr	clr	clr	a-c	clr	clr	clr	clr	clr	clr	a-c	clr	clr	clr	clr	clr
Apr	av	av	a-c	av	av	av	a-w	wmr	av	av	clr	a-c	av	a-c	av	av	av	av	clr	a-c	clr	av	av	a-c	av
May	a-w	a-w	av	wmr	av	a-w	wmr	wmr	av	a-w	a-c	av	a-w	clr	av	av	a-w	av	av	av	av	wmr	av	av	av
Jun	wmr	wmr	wmr	wmr	av	wmr	wmr	wmr	a-w	wmr	av	wmr	wmr	av	wmr	wmr	wmr	a-w	a-w	wmr	wmr	wmr	a-w	wmr	wmr
Jul	wmr	wmr	wmr	wmr	wmr	wmr	wmr	wmr	wmr	wmr	av	wmr	wmr	av	wmr	wmr	wmr	wmr	wmr	wmr	wmr	wmr	wmr	wmr	wmr
Aug	wmr	wmr	wmr	wmr	wmr	wmr	a-w	wmr	wmr	wmr	a-w	a-w	wmr	a-c	wmr	wmr	wmr	wmr	a-w	a-w	wmr	wmr	wmr	wmr	wmr
Sep	wmr	wmr	wmr	wmr	wmr	wmr	wmr	wmr	wmr	wmr	wmr	wmr	wmr	wmr	wmr	wmr	wmr	wmr	wmr	wmr	wmr	wmr	wmr	wmr	wmr
Oct	wmr	wmr	wmr	wmr	wmr	wmr	wmr	wmr	wmr	wmr	wmr	wmr	wmr	wmr	wmr	wmr	wmr	wmr	wmr	wmr	wmr	wmr	wmr	wmr	wmr
Nov	wmr	wmr	wmr	wmr	wmr	wmr	wmr	wmr	wmr	wmr	wmr	wmr	wmr	wmr	wmr	wmr	wmr	wmr	wmr	wmr	av	wmr	wmr	wmr	wmr
Dec	wmr	wmr	wmr	wmr	wmr	wmr	wmr	wmr	wmr	wmr	wmr	wmr	wmr	wmr	wmr	wmr	wmr	wmr	wmr	wmr	a-c	wmr	wmr	wmr	wmr
Year	wmr	wmr	wmr	wmr	wmr	wmr	wmr	wmr	wmr	wmr	a-w	wmr	wmr	a-w	wmr	wmr	wmr	wmr	a-w	wmr	av	wmr	a-w	a-w	wmr

wmr = 1.0> degrees above average | a-w = 0.6-1.0 degrees above average | av = 0.0-0.5 degrees from average | a-c = 0.6-1.0 degrees below average | clr = 1.0 > degrees below average

SEASONAL

	Ulster Province						Connaught Province							Leinster Province					MUNSTER PROVINCE						
	Aldergrove	Hillsborough	Armagh	Lough Navar	Malin Head	Ulster Overall	Sligo	Ardtarmon	Belmullet	Drumsna	Galway	Roscommon	Connaught Overall	Markree Castle	Edenderry	Kilkenny	Dublin	Leinster Overall	Shannon	Newport	Tralee	Killarney	Cork	Munster Overall	Overall
Winter	wmr	wmr	a-w	wmr	wmr	wmr	wmr	a-w	wmr	wmr	a-w	wmr	wmr	wmr	a-w	a-w	a-w	wmr	a-w	wmr	a-c	a-w	a-w	a-w	wmr
Spring	av	av	a-c	av	av	av	a-w	wmr	av	av	clr	a-c	av	clr	a-c	av	av	a-c	clr	av	clr	av	a-c	a-c	av
Summer	wmr	wmr	wmr	wmr	wmr	wmr	wmr	wmr	wmr	wmr	a-w	wmr	wmr	av	wmr	wmr	wmr	wmr	wmr	a-w	wmr	wmr	wmr	wmr	wmr
Autumn	wmr	wmr	wmr	wmr	wmr	wmr	wmr	wmr	wmr	wmr	wmr	wmr	wmr	wmr	wmr	wmr	wmr	wmr	wmr	wmr	wmr	wmr	wmr	wmr	wmr
December	wmr	wmr	wmr	wmr	wmr	wmr	wmr	wmr	wmr	wmr	wmr	wmr	wmr	wmr	wmr	wmr	wmr	wmr	wmr	wmr	a-c	wmr	wmr	wmr	wmr
Year	wmr	wmr	wmr	wmr	wmr	wmr	wmr	wmr	wmr	wmr	a-w	wmr	wmr	a-w	wmr	wmr	wmr	wmr	a-w	wmr	av	wmr	a-w	a-w	wmr

wmr = 1.0> degrees above average | a-w = 0.6-1.0 degrees above average | av = 0.0-0.5 degrees from average | a-c = 0.6-1.0 degrees below average | clr = 1.0 > degrees below average

ESTIMATED MINIMUM AVERAGE TEMPERATURES – QUICK REFERENCE

MONTHLY

	Ulster Province						Connaught Province							Leinster Province					MUNSTER PROVINCE						
	Aldergrove	Hillsborough	Armagh	Lough Navar	Malin Head	Ulster Overall	Sligo	Ardtarmon	Belmullet	Drumsna	Galway	Roscommon	Connaught Overall	Markree Castle	Edenderry	Kilkenny	Dublin	Leinster Overall	Shannon	Newport	Tralee	Killarney	Cork	Munster Overall	Overall
Jan	wmr	wmr	wmr	a-w	av	wmr	wmr	wmr	wmr	a-w	av	wmr	wmr	wmr	a-w	wmr	a-w	wmr	wmr	wmr	wmr	wmr	wmr	wmr	wmr
Feb	wmr	av	av	av	av	av	wmr	wmr	a-w	av	av	a-w	a-w	wmr	av	av	av	av	av	wmr	wmr	wmr	av	wmr	a-w
Mar	av	clr	clr	clr	a-c	clr	av	av	clr	clr	clr	clr	clr	av	clr	clr	clr	clr	clr	av	a-w	a-c	clr	clr	clr
Apr	a-w	av	av	clr	a-c	av	a-w	a-w	a-c	av	clr	av	av	a-w	av	av	clr	av	a-c	a-w	wmr	wmr	av	a-w	av
May	a-w	a-w	av	clr	av	av	wmr	wmr	av	av	a-c	a-w	av	wmr	a-w	av	av	a-w	a-c	wmr	wmr	av	av	a-w	av
Jun	av	av	av	clr	av	a-c	a-w	a-w	a-c	av	clr	av	av	a-w	av	clr	clr	av	clr	a-w	a-w	av	a-c	av	av
Jul	a-w	a-w	av	clr	av	av	wmr	wmr	a-c	av	clr	a-w	av	wmr	av	av	a-c	av	a-c	av	wmr	a-w	av	av	av
Aug	wmr	wmr	wmr	wmr	a-w	wmr	wmr	wmr	wmr	wmr	av	wmr	wmr	wmr	wmr	a-w	av	wmr	av	wmr	wmr	wmr	av	wmr	wmr
Sep	wmr	wmr	wmr	a-c	a-w	wmr	wmr	wmr	wmr	wmr	av	wmr	wmr	wmr	wmr	wmr	a-w	wmr	wmr	wmr	wmr	wmr	wmr	wmr	wmr
Oct	wmr	wmr	wmr	wmr	wmr	wmr	wmr	wmr	wmr	wmr	a-w	wmr	wmr	wmr	wmr	wmr	wmr	wmr	wmr	wmr	wmr	wmr	wmr	wmr	wmr
Nov	wmr	wmr	wmr	av	wmr	wmr	wmr	wmr	wmr	a-w	av	wmr	wmr	wmr	av	av	av	wmr	av	wmr	wmr	wmr	av	wmr	wmr
Dec	wmr	wmr	wmr	av	wmr	wmr	wmr	wmr	wmr	a-w	av	a-w	wmr	wmr	a-w	a-w	wmr	wmr	wmr	wmr	wmr	wmr	a-w	wmr	wmr
Year	wmr	wmr	a-w	a-c	av	a-w	wmr	wmr	a-w	a-w	a-c	a-w	a-w	wmr	a-w	av	av	a-w	av	wmr	wmr	wmr	av	wmr	a-w

wmr = 1.0> degrees above average | a-w = 0.6-1.0 degrees above average | av = 0.0-0.5 degrees from average | a-c = 0.6-1.0 degrees below average | clr = 1.0 > degrees below average

SEASONAL

	Ulster Province						Connaught Province							Leinster Province					MUNSTER PROVINCE						
	Aldergrove	Hillsborough	Armagh	Lough Navar	Malin Head	Ulster Overall	Sligo	Ardtarmon	Belmullet	Drumsna	Galway	Roscommon	Connaught Overall	Markree Castle	Edenderry	Kilkenny	Dublin	Leinster Overall	Shannon	Newport	Tralee	Killarney	Cork	Munster Overall	Overall
Winter	wmr	wmr	a-w	av	a-w	a-w	wmr	wmr	wmr	av	av	a-w	wmr	wmr	av	av	av	a-w	av	wmr	wmr	wmr	a-w	wmr	wmr
Spring	av	av	av	clr	a-c	a-c	av	av	a-c	a-c	clr	av	av	a-w	av	av	clr	av	clr	a-w	wmr	av	clr	av	av
Summer	a-w	a-w	a-w	clr	av	av	wmr	wmr	av	a-w	clr	wmr	a-w	wmr	a-w	av	a-c	av	a-c	wmr	wmr	a-w	av	a-w	av
Autumn	wmr	wmr	wmr	av	wmr	wmr	wmr	wmr	wmr	wmr	av	wmr	wmr	wmr	wmr	wmr	a-w	wmr	wmr	wmr	wmr	wmr	wmr	wmr	wmr
December	wmr	wmr	wmr	av	wmr	wmr	wmr	wmr	wmr	a-w	av	a-w	wmr	wmr	a-w	a-w	wmr	wmr	wmr	wmr	wmr	wmr	a-w	wmr	wmr
Year	wmr	wmr	a-w	a-c	av	a-w	wmr	wmr	a-w	a-w	a-c	a-w	a-w	wmr	a-w	av	av	a-w	av	wmr	wmr	wmr	av	wmr	a-w

wmr = 1.0> degrees above average | a-w = 0.6-1.0 degrees above average | av = 0.0-0.5 degrees from average | a-c = 0.6-1.0 degrees below average | clr = 1.0 > degrees below average

GARDENING GUIDE 2024

REST PERIOD:
The Rest Period in the garden is as important as the activity times. It's a time when the Moon is Void of Course, that is, traveling between one sign and another and during this time the Moon is deemed to have "no energizing power", also Rest Periods on the day of the Full Moon, New Moon and when the moon is at Southern and Northern Declinations. Hence it is a good time to sit and enjoy your garden or to attend to non-plant activities such as tidy your shed, repair or maintain outdoor structures or even to sit and plan your next "to do" list when the Moon is once more energized. Eclipse days and the day after the eclipse are also deemed days to rest in the garden.

ALL ABOVE-GROUND ACTIVITY SHOULD TAKE PLACE when the Moon is Waxing – i.e. building from New Moon through the 1st Quarter and up to the Full Moon. This is a time of increasing light and with the Moon pulling the sap upwards into the leaves and flowers.

ALL BELOW-GROUND ACTIVITY SHOULD TAKE PLACE when the Moon is Waning – i.e. from the Full Moon through the 3rd Quarter and up to the New Moon. This is a phase of decreasing light and the general trend for sap to be pulled downwards and into the root and tuber systems.

THE NEW MOON PHASE:
Activities can begin in the garden once the crescent of the New Moon begins to be visible.

At this time:

DO start sowing, transplanting, grafting or strike cuttings of all ABOVE-GROUND producers. Flowering annuals, brassicas (cabbage, broccoli etc) cover crops (mustard, lupins etc) and annual fruiting crops (cucumber, pumpkin, tomatoes etc) and house-plants all benefit from attention during this phase. Also, mow lawns should you wish to encourage growth or sow any large areas such as lawns.

DON'T plant root crops (as they will only produce plenty of leaves and little roots, or go to seed too early), nor prune at this time – your plants, and trees in particular, will be prone to excessive sap loss and suffer from die-back.

THE 1ST QUARTER PHASE:
Continue on with the New Moon phase activities and especially sow any seeds two days before the Full Moon to yield optimum results. During drought times, sow during this phase when the Moon is in Capricorn or Taurus for the strongest of survival qualities.

At this time:

DO sow and transplant all legume, grain, flowering plants (especially roses) and pot plants and start harvesting any above-ground produce, with 2-3 days before the Full Moon being the most beneficial time (especially for herbs and grape harvesting).

DO give plants a food boost if required, but do water it down and apply as close to the Full Moon as possible, particularly for those plants that require phosphorus.

DON'T harvest with the Moon in a water sign (Cancer, Scorpio or Pisces) unless for immediate use, nor prune or plant out root crops.

THE FULL MOON PHASE:
The Moon will have reached its full potency by the time of the Full Moon, and from this point to the New Moon, there is a gradual decrease in its vitality and the sap flow is now downwards into the roots – it is a time when the plants build up their root systems and store their goodness. However, you can continue to harvest above-ground producers for a few more days after the Full Moon. Leeks, onions, garlic etc are at their best potency when harvested with the Full Moon in Leo.

At this time:

DO sow or plant out any root crops (potatoes, kumaras, carrots etc) especially during the two days in the middle of the Full Moon phase. Also sow or plant out fruting plants (e.g. strawberries, passion fruit etc) as well as perennial flowering plants, bulbs and tubers (e.g. daffodils, dahlias, crocus etc) and this is also a good time to spray, prune and trim hedges.

DO perform drying activities of herbs, flowers and fruits.

DO dig herb roots or harvest leaves and bark intended for medicinal teas.

DO spray fruit trees, mulch your garden and kill weeds.

THE 3RD QUARTER PHASE:

This is considered to be the resting period of the Lunar cycle, the time when the sap in the plants runs at its slowest.

DO harvest all root crops that you wish to keep for long term storage (e.g. potatoes, garlic, onions and root herbs etc), especially if the Moon is in Taurus or Capricorn. Also lift your bulbs and tubers for their seasonal storage. You can also use this time to build up your compost, especially when the Moon is in Scorpio and attend to maintenance chores such as spreading compost, mulching, staking taller plants etc, as well as attend to all pruning and trimming of trees, vines and bushes, as there is less chance of die back and excessive sap bleeding. Cut your lawns during this phase to discourage growth.

DO plant trees and saplings during this phase for good strong foundations of root growth and bark development as well as plant strawberries or their runners so their growth is well developed before fruiting.

DO fertilize your crops for potassium if needed, as potassium absorption is at its peak at this phase.

When the Moon is in Cancer, Scorpio or Pisces:

This is a beneficial time for attending to vegetables such as lettuce, cabbage, herbs, broccoli, leafy pot plants, etc. Avoid any harvesting except for immediate use during these Moon signs as anything harvested for long term storage will deteriorate faster and be prone to mildew, except of course, unless you are attempting to build compost – choose when the Moon is in Scorpio for best results, as vegetation decomposes fastest under this influence.

When the Moon is in Taurus, Virgo or Capricorn:

This is a beneficial time for attending to vegetables such as potatoes, carrots, leeks etc. as well as to non-flowering shrubs and trees and ground covers such as lawns all respond best attended to under these signs. A Virgo Moon (particularly in the 3rd Quarter phase) is best for harvesting any medicinal leafy herbs due to the usual desire for long term storage.

When the Moon is in Gemini, Libra, Aquarius:

This is a beneficial time to attend to vegetables such as globe artichokes, flowers for medicinal purposes, and flowers in general. (Brocolli and cauliflower etc are considered "leafy" as if planted during a Flower phase they are more likely to bolt straight to seed). The Moon's focus is more directed towards developing perfume, fragrance, flower vibrancy and beauty as well as medicinal properties with the Full Moon in one of these signs.

When the Moon is in Aries, Leo or Sagittarius:

This is a beneficial time to attend to any fruit, berry, seed or grain crops such as tomatoes, peas, beans, peppers, wheat or corn, pumpkin, apples, strawberries, raspberries etc. Any seeds you wish to store for next year's planting, or for long term storage are best attended to during these Moon signs, as well as harvesting crops such as onions and garlic for their most potent flavour.

Above-ground - Leafy, Fruits and Seeds examples:

Vegetables	beans, broccoli, brussels sprouts, cabbages, cauliflower, chicory, celery, cress, cucumber, endive, lettuces, marrow, mustard, parsley, rhubarb, sage, silverbeet, corn, etc.
Flowers	cornflowers, forget-me-nots, marigolds, pansies, poppies, roses etc.
Fruits and seeds	apples, blue-berries, grapes, passion-fruit, pears, peas, peppers, raspberries, strawberries, tomatoes, wheat

Below-ground - Roots, Bulbs and Non-Flowering shrubs examples:

Vegetables	beetroot, carrots, garlic, leeks, onions, parsnips, radish, salsify, shallots, swedes, turnips etc.
Flowers	agapanthus, anemone, crocus, daffodils, dahlias, freesias, gladioli, hyacinth, ranunculus, red-hot pokers, tuberose, tulips
Fruits and Seeds	garlic, onions, peanuts

BASIC MOON IN SIGN GUIDE

Moon in Aries – Harvest root and fruit for storage. Good for destroying weeds and pests.

Moon in Taurus – Excellent time for planting and transplanting, particularly root crops. Also good for leaf crops.

Moon in Gemini – Good for harvesting root and fruit for storage. Good for destroying weeds and pests. Melon seeds respond well.

Moon in Cancer – Best time for all planting and transplanting. Good for grafting and irrigation.

Moon in Leo – Good for harvesting root and fruit for storage. Excellent for destroying weeds and pest.

Moon in Virgo – Good for destroying weeds and pests. Vines and some flowers respond well.

Moon in Libra – Best time for planting for fragrance and beauty of flowers, vines and herbs.

Moon in Scorpio – Best time for planting sturdy plants and vines. Especially favoured by tomatoes, corn and squash. Good for irrigation and transplanting.

Moon in Sagittarius – Good for harvesting roots. Especially favoured by onions and fruit trees. Good time to cultivate the soil.

Moon in Capricorn – Plant potatoes and root crops and encouraging hardy growth. Good for grafting and pruning to promote healing and applying organic fertilizers.

Moon in Aquarius – Harvest root and fruit for storage. Good for destroying weeds and pest. Good for planting onions.

Moon in Pisces – Good for planting and transplanting, especially for root growth. Good for irrigation.

JANUARY		2024		Hour: 1am – Midnight
A	1st	Mon	Below	Virgo
	2nd	Tue	Below	Virgo / REST
	3rd	Wed	Below	REST / Libra
3Q XhS	4th	Thur	Below	Libra
	5th	Fri	Below	Libra / REST / Scorpio
	6th	Sat	Below	Scorpio
	7th	Sun	Below	Scorpio / REST / Sagittarius
	8th	Mon	Below	Sagittarius
	9th	Tue	Below	Sagittarius / REST
V	10th	Wed	Below	REST DAY - Southern Declination
N	11th	Thur	Above	REST DAY - New Moon
	12th	Fri	Above	Capricorn / REST / Aquarius
P8	13th	Sat	Above	Aquarius / REST
	14th	Sun	Above	REST / Pisces
	15th	Mon	Above	Pisces
	16th	Tue	Above	Pisces / RE / Aries
XhN	17th	Wed	Above	Aries
1Q	18th	Thur	Above	Aries / REST / Taurus
	19th	Fri	Above	Taurus
	20th	Sat	Above	Taurus / RE / Gemini
	21st	Sun	Above	Gemini
	22nd	Mon	Above	Gemini / REST / Cancer
^	23rd	Tue	Above	REST DAY - Northern Declination
	24th	Wed	Above	Cancer / REST
F	25th	Thur	Below	REST DAY - Full Moon
	26th	Fri	Below	Leo / REST
	27th	Sat	Below	REST / Virgo
	28th	Sun	Below	Virgo
A	29th	Mon	Below	Virgo / REST
	30th	Tue	Below	REST / Libra
XhS	31st	Wed	Below	Libra

FEBRUARY		2024		Hour: 1am – Midnight
	1st	Thur	Below	Libra / REST / Scorpio
3Q	2nd	Fri	Below	Scorpio
	3rd	Sat	Below	Scorpio
	4th	Sun	Below	Scorpio / REST / Sagittarius
	5th	Mon	Below	Sagittarius
V	6th	Tue	Below	REST DAY - Southern Declination
	7th	Wed	Below	Capricorn
	8th	Thur	Below	Capricorn / REST / Aquarius
N	9th	Fri	Above	REST DAY - New Moon
P4	10th	Sat	Above	REST / Pisces
	11th	Sun	Above	Pisces
	12th	Mon	Above	Pisces / REST / Aries
XhN	13th	Tue	Above	Aries
	14th	Wed	Above	Aries / REST / Taurus
	15th	Thur	Above	Taurus

FEBRUARY		2024	Hour:	1am – Midnight
1Q	16th	Fri	Above	Taurus / REST / Gemini
	17th	Sat		Gemini
	18th	Sun		Gemini
^	19th	Mon		REST DAY - Northern Declination
	20th	Tue		Cancer
	21st	Wed		Cancer / REST / Leo
	22nd	Thur		Leo
	23rd	Fri		Leo / REST
F	24th	Sat	Below	REST DAY - Full Moon
A	25th	Sun		Virgo
	26th	Mon		Virgo / REST / Libra
XhS	27th	Tue		Libra / REST
	28th	Wed		REST DAY - Moon Void of Course all day
	29th	Thur		REST / Scorpio

MARCH		2024	Hour:	1am – Midnight
	1st	Fri	Below	Scorpio
	2nd	Sat		Scorpio / REST / Sagittarius
3Q	3rd	Sun		Sagittarius
	4th	Mon		Sagittarius / REST / Capricorn
V	5th	Tue		REST DAY - Southern Declination
	6th	Wed		Capricorn / REST
	7th	Thur		RE / Aquarius
	8th	Fri		Aquarius / REST
	9th	Sat		REST / Pisces
N P1	10th	Sun	Above	REST DAY - New Moon
	11th	Mon		RE / Aries
XhN	12th	Tue		Aries / REST
	13th	Wed		RE / Taurus
	14th	Thur		Taurus / REST
	15th	Fri		REST / Gemini
	16th	Sat		Gemini
1Q ^	17th	Sun		REST DAY - Northern Declination
	18th	Mon		Cancer
	19th	Tue		Cancer / RE / Leo
	20th	Wed		Leo
	21st	Thur		Leo
	22nd	Fri		Leo / REST / Virgo
A	23rd	Sat		Virgo
	24th	Sun		Virgo / REST / Libra
F XhS	25th	Mon	Below	REST DAY - Full Moon
	26th	Tue		Libra / REST
	27th	Wed		REST / Scorpio
	28th	Thur		Scorpio
	29th	Fri		Scorpio / REST / Sagittarius
	30th	Sat		Sagittarius
DLS Starts	31st	Sun		Sagittarius

APRIL		2024	Hour:	1am–Midnight
V	1st	Mon	Below	REST DAY - Southern Declination
3Q	2nd	Tue		Capricorn
	3rd	Wed		Capricorn / REST / Aquarius
	4th	Thur		Aquarius
	5th	Fri		Aquarius / REST / Pisces
	6th	Sat		Pisces
P5	7th	Sun		Pisces / REST / Aries
N XhN	8th	Mon	Above	REST DAY - New Moon - Total Solar Eclipse
	9th	Tue		REST DAY - Day after Eclipse Day
	10th	Wed		Taurus
	11th	Thur		Taurus / REST / Gemini
	12th	Fri		Gemini
^	13th	Sat		REST DAY - Northern Declination
	14th	Sun		Cancer
1Q	15th	Mon		Cancer
	16th	Tue		REST / Leo
	17th	Wed		Leo
	18th	Thur		Leo / REST / Virgo
	19th	Fri		Virgo
A	20th	Sat		Virgo
	21st	Sun		Virgo / REST / Libra
XhS	22nd	Mon		Libra
	23rd	Tue		Lib / REST / Scorpio
F	24th	Wed	Below	REST DAY - Full Moon
	25th	Thur		Scorpio
	26th	Fri		REST / Sagittarius
	27th	Sat		Sagittarius
V	28th	Sun		REST DAY - Southern Declination
	29th	Mon		Capricorn
	30th	Tue		Capricorn / RE / Aquarius

MAY		2024	Hour:	1am–Midnight
3Q	1st	Wed	Below	Aquarius
	2nd	Thur		Aquarius / REST / Pisces
	3rd	Fri		Pisces
	4th	Sat		Pisces / REST / Aries
P9	5th	Sun		Aries
XhN	6th	Mon		Aries / REST / Taurus
	7th	Tue		Taurus
N	8th	Wed	Above	REST DAY - New Moon
	9th	Thur		RE / Gemini
	10th	Fri		Gemini
^	11th	Sat		REST DAY - Northern Declination
	12th	Sun		Cancer
	13th	Mon		Cancer / REST / Leo
	14th	Tue		Leo
1Q	15th	Wed		Leo / REST / Virgo

Hours: 1am 2am 3am 4am 5am 6am 7am 8am 9am 10am 11am Noon 1pm 2pm 3pm 4pm 5pm 6pm 7pm 8pm 9pm 10pm 11pm Midnight

MAY		2024	Hour:	1am – 11pm Midnight
	16th	Thur	Above	Virgo
A	17th	Fri	Above	Virgo
	18th	Sat	Above	Virgo · REST · Libra
XhS	19th	Sun	Above	Libra · REST
	20th	Mon	Above	REST DAY - Moon Void of Course all day
	21st	Tue	Above	Scorpio
	22nd	Wed	Above	Scorpio
F	23rd	Thur	Below	REST DAY - Full Moon
	24th	Fri	Below	Sagittarius
V	25th	Sat	Below	REST DAY - Southern Declination
	26th	Sun	Below	Capricorn
	27th	Mon	Below	Capricorn · REST · Aquarius
	28th	Tue	Below	Aquarius
	29th	Wed	Below	Aquarius · REST
3Q	30th	Thur	Below	REST · Pisces
	31st	Fri	Below	Pisces

JUNE		2024	Hour:	1am – 11pm Midnight
	1st	Sat	Below	Pisces · REST · Aries
P12 XhN	2nd	Sun	Below	Aries · REST
	3rd	Mon	Below	REST · Taurus
	4th	Tue	Below	Taurus
	5th	Wed	Below	Taurus · RE · Gemini
N	6th	Thur	Above	REST DAY - New Moon
^	7th	Fri	Above	REST DAY - Northern Declination
	8th	Sat	Above	Pisces
	9th	Sun	Above	Pisces · RE · Leo
	10th	Mon	Above	Leo
	11th	Tue	Above	Leo · REST
	12th	Wed	Above	REST · Virgo
	13th	Thur	Above	Virgo
1Q A	14th	Fri	Above	Virgo · REST · Libra
XhS	15th	Sat	Above	Libra
	16th	Sun	Above	Libra
	17th	Mon	Above	Libra · RE · Scorpio
	18th	Tue	Above	Scorpio
	19th	Wed	Above	Scorpio · RE · Sagittarius
	20th	Thur	Above	Sagittarius
	21st	Fri	Above	Sagittarius
F V	22nd	Sat	Below	REST DAY - Full Moon / Southern Declination
	23rd	Sun	Below	Capricorn
	24th	Mon	Below	Capricorn · RE · Aquarius
	25th	Tue	Below	Aquarius · RE
	26th	Wed	Below	REST · Pisces
P13	27th	Thur	Below	Pisces
3Q	28th	Fri	Below	Pisces · RE · Aries
XhN	29th	Sat	Below	Aries
	30th	Sun	Below	Aries · REST · Taurus

JULY		2024	Hour:	1am – 11pm Midnight
	1st	Mon	Below	Taurus
	2nd	Tue	Below	Taurus · RE · Gemini
	3rd	Wed	Below	Gemini
	4th	Thur	Below	Gemini · RE · Cancer
N ^	5th	Fri	Above	REST DAY - New Moon / Nothern Declination
	6th	Sat	Above	Cancer
	7th	Sun	Above	Cancer · RE · Leo
	8th	Mon	Above	Leo
	9th	Tue	Above	Leo · REST · Virgo
	10th	Wed	Above	Virgo
	11th	Thur	Above	Virgo
A	12th	Fri	Above	Virgo · RE · Libra
1Q XhS	13th	Sat	Above	Libra · RE
	14th	Sun	Above	REST · Scorpio
	15th	Mon	Above	Scorpio
	16th	Tue	Above	Scorpio
	17th	Wed	Above	Scorpio · RE · Sagittarius
	18th	Thur	Above	Sagittarius
V	19th	Fri	Above	REST DAY - Southern Declination
	20th	Sat	Above	Capricorn
F	21st	Sun	Below	REST DAY - Full Moon
	22nd	Mon	Below	Aquarius
	23rd	Tue	Below	Aquarius · REST · Pisces
P10	24th	Wed	Below	Pisces
	25th	Thur	Below	Pisces · RE · Aries
XhN	26th	Fri	Below	Aries · REST
	27th	Sat	Below	REST · Taurus
3Q	28th	Sun	Below	Taurus
	29th	Mon	Below	Taurus · RE · Gemini
	30th	Tue	Below	Gemini
	31st	Wed	Below	Gemini

AUGUST		2024	Hour:	1am – 11pm Midnight
^	1st	Thur	Below	REST DAY - Northern Declination
	2nd	Fri	Below	Cancer
	3rd	Sat	Below	Cancer · RE · Leo
N	4th	Sun	Above	REST DAY - New Moon
	5th	Mon	Above	Leo · REST · Virgo
	6th	Tue	Above	Virgo
	7th	Wed	Above	Virgo
	8th	Thur	Above	Virgo · REST · Libra
A XhS	9th	Fri	Above	Libra · REST
	10th	Sat	Above	REST DAY - Moon Void of Course all day
	11th	Sun	Above	Scorpio
1Q	12th	Mon	Above	Scorpio
	13th	Tue	Above	Scorpio · REST · Sagittarius
	14th	Wed	Above	Sagittarius
V	15th	Thur	Above	REST DAY - Southern Declination

AUGUST		2024	Hour:	1am 2am 3am 4am 5am 6am 7am 8am 9am 10am 11am Noon 1pm 2pm 3pm 4pm 5pm 6pm 7pm 8pm 9pm 10pm 11pm Midnight
	16th	Fri	Above	Capricorn
	17th	Sat	Above	Capricorn / REST / Aquarius
	18th	Sun	Above	Aquarius
F	19th	Mon	Below	REST DAY - Full Moon
	20th	Tue	Below	Pisces
P7	21st	Wed	Below	Pisces / REST
	22nd	Thur	Below	RE / Aries
XhN	23rd	Fri	Below	Aries / REST
	24th	Sat	Below	REST / Taurus
	25th	Sun	Below	Taurus
3Q	26th	Mon	Below	Taurus / REST / Gemini
	27th	Tue	Below	Gemini
^	28th	Wed	Below	REST DAY - Northern Declination
	29th	Thur	Below	Cancer
	30th	Fri	Below	Cancer / REST / Leo
	31st	Sat	Below	Leo

SEPTEMBER		2024	Hour:	1am 2am 3am 4am 5am 6am 7am 8am 9am 10am 11am Noon 1pm 2pm 3pm 4pm 5pm 6pm 7pm 8pm 9pm 10pm 11pm Midnight
	1st	Sun	Belo	Leo
	2nd	Mon	Belo	Leo / REST / Virgo
N	3rd	Tue	Above	REST DAY - New Moon
	4th	Wed	Above	Virgo / RE / Libra
A XhS	5th	Thur	Above	Libra
	6th	Fri	Above	Libra
	7th	Sat	Above	Libra / RE / Scorpio
	8th	Sun	Above	Scorpio
	9th	Mon	Above	Scorpio / RE / Sagittarius
	10th	Tue	Above	Sagittarius
1Q	11th	Wed	Above	Sagittarius
V	12th	Thur	Above	REST DAY - Southern Declination
	13th	Fri	Above	Capricorn
	14th	Sat	Above	Capricorn / RE / Aquarius
	15th	Sun	Above	Aquarius
	16th	Mon	Above	Aquarius / REST / Pisces
	17th	Tue	Above	Pisces
F P3	18th	Wed	Below	REST DAY - Full Moon
XhN	19th	Thur	Below	Aries
	20th	Fri	Below	Aries / RE / Taurus
	21st	Sat	Below	Taurus
	22nd	Sun	Below	Taurus / RE / Gemini
	23rd	Mon	Below	Gemini
3Q ^	24th	Tue	Below	REST DAY - Northern Declination
	25th	Wed	Below	Cancer
	26th	Thur	Below	Cancer / RE / Le
	27th	Fri	Below	Leo
	28th	Sat	Below	Leo
	29th	Sun	Below	Leo / REST / Virgo
	30th	Mon	Below	Virgo

OCTOBER		2024	Hour:	1am – Midnight
	1st	Tue	B	Virgo / RE / Libra
N A XhS	2nd	Wed	Above	REST DAY - New Moon - Annular Solar Eclipse
	3rd	Thur		REST DAY - Day after Eclipse Day
	4th	Fri		Libra / RE / Scorpio
	5th	Sat		Scorpio
	6th	Sun		Scorpio / RE
	7th	Mon		RE / Sagittarius
	8th	Tue		Sagittarius
V	9th	Wed		REST DAY - Southern Declination
1Q	10th	Thur		Capricorn
	11th	Fri		Capricorn / REST / Aquarius
	12th	Sat		Aquarius
	13th	Sun		Aquarius / REST / Pisces
	14th	Mon		Pisces
	15th	Tue		Pisces / REST / Aries
	16th	Wed		Aries
F P2 XhN	17th	Thur	Below	REST DAY - Full Moon
	18th	Fri		Taurus
	19th	Sat		Taurus / RE / Gemini
	20th	Sun		Gemini
	21st	Mon		Gemini / REST / Ca
^	22nd	Tue		REST DAY - Northern Declination
	23rd	Wed		Cancer
3Q	24th	Thur		Cancer / REST / Leo
	25th	Fri		Leo
	26th	Sat		Leo / REST / Virgo
DLS Ends	27th	Sun		Virgo
	28th	Mon		Virgo
A	29th	Tue		Virgo / RE / Libra
XhS	30th	Wed		Libra
	31st	Thur		Libra / RE / Scorpio

NOVEMBER		2024	Hour:	1am – Midnight
	1st	Fri	B	Scorpio
N	2nd	Sat	Above	REST DAY - New Moon
	3rd	Sun		Scorpio / RE / Sagittarius
	4th	Mon		Sagittarius
V	5th	Tue		REST DAY - Southern Declination
	6th	Wed		Capricorn
	7th	Thur		Capricorn / RE / Aquarius
	8th	Fri		Aquarius
1Q	9th	Sat		Aquarius
	10th	Sun		Aq / REST / Pisces
	11th	Mon		Pisces
	12th	Tue		Pisces / RE / Aries
XhN	13th	Wed		Aries
P6	14th	Thur		Aries / RE / Taurus
F	15th	Fri		REST DAY - Full Moon
	16th	Sat		Taurus / RE / Gemini

NOVEMBER		2024	Hour:	Schedule
	17th	Sun	Below	Gemini
^	18th	Mon	Below	REST DAY - Northern Declination
	19th	Tue	Below	Cancer
	20th	Wed	Below	Cancer / REST / Leo
	21st	Thur	Below	Leo
	22nd	Fri	Below	Leo / REST / Virgo
3Q	23rd	Sat	Below	Virgo
	24th	Sun	Below	Virgo
	25th	Mon	Below	Virgo / REST / Libra
A XhS	26th	Tue	Below	Libra
	27th	Wed	Below	Libra / REST
	28th	Thur	Below	RE / Scorpio
	29th	Fri	Below	Scorpio
	30th	Sat	Below	Scorpio / REST / Sagittarius

DECEMBER		2024	Hour:	Schedule
N	1st	Sun	Above	REST DAY - New Moon
V	2nd	Mon	Above	REST DAY - Southern Declination
	3rd	Tue	Above	Capricorn
	4th	Wed	Above	Capricorn / RE
	5th	Thur	Above	REST / Aquarius
	6th	Fri	Above	Aquarius
	7th	Sat	Above	REST / Pisces
1Q	8th	Sun	Above	Pisces
	9th	Mon	Above	Pisces / REST / Aries
XhN	10th	Tue	Above	Aries / REST
	11th	Wed	Above	REST / Taurus
P11	12th	Thur	Above	Taurus
	13th	Fri	Above	Taurus / REST / Gemini
	14th	Sat	Above	Gemini
F ^	15th	Sun	Below	REST DAY - Full Moon / Northern Declination
	16th	Mon	Below	Cancer
	17th	Tue	Below	Cancer / REST / Leo
	18th	Wed	Below	Leo
	19th	Thur	Below	Leo
	20th	Fri	Below	Leo / REST / Virgo
	21st	Sat	Below	Virgo
3Q	22nd	Sun	Below	Virgo / REST / Libra
Xhs	23rd	Mon	Below	Libra
A	24th	Tue	Below	Libra / REST
	25th	Wed	Below	REST / Scorpio
	26th	Thur	Below	Libra
	27th	Fri	Below	Scorpio / REST / Sagittarius
	28th	Sat	Below	Sagittarius
	29th	Sun	Below	Sagittarius / REST
N V	30th	Mon	Abo	REST DAY - New Moon / Northern Declination
	31st	Tue	Abo	Capricorn

FISHING CALENDAR 2024

AV	VG	EX
🐟	🐟🐟	🐟🐟🐟

JANUARY	1st	2nd	3rd	4th	5th	6th	7th	8th	9th	10th	11th	12th	13th	14th	15th	16th	17th	18th	19th	20th	21st	22nd	23rd	24th	25th	26th	27th	28th	29th	30th	31st
MOON	Ap			Third Q							New Moon		P8					First Q							Full Moon				Ap		
Best Bite Time am/pm	4.30	5.00	5.30	6.00	7.00	8.00	9.00	10.00	11.00	11.30	12.00	1.00	2.00	3.00	4.00	5.00	5.30	6.00	7.00	8.00	9.00	10.00	10.30	11.00	11.30	12.00	12.30	1.00	2.00	3.00	4.00
Next Best Bite Time am/pm	10.30	11.00	11.30	12.00	1.00	2.00	3.00	4.00	5.00	5.30	6.00	7.00	8.00	9.00	10.00	11.00	11.30	12.00	1.00	2.00	3.00	4.00	4.30	5.00	5.30	6.00	6.30	7.00	8.00	9.00	10.00
RATING	🐟	🐟🐟	🐟🐟	🐟🐟	🐟	🐟	🐟	🐟	🐟🐟🐟	🐟🐟🐟	🐟🐟🐟	🐟	🐟	🐟	🐟	🐟🐟	🐟🐟	🐟🐟	🐟🐟	🐟	🐟	🐟	🐟	🐟🐟🐟	🐟🐟🐟	🐟🐟🐟	🐟🐟🐟	🐟	🐟	🐟	🐟

FEBRUARY	1st	2nd	3rd	4th	5th	6th	7th	8th	9th	10th	11th	12th	13th	14th	15th	16th	17th	18th	19th	20th	21st	22nd	23rd	24th	25th	26th	27th	28th	29th
MOON		Third Q							New Moon	P4						First Q								Full Moon	Ap				
Best Bite Time am/pm	5.00	5.30	6.00	7.00	8.00	9.00	10.00	11.00	11.30	12.00	1.00	2.00	2.30	3.00	4.00	5.00	6.00	7.00	7.30	8.00	9.00	10.00	11.00	12.00	12.30	1.00	2.00	3.00	4.00
Next Best Bite Time am/pm	11.00	11.30	12.00	1.00	2.00	3.00	4.00	5.00	5.30	6.00	7.00	8.00	8.30	9.00	10.00	11.00	12.00	1.00	1.30	2.00	3.00	4.00	5.00	6.00	6.30	7.00	8.00	9.00	10.00
RATING	🐟🐟	🐟🐟	🐟🐟	🐟	🐟	🐟	🐟	🐟🐟🐟	🐟🐟🐟	🐟🐟🐟	🐟🐟🐟	🐟	🐟	🐟	🐟	🐟🐟	🐟🐟	🐟🐟	🐟	🐟	🐟	🐟	🐟🐟🐟	🐟🐟🐟	🐟🐟🐟	🐟🐟🐟	🐟	🐟	🐟

MARCH	1st	2nd	3rd	4th	5th	6th	7th	8th	9th	10th	11th	12th	13th	14th	15th	16th	17th	18th	19th	20th	21st	22nd	23rd	24th	25th	26th	27th	28th	29th	30th	31st
MOON			Third Q							New Moon P1							First Q						Ap		Full Moon						
Best Bite Time am/pm	4.30	5.00	5.30	6.00	7.00	8.00	9.00	10.00	11.00	12.00	1.00	2.00	3.00	4.00	5.00	5.30	6.00	7.00	8.00	9.00	9.30	10.00	10.30	11.00	12.00	1.00	2.00	3.00	4.00	4.30	5.00
Next Best Bite Time am/pm	10.30	11.00	11.30	12.00	1.00	2.00	3.00	4.00	5.00	6.00	7.00	8.00	9.00	10.00	11.00	11.30	12.00	1.00	2.00	3.00	3.30	4.00	4.30	5.00	6.00	7.00	8.00	9.00	10.00	10.30	11.00
RATING	🐟	🐟🐟	🐟🐟	🐟🐟	🐟	🐟	🐟	🐟	🐟🐟🐟	🐟🐟🐟	🐟🐟🐟	🐟	🐟	🐟	🐟🐟	🐟🐟	🐟🐟	🐟🐟	🐟	🐟	🐟	🐟	🐟🐟🐟	🐟🐟🐟	🐟🐟🐟	🐟🐟🐟	🐟	🐟	🐟	🐟	🐟🐟

APRIL	1st	2nd	3rd	4th	5th	6th	7th	8th	9th	10th	11th	12th	13th	14th	15th	16th	17th	18th	19th	20th	21st	22nd	23rd	24th	25th	26th	27th	28th	29th	30th
MOON		Third Q					P5	New Moon							First Q					Ap			Full Moon							
Best Bite Time am/pm	5.30	6.00	7.00	8.00	9.00	10.00	11.00	11.30	12.00	1.00	2.00	3.00	4.00	5.00	5.30	6.00	7.00	8.00	9.00	10.00	10.30	11.00	11.30	12.00	1.00	2.00	3.00	4.00	5.00	5.30
Next Best Bite Time am/pm	11.30	12.00	1.00	2.00	3.00	4.00	5.00	5.30	6.00	7.00	8.00	9.00	10.00	11.00	11.30	12.00	1.00	2.00	3.00	4.00	4.30	5.00	5.30	6.00	7.00	8.00	9.00	10.00	11.00	11.30
RATING	🐟🐟	🐟🐟	🐟	🐟	🐟	🐟	🐟🐟🐟	🐟🐟🐟	🐟🐟🐟	🐟🐟🐟	🐟	🐟	🐟	🐟🐟	🐟🐟	🐟🐟	🐟🐟	🐟	🐟	🐟	🐟	🐟🐟🐟	🐟🐟🐟	🐟🐟🐟	🐟🐟🐟	🐟	🐟	🐟	🐟🐟	🐟🐟

MAY	1st	2nd	3rd	4th	5th	6th	7th	8th	9th	10th	11th	12th	13th	14th	15th	16th	17th	18th	19th	20th	21st	22nd	23rd	24th	25th	26th	27th	28th	29th	30th	31st
MOON	Third Q				P9			New Moon							First Q		Ap						Full Moon							Third Q	
Best Bite Time am/pm	6.00	7.00	8.00	9.00	10.00	11.00	11.30	12.00	1.00	2.00	3.00	4.00	5.00	5.30	6.00	7.00	8.00	9.00	9.30	10.00	10.30	11.00	11.30	12.00	1.00	2.00	3.00	4.00	5.00	5.30	6.00
Next Best Bite Time am/pm	12.00	1.00	2.00	3.00	4.00	5.00	5.30	6.00	7.00	8.00	9.00	10.00	11.00	11.30	12.00	1.00	2.00	3.00	3.30	4.00	4.30	5.00	5.30	6.00	7.00	8.00	9.00	10.00	11.00	11.30	12.00
RATING	🐟🐟	🐟	🐟	🐟	🐟	🐟🐟🐟	🐟🐟🐟	🐟🐟🐟	🐟🐟🐟	🐟	🐟	🐟	🐟	🐟🐟	🐟🐟	🐟🐟	🐟	🐟	🐟	🐟	🐟🐟🐟	🐟🐟🐟	🐟🐟🐟	🐟🐟🐟	🐟	🐟	🐟	🐟	🐟🐟	🐟🐟	🐟🐟

JUNE	1st	2nd	3rd	4th	5th	6th	7th	8th	9th	10th	11th	12th	13th	14th	15th	16th	17th	18th	19th	20th	21st	22nd	23rd	24th	25th	26th	27th	28th	29th	30th
MOON		P12				New Moon								First Q Ap								Full Moon					P13	Third Q		
Best Bite Time am/pm	7.00	8.00	9.00	10.00	11.00	11.30	12.00	1.00	2.00	3.00	4.00	5.00	5.30	6.00	7.00	8.00	9.00	10.00	10.30	11.00	11.30	12.00	1.00	2.00	3.00	4.00	5.00	5.30	6.00	7.00
Next Best Bite Time am/pm	1.00	2.00	3.00	4.00	5.00	5.30	6.00	7.00	8.00	9.00	10.00	11.00	11.30	12.00	1.00	2.00	3.00	4.00	4.30	5.00	5.30	6.00	7.00	8.00	9.00	10.00	11.00	11.30	12.00	1.00
RATING	🐟	🐟	🐟	🐟	🐟🐟🐟	🐟🐟🐟	🐟🐟🐟	🐟🐟🐟	🐟	🐟	🐟	🐟	🐟🐟	🐟🐟	🐟🐟	🐟	🐟	🐟	🐟	🐟🐟🐟	🐟🐟🐟	🐟🐟🐟	🐟	🐟	🐟	🐟	🐟🐟	🐟🐟	🐟🐟	🐟

AV	VG	EX
1 fish	2 fish	3 fish

JULY	1st	2nd	3rd	4th	5th	6th	7th	8th	9th	10th	11th	12th	13th	14th	15th	16th	17th	18th	19th	20th	21st	22nd	23rd	24th	25th	26th	27th	28th	29th	30th	31st
MOON					New Moon							Ap	First Q								Full Moon			P10				Third Q			
Best Bite Time am/pm	8.00	9.00	10.00	11.00	11.30	12.00	1.00	2.00	3.00	4.00	4.30	5.00	5.30	6.00	7.00	8.00	9.00	10.00	11.00	11.30	12.00	1.00	2.00	3.00	4.00	5.00	5.30	6.00	7.00	8.00	9.00
Next Best Bite Time am/pm	2.00	3.00	4.00	5.00	5.30	6.00	7.00	8.00	9.00	10.00	10.30	11.00	11.30	12.00	1.00	2.00	3.00	4.00	5.00	5.30	6.00	7.00	8.00	9.00	10.00	11.00	11.30	12.00	1.00	2.00	3.00
RATING	AV	AV	AV	EX	EX	EX	EX	AV	AV	AV	AV	VG	VG	VG	AV	AV	AV	AV	EX	EX	EX	AV	AV	AV	AV	VG	VG	VG	AV	AV	AV

AUGUST	1st	2nd	3rd	4th	5th	6th	7th	8th	9th	10th	11th	12th	13th	14th	15th	16th	17th	18th	19th	20th	21st	22nd	23rd	24th	25th	26th	27th	28th	29th	30th	31st
MOON				New Moon					Ap			First Q							Full Moon		P7					Third Q					
Best Bite Time am/pm	10.00	11.00	11.30	12.00	1.00	2.00	3.00	3.30	4.00	4.30	5.00	5.30	6.00	7.00	8.00	9.00	10.00	11.00	11.30	12.00	1.00	2.00	3.00	4.00	5.00	6.00	7.00	8.00	9.00	10.00	10.30
Next Best Bite Time am/pm	4.00	5.00	5.30	6.00	7.00	8.00	9.00	9.30	10.00	10.30	11.00	11.30	12.00	1.00	2.00	3.00	4.00	5.00	5.30	6.00	7.00	8.00	9.00	10.00	11.00	12.00	1.00	2.00	3.00	4.00	4.30
RATING	AV	EX	EX	EX	EX	AV	AV	AV	AV	AV	VG	VG	VG	AV	AV	AV	AV	EX	EX	EX	AV	AV	AV	AV	VG	VG	VG	AV	AV	AV	AV

SEPTEMBER	1st	2nd	3rd	4th	5th	6th	7th	8th	9th	10th	11th	12th	13th	14th	15th	16th	17th	18th	19th	20th	21st	22nd	23rd	24th	25th	26th	27th	28th	29th	30th
MOON			New Moon		Ap						First Q							Full Moon P3						Third Q						
Best Bite Time am/pm	11.00	11.30	12.00	1.00	2.00	3.00	4.00	4.30	5.00	5.30	6.00	7.00	8.00	9.00	10.00	11.00	11.30	12.00	1.00	2.00	3.00	4.00	5.00	5.30	6.00	7.00	8.00	9.00	10.00	10.30
Next Best Bite Time am/pm	5.00	5.30	6.00	7.00	8.00	9.00	10.00	10.30	11.00	11.30	12.00	1.00	2.00	3.00	4.00	5.00	5.30	6.00	7.00	8.00	9.00	10.00	11.00	11.30	12.00	1.00	2.00	3.00	4.00	4.30
RATING	EX	EX	EX	EX	AV	AV	AV	AV	VG	VG	VG	AV	AV	AV	AV	EX	EX	EX	AV	AV	AV	AV	AV	VG	VG	VG	AV	AV	AV	AV

OCTOBER	1st	2nd	3rd	4th	5th	6th	7th	8th	9th	10th	11th	12th	13th	14th	15th	16th	17th	18th	19th	20th	21st	22nd	23rd	24th	25th	26th	27th	28th	29th	30th	31st
MOON		New Moon Ap								First Q							Full Moon P2							Third Q					Ap		
Best Bite Time am/pm	11.00	11.30	12.00	1.00	2.00	3.00	4.00	4.30	5.00	5.30	6.00	7.00	8.00	9.00	10.00	11.00	12.00	1.00	2.00	3.00	4.00	5.00	5.30	6.00	7.00	8.00	9.00	9.30	10.00	10.30	11.00
Next Best Bite Time am/pm	5.00	5.30	6.00	7.00	8.00	9.00	10.00	10.30	11.00	11.30	12.00	1.00	2.00	3.00	4.00	5.00	6.00	7.00	8.00	9.00	10.00	11.00	11.30	12.00	1.00	2.00	3.00	3.30	4.00	4.30	5.00
RATING	AV	EX	EX	EX	AV	AV	AV	AV	VG	VG	VG	AV	AV	AV	AV	EX	EX	EX	AV	AV	AV	AV	VG	VG	VG	AV	AV	AV	AV	AV	EX

NOVEMBER	1st	2nd	3rd	4th	5th	6th	7th	8th	9th	10th	11th	12th	13th	14th	15th	16th	17th	18th	19th	20th	21st	22nd	23rd	24th	25th	26th	27th	28th	29th	30th
MOON		New Moon							First Q					P6	Full Moon								Third Q			Ap				
Best Bite Time am/pm	11.30	12.00	1.00	2.00	3.00	4.00	5.00	5.30	6.00	7.00	8.00	9.00	10.00	11.00	11.30	12.00	1.00	2.00	3.00	4.00	5.00	5.30	6.00	7.00	8.00	9.00	10.00	10.30	11.00	11.30
Next Best Bite Time am/pm	5.30	6.00	7.00	8.00	9.00	10.00	11.00	11.30	12.00	1.00	2.00	3.00	4.00	5.00	5.30	6.00	7.00	8.00	9.00	10.00	11.00	11.30	12.00	1.00	2.00	3.00	4.00	4.30	5.00	5.30
RATING	EX	EX	AV	AV	AV	AV	VG	VG	VG	AV	AV	AV	AV	EX	EX	EX	AV	AV	AV	AV	AV	VG	VG	VG	AV	AV	AV	AV	EX	EX

DECEMBER	1st	2nd	3rd	4th	5th	6th	7th	8th	9th	10th	11th	12th	13th	14th	15th	16th	17th	18th	19th	20th	21st	22nd	23rd	24th	25th	26th	27th	28th	29th	30th	31st
MOON	New Moon							First Q				P11			Full Moon							Third Q		Ap						New Moon	
Best Bite Time am/pm	12.00	1.00	2.00	3.00	4.00	4.30	5.00	5.30	6.00	7.00	8.00	9.00	10.00	11.00	12.00	1.00	2.00	3.00	4.00	4.30	5.00	5.30	6.00	7.00	8.00	9.00	10.00	10.30	11.00	11.30	12.00
Next Best Bite Time am/pm	6.00	7.00	8.00	9.00	10.00	10.30	11.00	11.30	12.00	1.00	2.00	3.00	4.00	5.00	6.00	7.00	8.00	9.00	10.00	10.30	11.00	11.30	12.00	1.00	2.00	3.00	4.00	4.30	5.00	5.30	6.00
RATING	EX	EX	AV	AV	AV	AV	VG	VG	VG	AV	AV	AV	AV	EX	EX	EX	AV	AV	AV	AV	AV	VG	VG	VG	AV	AV	AV	AV	EX	EX	EX

Printed in Australia
Ingram Content Group Australia Pty Ltd
AUHW011722160124
389174AU00014B/157

9 781738 623013